Futuring Human Empowerment

Cover image: Cellular reprogramming at the human synthetic biology enterprise bit.bio in Cambridge, UK. Cells are defined by distinct transcription factor programs, and every cell type has a unique code. This image provides a glimpse at the discovery process of liver cells. The cells in green are successfully reprogrammed cells while the blue cells are not. In one experiment thousands of different programs are tested. (See interview with founder Mark Kotter on page 135.)

Hans Georg Näder

Futuring Human Empowerment

Concept, realization, and interviews by
Thomas Huber

Photographs by Christoph Neumann
with Sascha Boldt

Steidl

This book takes readers on a journey to the remaining secrets of the human body, offering an exclusive look at laboratories and think tanks worldwide—normally closed to the public—where the most innovative researchers, engineers, and start-ups are working to expand the capabilities of the human body with artificial intelligence and progressive technologies. In exclusive, in-depth conversations experts explain how they are using biotechnologies in the fight against pandemics and cancer, mobilizing the human body's hidden self-healing powers, deciphering the mathematical algorithm of life, cultivating stem cells on an industrial scale, and printing artificial replacements of vital organs on 3D bioprinters. Another focus of this book is the digitization that determines the future of every company with aspirations to be a leader in technology and innovation. The issue at stake in orthopedic technology is nothing less than the complete digitization of a century-old craft (see also the essay "The Fourth Revolution," page 323).

An additional focus of this book is the design of the human-machine interface. While humanity entered the age of robots and cyborgs long ago, robot "coworkers" are increasingly taking on more of the workload in modern industry and numerous technologies are seamlessly incorporated into the human body and expanding its capabilities. As the human-machine interface becomes smoother in the future, Ottobock's protheses and orthoses will become ever better. This opens up new opportunities for our company, a world leader in medical technologies. With wearable robots, we are entering new markets for users who wish to boost their power at work and in leisure activities.

This positions Ottobock at the intersection of the most important disciplines of the future that, in each chapter of this book, are presented by visionaries and progressive thinkers and explained by engineers and start-up founders in specific application scenarios. I have known some of the researchers and protagonists for many years. As experts in their fields, they are able to very precisely predict the future of the human body. While many new technologies that are discussed in this book are practically within our reach, others are still in the distant future. This is why we have placed a discussion between the researchers and developers at Ottobock at the end of each chapter to focus on how the various technological advances can be implemented in and on the human body. Through our network of laboratories around the globe, we intend to explore new areas of growth with this knowledge and to continue our pioneering role in translating intriguing technologies into solutions that enable people to transcend themselves.

I wish you many exciting discoveries on your personal journey to the future of the human body!

Hans Georg Näder

Digitization

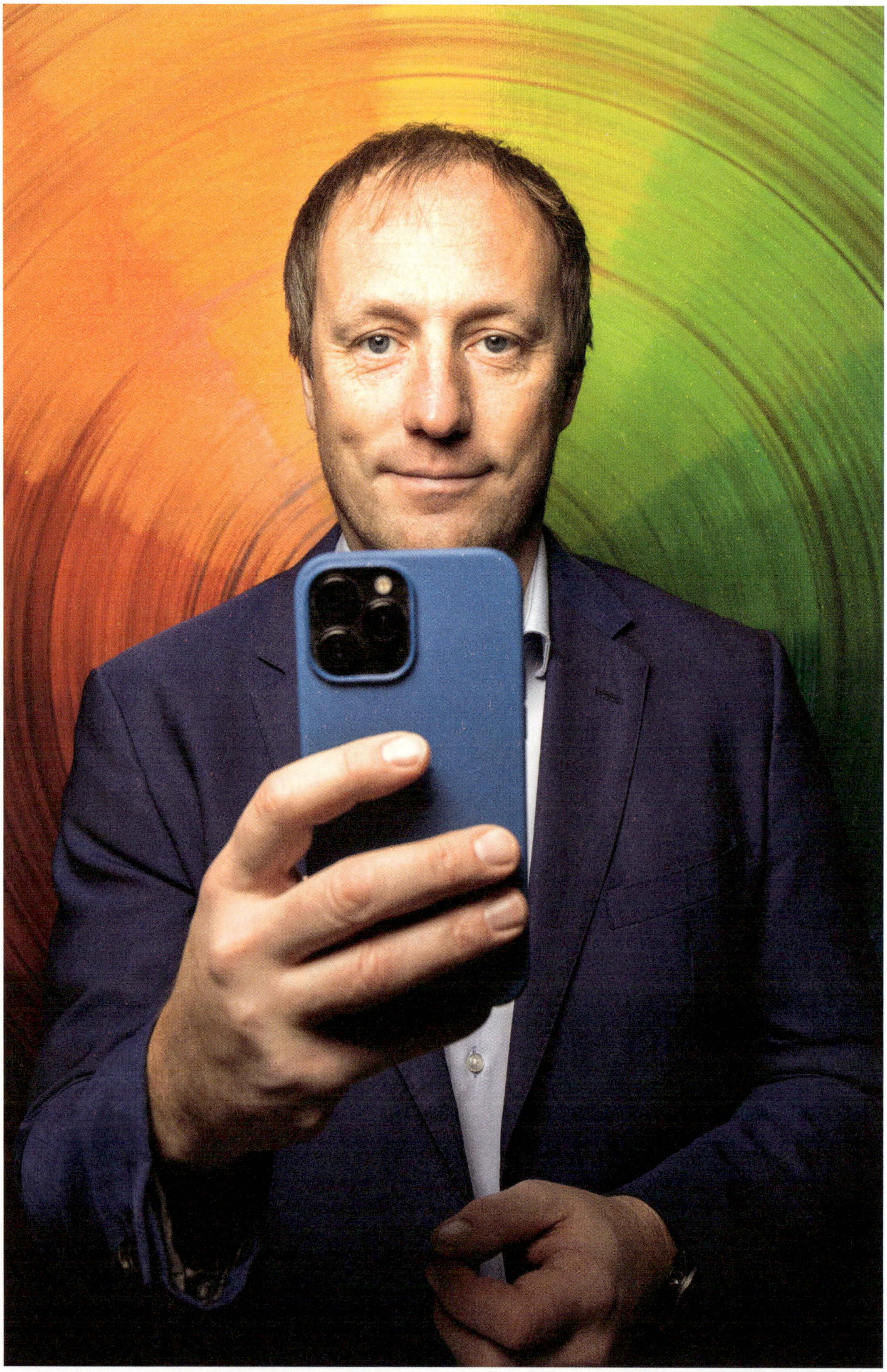

TAAVI KOTKA

"Listen to Your Engineers!"

Silicon Valley didn't start in 1970; it started in Estonia in the 1930s. The former republic of the Soviet Union was extremely adept at software engineering and computer science. After 1991 they built up a digital infrastructure. As the Chief Information Officer of Estonia Taavi Kotka designed software for this infrastructure and digitized the entire society. The data is stored securely in a cloud—secure not only against abuse but also against Russia, which has started an aggressive war against Ukraine in pursuit of its military interests in the region. Kotka is now digitizing India, which is more than a hundred times the size of the population of Estonia. However, the principles he applies are always the same, and many companies could apply them to their organizations as well, if only they trusted their engineers more.

You were responsible for introducing e-government in Estonia a decade ago. Is Estonia still the leader in this development?

Estonia started this change at the end of the 1990s because it was the only way to efficiently manage this country. Our country has a small population and a huge amount of land—we are actually bigger than Switzerland or Denmark. If you want to manage this kind of country, you need to automate and push people toward self-service, because there aren't enough people to serve the others. When the internet came, we instantly recognized that it could be our savior. Instead of having a small bank or government office in every small town, we decided to build proper tools for people to interact digitally, which also means independent of location, so it doesn't matter where you are in Estonia. Thirty years ago I was still a kid, so I'm not the author of Estonian e-government, but I am definitely an author of the second wave. I wasn't the one who actually put the infrastructure in place. I was the guy who built the software on top of the infrastructure. You know, I don't like the term *e-government*.

Do you prefer the word e-society*?*

Yes, because if you think about it, how many times do you interact with the government during the year?

Two or three times maybe?

It depends on your profession, but for the average person it's 2.6, not including healthcare and education. So there is no point in making e-government fully digital if your private sector is still lagging behind without any digital tools or anything. The goal in Estonia was to build a digital society, not e-government. E-government is just a small part. We defined seven steps. The first is obvious: you have to have an infrastructure. If you don't have internet, you can't build a digital society. You have to get the data moving, which from an engineering perspective means that certain elements need to be put in place. That's where the whole story ends for Germany, the UK, the US, and Canada—because you're not willing to connect the data. You have a wrong understanding of privacy. We'll come back to that. But if you want to push people to the internet, they have to be able to authenticate themselves. How can I be sure who is using a device? It has to be a combination of entering a PIN and showing my face—this combination is actually secure. Another thing is free will. Let's say I sell my apartment or vote in an election—how can you be sure that it was actually my free will to do that? You need some kind of digital signature. The private sector needs that every day for signing documents, bank transfers, or anything. We are constantly signing things. If you have those core elements in place, everything starts to work. There are not too many ways to do that: northern Europe, China, and Singapore have understood that. The digital societies in Sweden and China are one hundred percent the same from the engineering perspective. The difference is: Swedes are careful about how they store data about citizens, and they protect privacy. In China, they use the data to manipulate the people's lives.

It's a cultural difference.

Scandinavian countries have understood the essence of the General Data Protection Regulation (GDPR), which is that people's rights and privacy need to be protected. But it doesn't mean that the data cannot be connected. In Estonia, for example, we share all medical data so doctors can see your whole medical history.

That's not possible in Germany.

But what about your hospitals? What do they see when they type your name in their system?

Nothing.

They can't see anything because you have removed everything.

Patients tell the doctor about their condition. Of course, patients can lie and say that they never drink alcohol or smoke cigarettes.

And every time you have to describe once again that you have had the measles. I had an operation in south Estonia, and later I was treated in north Estonia. The doctors shared all the information about my therapy, and they didn't have to send it back and forth. It was in front of them because the systems are all combined.

It's connected to your personal identification number.

Our hospital system makes the data public. But in Germany, even if the data has stayed in the hospital, you have no understanding who has access to that data. The GDPR now protects the data. But our view is that the data should be available to every doctor. In Estonia I can also see who has accessed my data. If you have accessed my data, I can ask you why. If you don't have a proper answer, you get fired and lose your license. All doctors know what can happen. You have to do a couple of hangings as examples. In a recent case, the person went to jail for one year. That's how you build the trust in society. So, yes, every policeman and doctor can look me up, but if you misuse the data ...

Do you get a notification?

No, it would be too many notifications. But we have a central portal that you can monitor. This actually protects privacy even more. In German hospitals, this information is digitized, but you can't see who has consulted the data. In Estonia we have digitized all the hospitals, plus the customer actually sees who has accessed and why. We built that before GDPR. That's how we got the data moving, which is essential for digital society. The processes became faster, and we can easily share information and case histories with specialists from all over the world. They know that I have a heart condition and that I also had cancer twenty years ago, so they can consider how the cancer influences my heart condition, or vice versa. If you don't have that link, this nuance would probably be missed. They should know

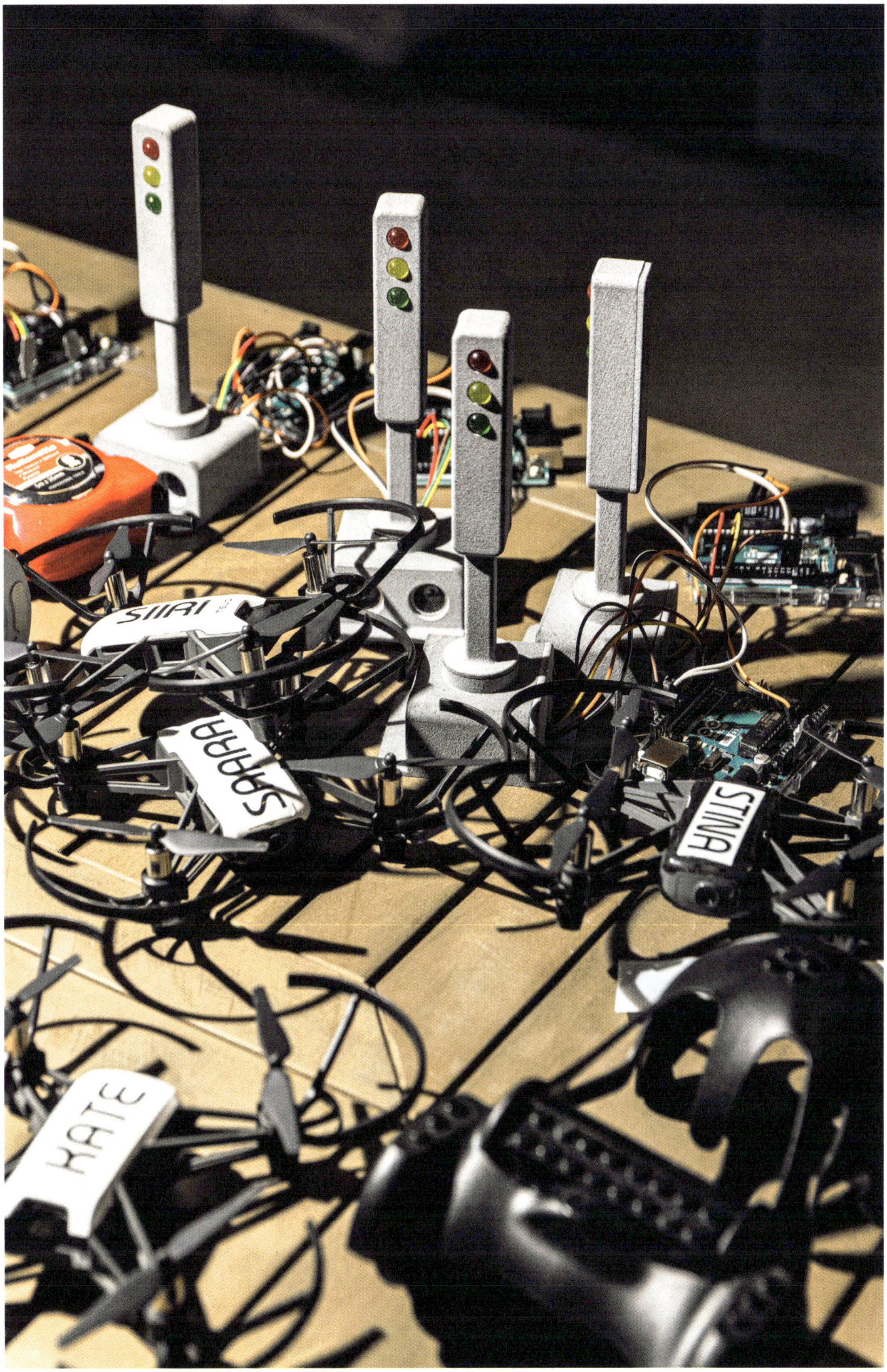
SIIRI
SAARA
STINA
KATE

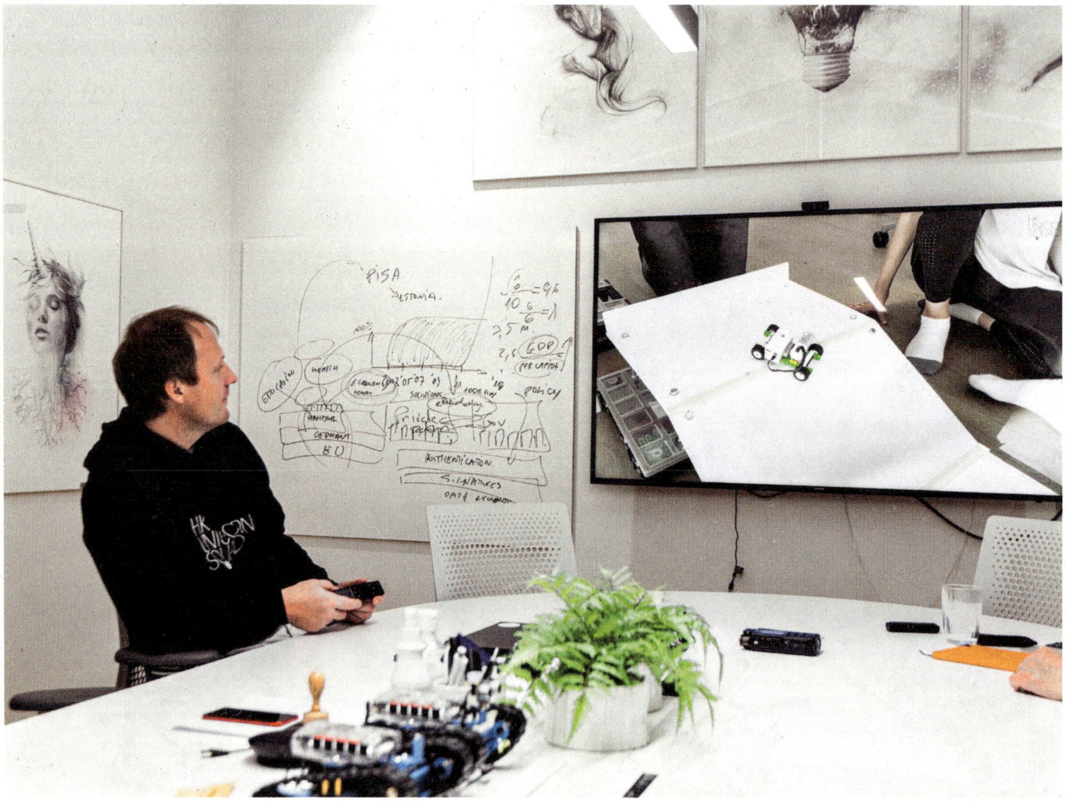

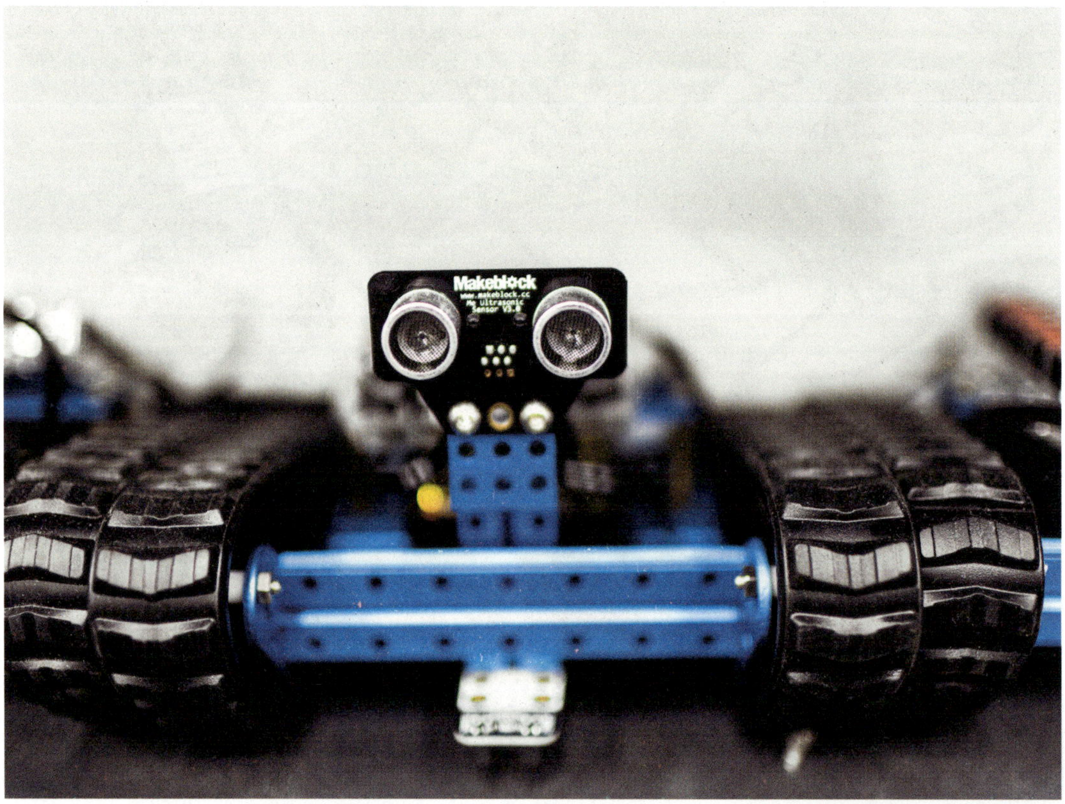
Makeblock
www.makeblock.cc
Me ultrasonic
Sensor V3.0

everything about you as a patient to provide you with the maximum benefit. There might be some funky diseases that you got in Thailand, or, let's say, a young lady had an abortion and wants to hide that fact. You can hide it, but statistically, it's still there, if the country wants to know how many abortions there were in 2016. Having data connected actually protects the data more, because you have control. It's not about data connections, it's about the control. That's one thing that everybody misses.

Why were the government and civil society open to your ideas? How could you convince them to build meaningful software for this infrastructure?

The idea of having personal identity numbers for every citizen was actually invented in Sweden in the 1940s. We copied a huge number of solutions from Finland and Sweden in Estonia. We have historical relationships with those countries. Finns are like brothers because we speak a similar language and emotionally we have the same values. The Swedes and the Germans have ruled us, but there was a moment when we actually passed them up, because we continued to innovate and try out new things.

Estonia's e-society—including you as the country's first Chief Information Officer—is still the benchmark in Europe and in the world. How did this come about?

There's a very simple answer to this. If you ask a German engineer to describe what a digital society looks like, he or she will explain to you exactly what has happened in Sweden, Estonia, and China. Because from an engineering perspective, you can't build it differently. I have studied all meaningful digital societies in the world, and I can describe what has happened in Chile, Japan, Australia, Canada, and the US. They all struggle with the same things. That's why we say it's a playbook—we wrote down the 126 steps.

Every country could do the same.

Exactly, but in Estonia we needed to become a digital society due to the fact that as a country consisting of a huge amount of land but not so many people, we have to be able to make it. There is no such need in Germany, where life was good before digitization. For me it's super hard to explain to a German why a digital signature is so useful and why this infrastructure is needed.

Life in Estonia is so much better now than before.

We started in 1991. In my generation, there is no person who can say, "My entrepreneurial father said ..." because there was no entrepreneurship in the Soviet Union. We basically had nothing in 1991, and now we are quickly approaching Finland and still growing. I think we are number twelve in the world in terms of cost of living and income. Our average salary is twice as high as in any other former republic of the Soviet Union. We all had the same start, but we focused on digitization.

Did you get any advice from McKinsey, Ernst & Young, or any consulting companies?

We did. And we said screw you—because McKinsey gets advice from us. The problem is that the national IT budget in the UK is twenty billion per year. It's two hundred million in Estonia, which is one hundred times less. But the functionality of our hospitals is ninety-five percent, the same as German hospitals. You may have more hospitals, but in the age of the cloud, it's just a question of scaling. We are much more efficient because of our core.

You built it from scratch.

What do you mean with "from scratch"? The Soviet Union was behind Europe, but it wasn't the Stone Age. The first man in space was a cosmonaut from the Soviet Union.

Yes, Yuri Gagarin!

Every Soviet Union republic contributed something to the Common Purpose. Latvia was extremely good in electronics, but software engineering and computer science were Estonia's contribution. Silicon Valley didn't start in 1970; it started here in the 1930s. We already had generations of good IT people because it was our job under communism. There were quotas in communism, and our job was software engineering and computer science. The first Soviet personal computers were built in Estonia. So it wasn't from scratch.

Estonia became independent in 1991?

Our country gained its independence in 1918. So it's two months younger than Finland. We were occupied for fifty years, from 1941 to 1991.

And then you flourished because you connected to the Soviet heritage of computing and software engineering.

It's very important that the government trust engineers. So the Estonian government said, "Go and do it." The German government says, "No, no, no, we have to ask the people."

I imagine that you get a lot of invitations from German institutions and companies.

I had a dream that I was in hell. My eternal punishment was that I had to explain how to build a digital society to Germans and Swiss for eternity.

You were also an advisor of the European Commission. What kind of advice do you have for German society?

Listen to your engineers. I mean, it's the same thing in the German car industry. I am trying to explain the benefits of self-driving cars for climate change, and you say, "They have a wheel, but in our society, the wheel is prohibited. You can't use a wheel." That's like asking an engineer to build a car without a wheel. That's my position. We have met with Merkel and all kinds of heads and presidents, and I have had hundreds of lectures in Germany. I have done consulting for Deutsche Telekom on a big scale in 2018, and they were also here last summer. They came to Estonia with their top management, but there is a huge difference between what should be built and what makes money. One of the key success factors in Estonia is that all these infrastructure elements, such as digital signature, authentication, and unique identifiers, are free of charge or quite inexpensive—like twenty-five euros for four years—which means that the entry barriers are only behavioral. It might be complicated or difficult to understand, but it's not because of money. You have to be patient with those free forums. Australia has tried twice to do what we have been doing. Both times after four years when the government changes, their new prime minister does away with everything that was done by his predecessor. They are rich country, so it's super simple for them to grow. I think we started working on their national ID card in 2002. Even though it was mandatory, it wasn't accepted. So, ten years later, this is what you call a CIO nightmare. You can imagine the blame we got from journalists every year. They didn't understand why we developed this plastic card that doesn't solve anything. Nobody used it, there were no services for it—a total nightmare. And then in 2018, we actually bought chips from Germany. One of the series of chips had mathematical flaws, which meant that we had to cancel all ID cards for that series. It was hundreds of thousands of people.

Germany failed again.

No, it was a private company that produced chips in Germany. So our prime minister immediately canceled an official trip to Poland and returned to Estonia for a press conference. The whole country was in the panic that they wouldn't have access to our digital society. It's the same situation as your bank cards not working. I mean, you can still use cash, but the bank cards are out.

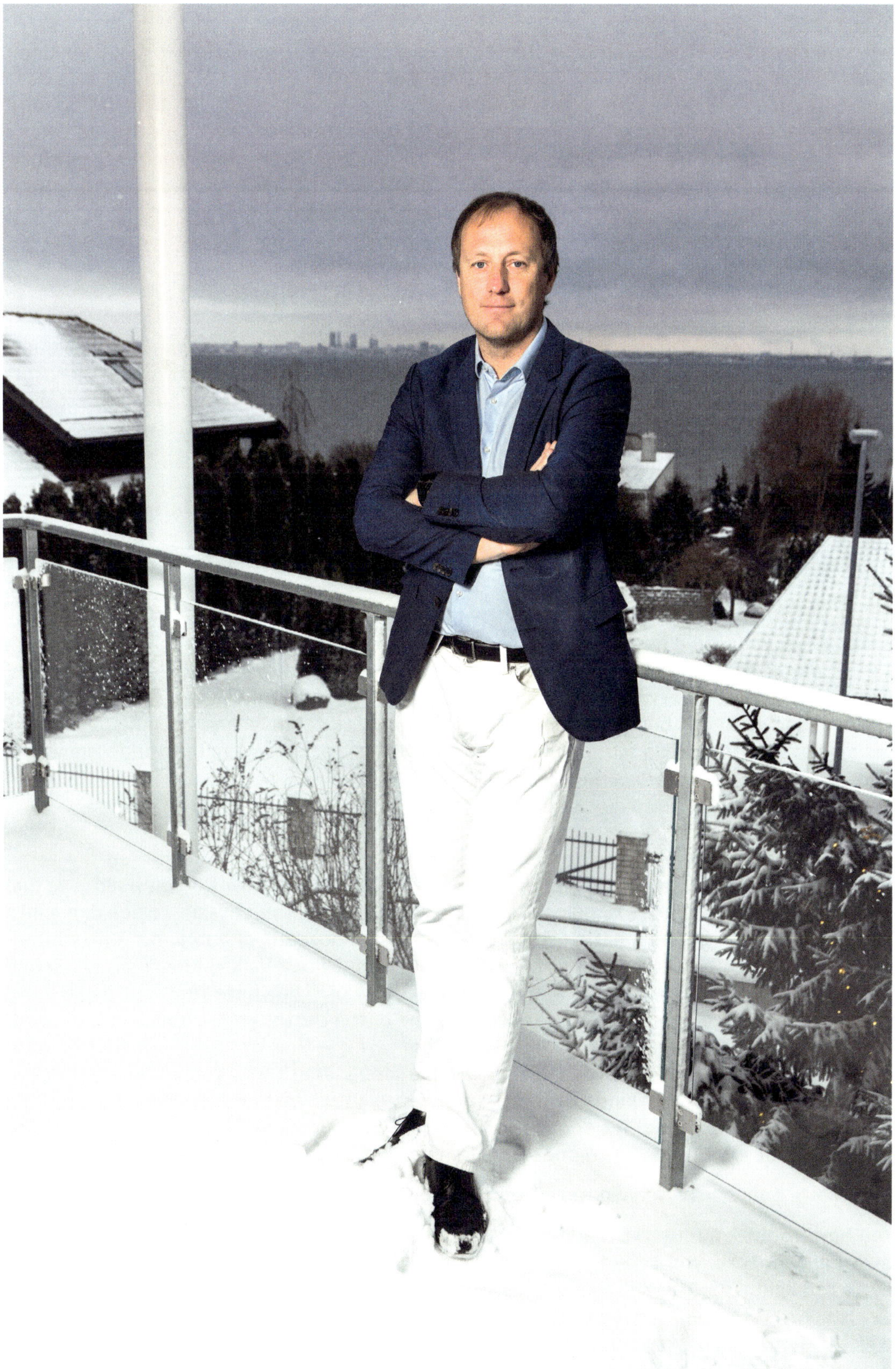

How did you fix it?

We cancelled those cards, and we started to change them. But it's like losing your debit card. You have other credit cards. We don't rely only on the national ID card. We also have smart ID, mobile ID, and other tools. Most people have several tools. So, even if one fails, they have a backup. We started to play with that idea, especially with the Crimea events in 2014. Estonia has been occupied every century, and we have had a war every century, even this century. It was a cyber war in 2007, but it was still a war. So we started to play with the idea of how we could survive as a nation if we lost control over our land again. It is not only built on Luxembourg. Luxembourg is the public news that signals to your opponent that you're prepared. We actually keep it in twenty places on the cloud.

Is it on an Amazon service or on Microsoft?

It's a national secret. There is a hugely complex problem with privately owned public clouds. It's exactly what you see in Russia at the moment. Putin tells Apple or Amazon or Microsoft that they can sell their stuff in Russia, if they give him the keys to the service of the Estonian data embassy. I mean, obviously Facebook and Microsoft will say no.

They did it in China. So they will do it in Estonia as well.

That's my point: it's a business decision. The same thing is happening in Germany now. Having gas becomes a business decision. You need to heat your home, so can you do without those Eastern European countries, or do you stand up and pay triple?

We could buy American gas.

Exactly, but you would have to pay triple. So we played with that idea. It actually goes well if you can reboot, but can you run it from there? If your country's occupied, how can you assure that events in an occupied country are actually free will? Let's say, Russia occupies Estonia, comes to this building where we are having this interview, and tells me to sign this building over to them. I do so digitally. How can you be sure that I wasn't pushed? The whole concept of virtual country is possible.

Only in freedom.

I have done some consulting for Catalonia. They want to be free. They asked me as an engineer, and I answered as an engineer. I'm not politician. So how do you build a virtual country? We have positive examples. If you think about Israel, the physical region came into existence in 1948. But as a nation, the Jews survived thousands of years. It's a very interesting concept. When you hear people, especially from Silicon Valley, talking about virtual countries and virtual states, they were all educated by us. They have read our papers and are familiar with the concept. Balaji Srinivasan, one of the best crypto experts in the world, is actually a very close friend of Estonia. What I am saying is that it is super simple to create a virtual country in the age of blockchain and the cloud. That's why you could say that the Metaverse and any other kind of virtual society are super interesting. Because creating country is not about land, it's about people agreeing or being willing to work together or believe or obey the rules. So you need certain people, but you don't need land.

But virtual countries are always a vision of billionaires to get rid of taxes and rules.

You have to think about a virtual country like a game. Even the game needs rules. If you're a Jew in New York, you obey the local rules, and you pay local taxes. There's a green light and a red light. They obey all those rules. But on top of that, they have another layer, called Jewish laws. So you have to see the world in this kind of mixture. You can add extra layers, which doesn't mean that it is a different part of that virtual country. Let's say we have our own education system, our own health care system, location independent, access to the best doctors in the world. But we follow the orders of the physical world, where we are physically located. If I am in Finland, I follow Finnish laws. In Estonia, I follow Estonian laws, and in Germany, German laws. You have to see the world in a more layered way. That's the outcome of data embassy project. That's how to build those extra layers. So the Catalonians can culturally and virtually define themselves as a nation, while still obeying the Spanish rules and paying Spanish taxes.

They want to have their own currency. So they can have a digital currency—

That's the beauty of blockchain. What happened is that there was a huge number of frauds. So a notary layer was developed to guarantee that it's not so easy to rewrite the contracts. Everybody knows their rights and stuff. And today, thirty years later, this site is so mature that it doesn't make sense anymore for everybody to bring their own lawyer and we visit a third lawyer. There are hundreds of people related to this sector, so it's very hard to remove it. We have already ditched notaries.

You were the Chief Information Officer of Estonia for several years, and then you stepped down to make a private company again. How are they running your baby?

My project is e-residency. You can become a digital Estonian, and many Germans have become e-residents of Estonia.

It costs only seventy-five euros.

It's slightly more, but one third of all companies created last year were by e-residents. So our virtual population is growing as those people connect their companies with us, and our economy is growing. It's basically solving the immigration problem. Those people want to come to Europe for a better life—although weather-wise, it's not better. The weather is actually better in many African countries, but they don't have the infrastructure. So what if we give them the infrastructure, but culturally and environmentally and climate-wise they can stay where they are? We just provide them with the infrastructure. If you provide them with access to the market and tools, this kind of digital society could make Europe very strong, and you could partially solve the immigration problem. We give them fish, but we also show them how to catch the fish. That's the thinking. My embassy project still depends on physical appearance. You have to give your fingerprints, and we have face recognition, but it's not progressing as rapidly as I had hoped. I also have a zero-bureaucracy project. Every company has an accountant who reports to the government, and the job of reporting doesn't add any value to your product or your service. It's something extra you need to do because of the government. Can we fully automate that? In this country, we can. We have seen that it's possible. Imagine running your company without having to report anything to the government, because the government automatically receives that information. That is how we killed VAT fraud in Estonia. Companies started to provide more data automatically, and the government will build AI robots on top of that. It was significant. We lost two to three percent of the national budget every year because of VAT fraud. The same fraud still exists in Lithuania, and they are losing billions. According to the Organization for Economic Cooperation and Development, Estonia is the best tax collector in the world. Everybody pays proportional taxes according to their income. We have flat-rate taxes, and the company tax is zero.

So in your vision, we will be global nomads. We will have the freedom to live wherever we want.

I still believe that states, borders, and different jurisdictions will remain. Let's take the city of Hamburg. Hamburg is part of Germany, which is part of Europe and the EU. So we have different layers already. There are some EU directives that you have to obey at the same

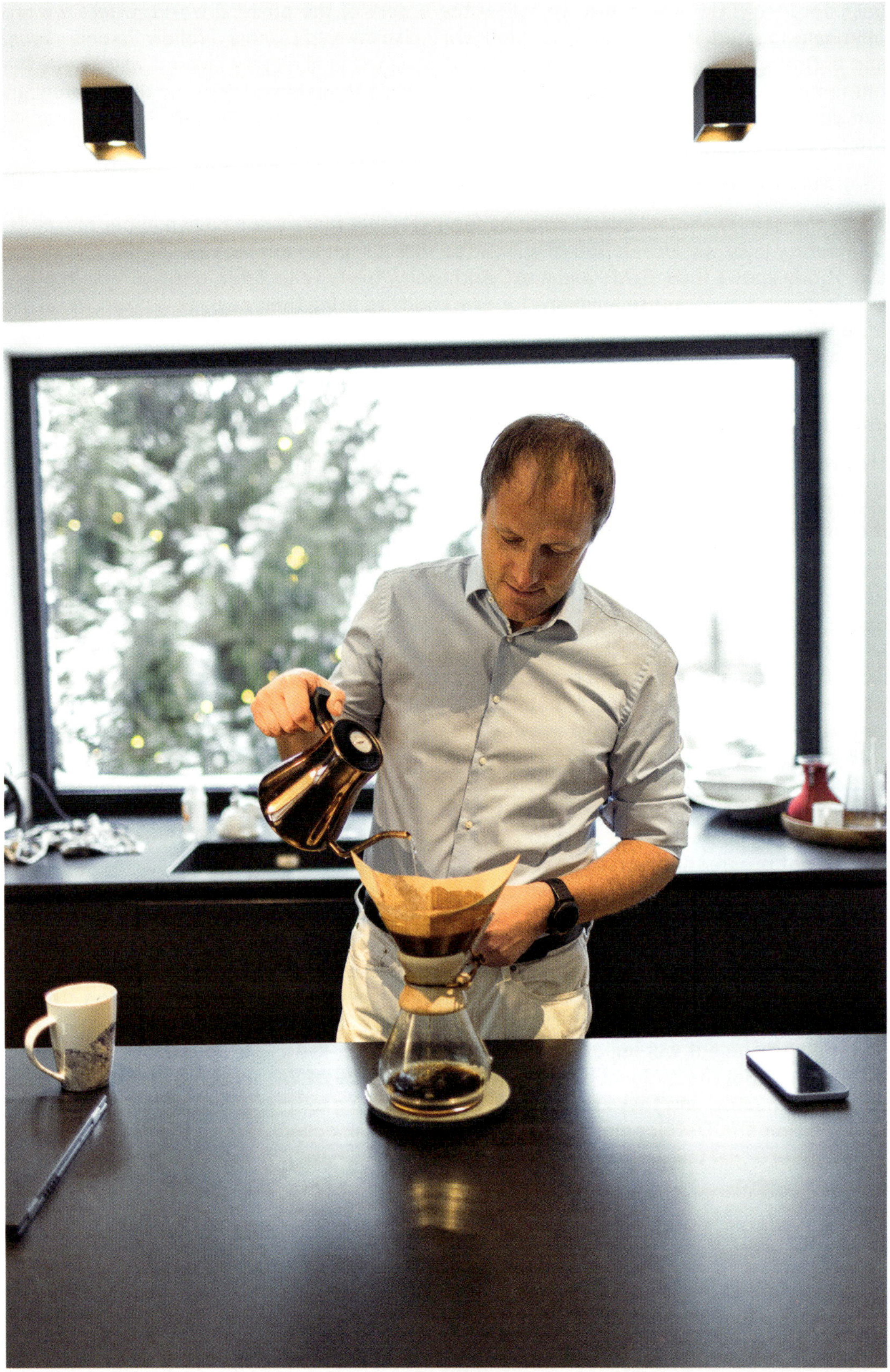

time as you have your local rules. And you have the local municipalities on the border. My point is that this will all remain. But what does this structure provide for me? Historically it has provided me with an economic environment, health care, and education. It's a monopoly. If you don't like the school, what do you do? It used to be that you had to move to go to another school. Today I can get an MIT degree without leaving Estonia. So if the education is better somewhere else, why should I buy it here? It's a service. What if I buy it there? Running a company in Hamburg costs more than ten thousand euros per year. Here it only costs one thousand euros per year.

I still have to pay taxes in Hamburg.

You also have to pay taxes here. But the difference is nine thousand. You still pay taxes, but accounting, salaries, and bureaucracy costs are so much cheaper. You can still have your German bank account; you don't have to have an Estonian one. Which is cheaper: Switzerland or Estonia? There will be more competition. Who provides what kind of service? You can look it up. It's a national PISA test. Who do you think is number one in Europe?

Estonia?

If you want your people to come back from other countries, you have to have top-notch education and health care. That's what they care about. You can run your business in Panama, but education and health care are important. If those things are not good, your immigrants won't come back. That's a problem in Latvia or Lithuania. Lithuania used to have a population of 3.5 million. Today it's 2.6 million—they've lost one Estonia in thirty years. If you ask me what I'm proud of, it is that we are basically changing the whole internal economy. Estonia used to be an agricultural country because St. Petersburg, the capital of Russia, is one hundred kilometers from our border. The capital was moved to Moscow only last century. Estonia was an agricultural country, and everything from vegetables to vodka and bread was from here. Today fifty percent of the country is covered with forest. It used to be less than thirty because of the fields. But when the Soviet Union occupation began, the rich farmers were actually in Siberia, so nobody took care of the fields anymore. Today our agriculture cannot compete with southern countries. That's why we expect to have around thirty percent of our economy connected with digital IT and start-ups by 2030. It's a total change.

It's generating real jobs here in Tallinn.

Yes. We have a shortage of people in Tallinn in the IT field. There's an international study that says that every high-salary job creates three to five low-salary jobs, such as hairdressers and baristas. This has happened because we have invested for the past thirty years. So we are on track with my baby. I'm very happy about that.

Now you have your own company and you're consulting. You are also creating an e-society in India with an Indian billionaire. So you went from one million to one billion e-residents.

He is also an e-resident, and he pays the heat and electricity of this building.

Tell me about your new vision.

We have already been doing it for three years, and we are starting to see the first changes. But there's one thing you need to understand. We've been using the internet for twenty years, but in India, the average person got access to the internet in 2016. That's only five or six years ago. So India has only just started to see what we have seen and experienced in the past twenty years. Mukesh Ambani basically built a countrywide mobile network. India is bigger than Europe. Imagine somebody building a new Deutsche Telekom and putting up masts not just in Germany, but all over Europe. They started to build that network in 2010, and they went live in early 2016. Today they have more than six hundred million customers. It's the largest telecommunications company in the world. There was only one idea: to give cheap access to everybody. So even very poor people have phones—not iPhones, but decent phones. And they have access to movies and entertainment in one payment.

It's so important for their business because they make all the payments over the mobile phone.

Their significant income still comes from the petrol business. Reliance—the largest company in India—has the world's largest refinery. But what is important is that it's a new growing economy. And India actually listens to engineers. So India has been doing all those changes that have not been possible for political reasons in Germany or the UK. They introduced a unique identifier ten years ago, and they have 1.3 billion people. The US didn't do unique ID, but India is doing a fabulous job. Money moves faster from account to account in India than in Europe. In many ways Europe lags behind. That was one of the problems when Estonia joined Europe. We had to accept SEPA payments, which increased the time from seconds to days. It used to be seconds from bank to bank in Estonia, but because of the European regulations, it took half a day or something. In India, because they listen to engineers, they're now solving the health-care field. There's no free health care in India. Everybody has to buy their own insurance—you grab your chicken and go to the doctor and pay the chicken for treatment.

But they are producing medicine for us.

Yes, and they don't have enough vaccines for themselves. It's a sad story. But Mukesh Ambani, the richest man in Asia, reached out to me.

How did you meet him?

It's a funny story. His team sent me an email about coming to India and discussing Digital India stuff. Everybody gets those emails from Nigeria that you have won the lottery, so I ignored it. And then one of my friends in Silicon Valley called me and said that there was a rich Indian guy who wanted to meet me. So I said that I would be in London the next week. I had Cisco back then, and when I finished with Cisco, I just stopped by. I didn't imagine that he would come to meet me, but he was there. We had a chat, and a week later I got an invitation by email to come to India with my wife. He has this house in Mumbai, which is the second-most expensive piece of real estate in the world after Buckingham Palace. It's a two-billion-dollar skyscraper. His father built up the Indian exchange and started this petroleum company. Reliance is actually an oil company. Now it's Reliance Jio, which also includes the telecom company. He had a very strong vision.

He is a philanthropist who wants to help his country.

If you ask people there, his image is quite negative. When the taxi driver realized that I was going to Mukesh Ambani's house, it wasn't very positive. But he has given back now in huge quantities. And for me and my team, the most important thing is that our work is influential for many people.

It's about making the world better.

That's his vision. But you can build a bridge over a small creek, and it can be a decent bridge. And then somebody gives you a chance to build the Golden Gate in San Francisco. Would you take the challenge?

It's the "from scratch" situation again.

Exactly. But there's a lot of legacy, especially cultural legacy, and twenty percent of the people are illiterate. At the same time, thanks to technology, poverty has been reduced a lot. I mean, it's a corrupt country. They share rice with poor people, but every layer takes some rice, so not much rice reaches the poor people. Instead of that, they gave everybody a number and a bank account. If you're illiterate, you can use your fingerprint. And they gave those people a small amount of money.This basically cut all the corrupt layers in between.

What a good idea!

Exactly. That's what brought about the whole change. In Estonia we say that you need huge pain to make change. Their pain in India was corruption, and now they have understood that technology is the way to change the country.

Is there a blockchain involved in that?

Not at the moment, but it will be more and more because there are different ways to get it. You need trust in society. One of the seven principles is to build trust. You can build trust if the government is popular with the people. It's like the Nordic people who are descended from the Viking state. When you have such extreme weather conditions, you have to rely on each other's contribution. When I run out of food, the community takes care of me. That's how we survive. That's why Nordic countries trust the government more, as opposed to Italians, who don't like anything to do with government. The same thing applies to India. They have states where there's high trust and states where it is lower. If there is no trust in the government, then the question is can you trust math? Blockchain is about trusting math.

What percentage of the population in India is already involved in this project?

The unique identifiers? Everybody. More than one billion.

Did you use the same tools as in Estonia?

The point is they started this way before we went to India. They probably observed different solutions around the world, and they probably also saw our solutions. But our role in India is to build products like the digital signature.

So you program meaningful software for them?

Yes, but our main reason there is to explain to Mr. Ambani how the economy and society will change.

So you are a kind of think tank for this guy.

The building we are in is a Jio Research center.

You recently did a big cooperation with the university. What is the project called?

"6G." We are hiring for Jio all around Europe.

6G will open up a completely new world with very low latency.

It is important. It basically opens up the world of virtual reality. That's the future, and I would guess that 6G will come to every country. There are two ways to go: you can wait for Huawei, Nokia, and Ericsson to build it, or if you have the money, you build your own. They have chosen to build their own.

It's also important for the defense industry.

They are covering the entire country of India. We are used with 2D maps from school, which show India as quite small. If you superimpose India on top of Europe, you'll see that it's bigger. Imagine that you're a telecommunications firm that needs to cover all of Europe with antennas. Would you buy in or build it yourself? That's a super simple question. The people from Deutsche Telekom helped them.

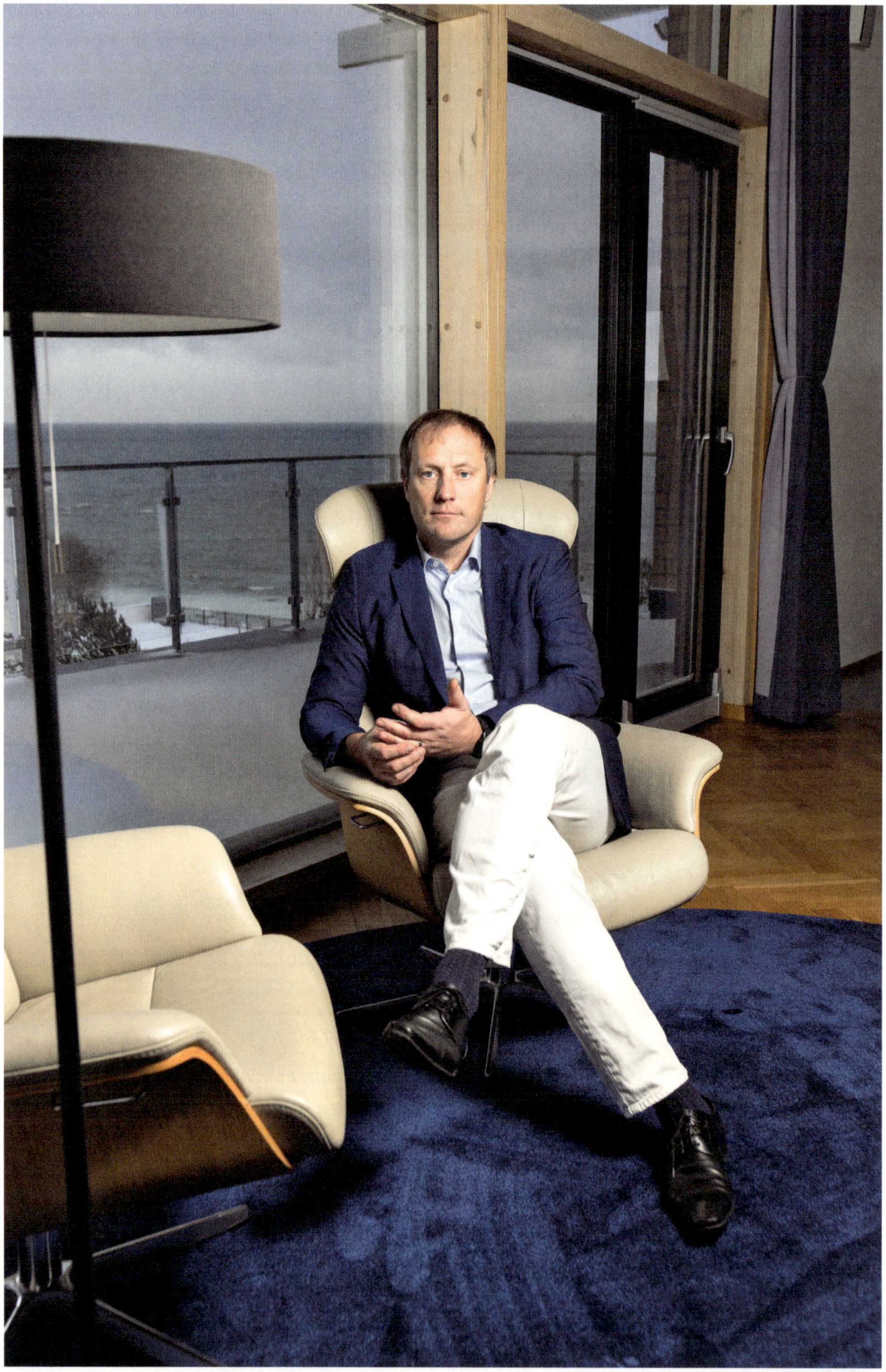

And Mukesh Ambani is planning that.

It's quite patriotic. He basically said to Apple, "Either you build your phones in India, or we won't sell Apple at Jio. If you want access to six hundred million customers ... You do. So build factories here. We will only sell phones that are produced in India. We will put up those antennas that are produced in India." That's the true patriotism. That's actually how China became big. So you can't blame them. This will cause everything—the economy, the brains, and all the significant markets—to shift to India and China, and Europe will be marginalized.

When will you become the CIO of the Republic of India?

Never. There is no need. I mean, India formed the team way before me. There are a couple of Google billionaires behind it. Just remember that all the technology companies in the world like Microsoft, Novartis, you name it, are all run by Indians. They're super smart. The first ones who made their billions more than ten years ago formed a similar think tank and basically started to build and use that money for making India better. They are the real CIOs of the country. We are like pioneers who can sometimes say that if you do this, that will happen. That's all. But the point is, they're building it exactly like we did. Unique identifiers, authentication, digital signature, data exchange, and pilot by government partnership.

So this is a scalable model for the whole planet.

Yes. We need communication experts who can explain this to Germans.

Perhaps we should give all the popular power to the engineers! You are already a consultant for companies such as Deutsche Telekom. What kind of mindset do you need to digitize an organization? What is your advice in general?

Again, there are different layers, if you see this in terms of Maslow's pyramid. You can't have a company nowadays without the internet. You have to have proper tools, proper infrastructure, proper cybersecurity, and privacy policies. You have to optimize your business, improve your sales, improve your production, everything. That's something that every company should do. But then there's another future layer, which is how easily you can embrace change, or like we said at the beginning that we don't want to produce our own body parts anymore. We will outsource it, or we want to have the best doctors join our network with their knowledge and experience. How rapidly can you make these changes happen? That's usually the problem of western European companies: everybody understands the cloud and everybody understands that change needs to be made, but then you look at how many years it took them to actually go to the cloud. I mean, I was already sick of the word *cloud* in 2008. How many years can you just talk about this? Just do it! And now thirteen or fourteen years later, people are starting to understand the cloud and use it to store their photos. Changes take time. Even a year ago, I think everybody laughed when someone said that Tesla would be more valued than Volkswagen. Nobody laughs now. So one company saw it, made a change, and is now number one in the world. The same happens in all industries. It first starts from more obvious ones like financial services and car industry, but it eventually happens everywhere. Also, your sponsors build. Then the question is, how ready are you? I mean, Kuehne+Nagel has a development center here in Tallinn. They're responsible for all of Amazon's delivery from the US to Europe. It's huge. They are very tech-savvy company.

Do you also consult for Kuehne+Nagel?

No, but they have visited us, and we've had these small face-to-face sessions here. They are huge Estonian fans. Kuehne+Nagel is also in Hamburg.

Maybe it has something to do with the Hanseatic League. They now have their headquarters in Zug, Switzerland. Mr. Kuehne lives there, and he is a Swiss citizen.

Zug now wants to become the center of crypto and blockchain. You can't be a pioneer if you don't test it out somewhere. One of the benefits we will have in Estonia is that all the software, all those processes, are made by Estonians. So basically we used our private sector and government project to train and coach our own developers.

So you are the world leaders because you have every software through the Estonian back-door?

Yes.

You can take control of everything.

Definitely not. I mean, Israel is way better at this. They start by thinking about what kind of backdoors they will build. In Estonia, it's different. NATO has the Cybersecurity Center here. So we don't do backdoors, although ten years ago, a Russian living in Estonia had the largest bot network in the world. Also, before Skype, there was a file-sharing platform called Kazaa, which was developed here. The same team built Skype, and Skype actually uses the same technology.

My last question is about the Unicorn squad, your project for girls.

We have invested quite a lot in Science, Technology Engineering, and Mathematics (STEM) education in Estonia. Robotics lessons are supposed to be held in mixed groups, but if you look more closely, you'll find ten boys and one girl per class. The concept doesn't work for the girls, because an eight-year-old boy is at least one and a half years more advanced than an eight-year-old girl. They don't want to work together due to the huge gap in knowledge. My wife and I got the idea for the club when our daughter and another girl were thrown out of their school's robotics club. That's how Unicorn Squad was born. Girls now have a better environment in which to learn about technology. They prefer classes with no boys, so we removed the boys. We have six courses, which explains how a gearbox works or how to build a car. Every girl in Estonia gets a hoodie for free. The exercises are also part of the national STEM competition. There's a national competition with the fifteen smartest children competing for number one in science. They use our exercises, and it's also used in a national TV show. A very famous Estonian physicist explains what is important, what are the mistakes to avoid in a design, how you should build the engine, and everything. We have trained thousands of children.

So you lowered the barriers for the STEM professions.

I developed the exercises myself. As a programmer, I know what they need to study. Everybody can repeat those exercises at home.

And you give something back to society.

Yes, but there are a lot of private initiatives like that. Great ideas and changes do not always happen thanks to great governance but often because we are fortunate to have people in our society who really care and want to give back. People who notice the trends and are willing to invest their time, money, and energy in finding the best, most efficient, and smartest solutions.

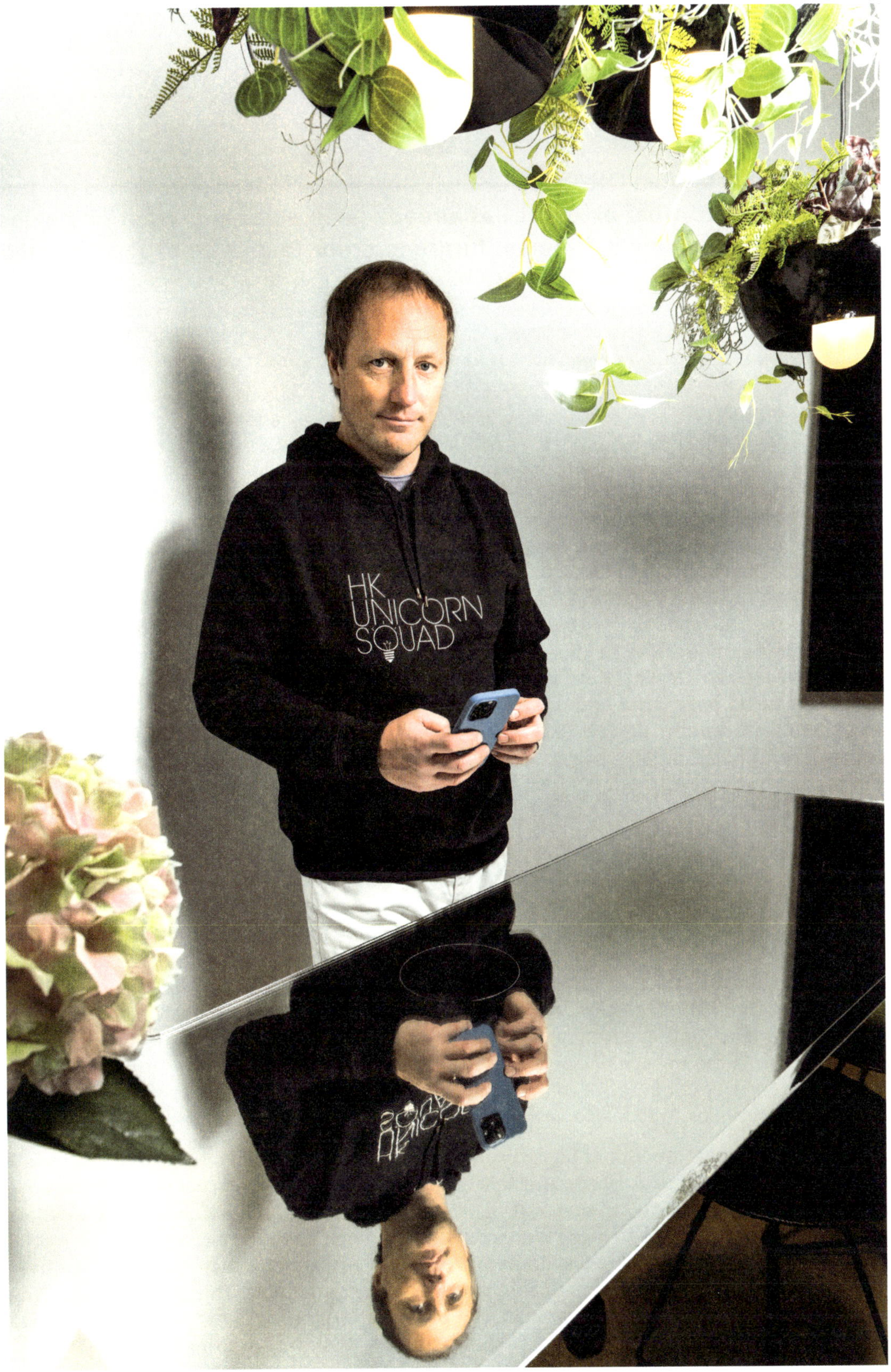
HK
UNICORN
SQUAD

MARTIN BÖHM, HANS DIETL, ANDREAS GOPPELT,
BERNHARD GRAIMANN, AND GÜNGÖR KARA

Reflections on Digitization

The five experts from Ottobock discuss how important digitization is for the company and to what extent it has already been achieved. What obstacles need to be overcome in the near future in order to digitize production and customer relations?

Andreas Goppelt: Hans Georg Näder's initiated a complete digitization of the entire company very early on. He was a pioneer in bringing the first microprocessor-controlled prostheses on the market. However, digitization does not only involve products and overall production to the greatest extent possible, even if it still contains elements of manufacturing, but also all processes including patient care. The basic principle of digitization also always includes getting to know customers and their needs better, speaking to them individually, and familiarizing them with the company and its technologies, and also including them as early as possible in the development and design of technical innovations. We've had a lot of experience with networking data. Since all our mechatronic products already have built-in connectivity, it was no problem for us at all to make a connection to the cloud. Our platform MyOttobock gave us a small sense of achievement. We were positively surprised at its popularity among our customers—there are already over five thousand users. We were amazed at how readily people share data with us, which we naturally take great care to protect. This is why Ottobock already has a considerable head start in the digital transformation.

Martin Böhm: In his interview Taavi Kotka graphically illustrated the cultural aversion in Germany with respect to digitization, and in particular data protection. At Ottobock there was also initial resistance, as is the case with every transformation in a complex organization. We demand new capabilities and a rethink from our employees. We are now beginning to think in terms of customer journeys, digital platforms, and ecosystems, and establish digital points of contact with our patient applications. We first need to lay the foundations for a dialogue with our users, whether via our website, social media channels, or sales platforms. Other high-tech companies are quite advanced in terms of customer centricity, but the customer journey often ends at the digital/analogue interface, for example, at the retail market in the consumer goods industry. The fantastic thing about Ottobock is that we cultivate a very cooperative relationship with the medical supply stores, hospitals, and rehabilitation clinics, which also have a great interest in a continuous patient journey.

Güngör Kara: In the process of digitizing Estonia, Taavi Kotka really changed all the structures, holistically and in a vigorous fashion. That can serve as a good role model for us. We first digitize our products, then our processes, and finally set about changing the whole network. For this we have to make numerous adjustments, so that the company is oriented to the digital future while opening up new paths. Like a pharmaceutical company, we have to have contact with the patients as early and as closely as possible. This patient integration and bonding ensures that we receive a lot of data and insights and that we will understand them a lot better as health customers. That is actually the most exciting thing.

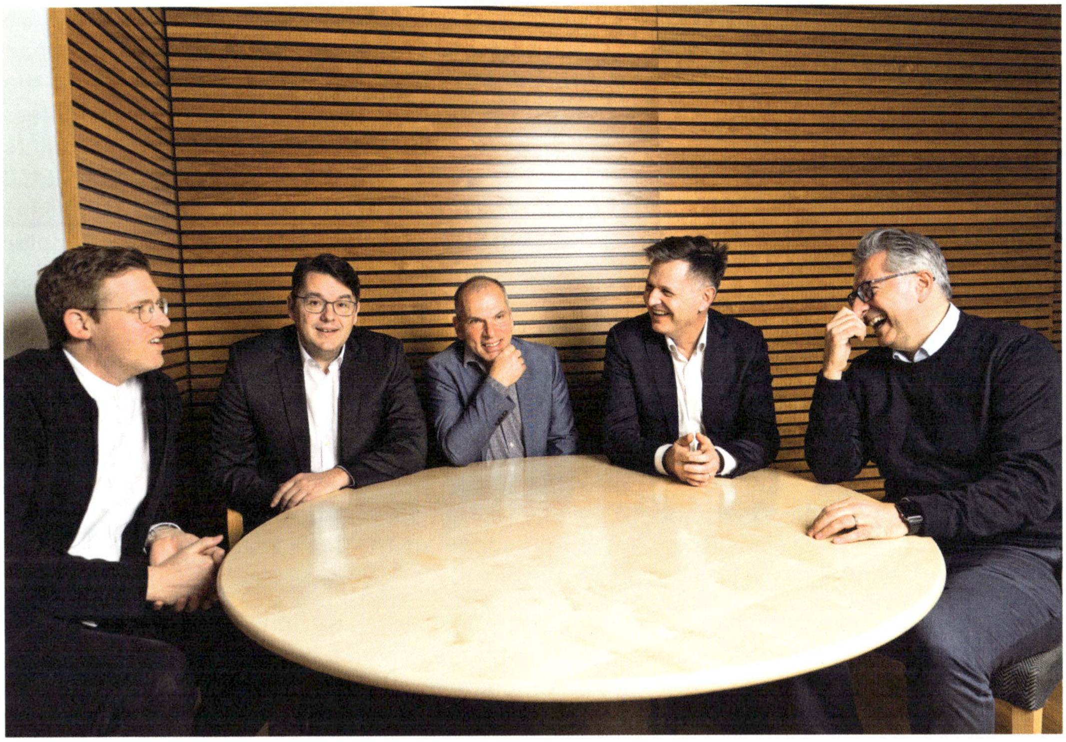

Martin Böhm, Bernhard Graimann, Hans Dietl, Andreas Goppelt, and Güngör Kara (from left to right)

Then things get quite complex. We have to use the customer data intelligently, in order to make the customer journeys into a unique experience with Ottobock. That begins with specific information that only we have. The next step is to offer services, products, and follow-up treatment as well as an exchange with like-minded people so that patients who are confronted with drastic restrictions in their mobility can gain new courage.

MB: This dialogue is very important. The interesting thing is that not just the people affected talk about their problems, but the entire community can also participate. How can one regenerate after an operation? How does it feel to walk on a test prosthesis? How should it feel? How far can one walk? What goals can I set myself? The discussion in the subgroups is very intense. However, instead of doing everything ourselves, we need to provide an ecosystem that develops into a platform with the relevant content and services for this dialogue. For example, we have started a fitness course with a trainer who is also a leg amputee—and this is very well received by the patients.

GK: Digitization is not just about a digitized customer relationship. The experiences that are gained through closer customer contact must be fed back into the company and product development. It is also about the digitization of production and indication-based product solutions. For example, for orthoses we set up a scan, modeling, and print-file chain with a connected 3D printer. Our iFab 4.0 is an innovation project through which we can run—like a conveyor belt—the entire chain of different indications and products. We have developed our own high-performance scanner the size of an electric shaver; we have our own modeling software and can use our own printers or those of other manufacturers. While other suppliers only offer partial solutions, we offer a complete solution in which everything is coordinated and ready to use. We have developed the software to generate a complete mesh, that is, a virtual wire mesh to provide a tailor-made image of the head region with high-resolution data. For example, with the children's helmets we have configured the scanner so that it doesn't just scan the deformation of the upper head form, but also carries out an automatic facial recognition at the same time, so that the data can always be unequivocally assigned to the respective child. If the child is restless during the scanning procedure and moves its head, the scanner always finds anchor points in the face for the digital "wire mesh model," enabling it to complement it correctly. Using artificial intelligence this can be dynamically recalculated. Orthopedic technicians need very expensive scanners and computers for this; with our solution it is a lot simpler for them launch digitization, and that at a good price-performance ratio. You won't find this anywhere else. Our goal is a network process from which we can extract data that we then bring together in a meta-AI. The central intelligence is in Duderstadt; the scan is globally independent.

AG: Digitization makes the regionalization of knowledge possible. That is essential for consistently achieving high-quality standards. In the various regions there are very great differences both in the training of technicians and in the service. Few countries operate at such a high level as Germany. Guaranteeing this consistent standard is a strong motivation for digitization, as is the availability of trained personnel or the cost-pressure in the healthcare system. In this light, everyone has to move.

GK: That is also an advantage of our system: the modeling can be done by people anywhere in the world. We repeatedly bring the data together, extracting and analyzing it—and optimizing the quality worldwide. If we have a complete digital twin of the supply process, we can, for example, improve the software on the scanner, when fitting, when modeling, and with respect to the individual 3D-printed prostheses or orthoses, so that we can print more individualized components and thus reduce the cost per component. To date we have global printing and manufacturing centers in Salt Lake City and Duderstadt. In the future we will position ourselves more regionally with production hubs, and in a next step we will be really present in the individual medical supply stores and shops on a decentralized basis, as soon as the costs for 3D printers and materials make it profitable. This will be a huge step toward mass customization. This is a decade of transformation, and many of our customers are going to increase their use of digital solutions while still using some analog processes for special cases. The consequence of this is that we will still have to offer some analog solutions so that our customers can sequentially integrate digital solutions. That offers our customers the chance to decide themselves at what point they would like to sequentially switch to the digital tools. That is the challenge with innovations: transforming existing successful business models into a completely different system.

Bernhard Graimann: Digitization as a consequence of the interconnection of digital infrastructures and communication technologies has been radically changing the markets. Amazon, Apple, and many other innovative companies have already impressively shown how the possibilities of digitization can be applied for the benefit of customers. We have to understand the possibilities of this interconnected digital world and use it to the advantage of our customers and patients. If we don't do it, somebody else will. It is therefore our job to advance digitization in our business sector. A particular challenge is the shift from digital to analog. The convergence of various technologies such as digitization and 3D printing helps us to do so.

MB: Our goal is to digitize this whole landscape of customer relations. First of all, an ID has to be assigned to each patient so they can be clearly identified, ideally in a comprehensive set of patient records so that all the collected data can be used by each partner. This makes a program such as Virtual Remote Assistance possible: if someone has problems with their prosthesis, they can show it to an expert virtually, who then takes pictures of it and passes

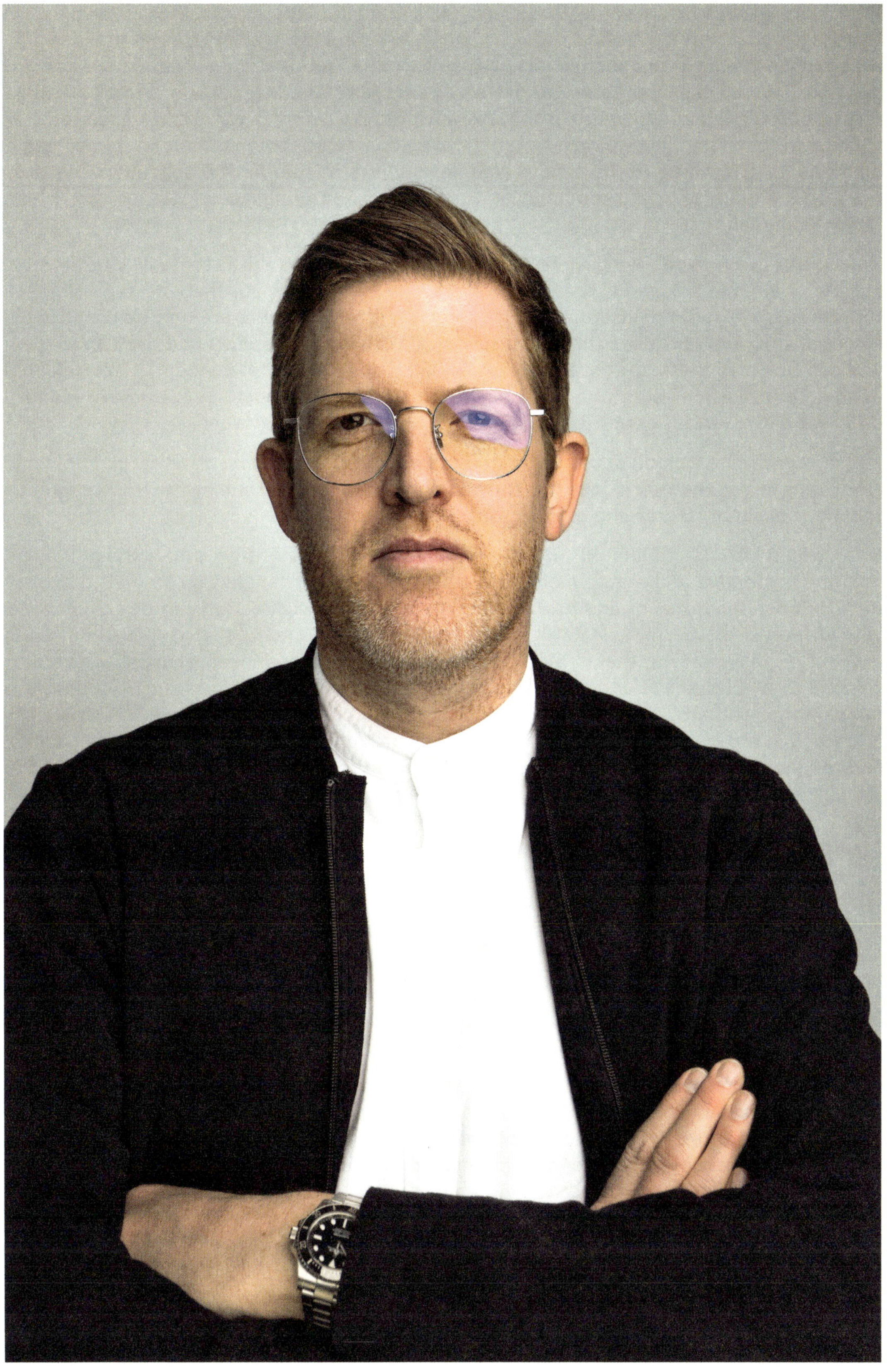

Martin Böhm, Chief Experience Officer, Ottobock

them on to a service team in real time. Everything remains in the system. We still have a ways to go: Ottobock tended to be product-centric, and now we want to be patient-centric, which means we have to build this dialogue in customer services too. Whether the patient turns to the medical supply store or us—we are always there for them. In order to digitize you first have to standardize the processes worldwide. Customer service is a huge topic at Ottobock; for a long time we didn't even have a global telephone number. So we still had a lot of cleaning up to do. To be able to digitize you have to standardize procedures all over the world. We are prone to reducing digitization to 3D printing, but this is just one piece in our whole mosaic.

Hans Dietl: In any event, we have to begin with the patients as this is the only interface we have in common throughout the company worldwide. Apart from that the supply structures are very different. In Germany, Austria, and Switzerland there are only medical supply stores. If you want to digitize them, you have to start with the patients and think about how you can optimize the respective value creation chain with digital concepts. This can't be done uniformly across the globe because the situation and the customer structure are very different in the various countries. The patient is the central interface.

MB: We can create a modular patient journey. It will be a combination of digital and physical. The strength is in the hybrid journey. The exciting thing is strengthening the involvement of the medical supply stores and orthotic and prosthetic shops.

BG: I would also like to link up with the patients by extending them into a triangle along with the customers and reimbursement systems. Using this triangle we can consider what digitization actually means and how it can generate value. Digitization can be summarized with three words: perfect, instant, and free. Digitization means that the information is always available, perfect, that is, always the same, and the marginal costs approach zero. If we have a comprehensive understanding of the jobs to be done and, deriving from that, the needs of our patients, the customers, and the reimbursement systems, we have the right approach in order to generate innovations. Digitization, in combination with other trends and technologies, offers a multitude of possibilities for this. Big data, algorithms, artificial intelligence—all these trends are merging together and will decide over competitive advantages. Operative excellence, privileged access to resources, economies of scale, strong branding, customer retention, and superior technologies—these are the six dimensions of competitive advantage. In the past it was primarily technology such as the C-Leg. With digitization we now have the possibility of not just improving the technology, but placing the customers center stage, intensifying customer loyalty, showing the brand in a new way, and developing operational excellence. With the exception of access to resources, we can address all dimensions of the competitive advantage.

HD: The most attractive thing about Ottobock is its global market access in the prosthetics sector.

MB: For a long time we didn't really recognize this. In addition, we have a highly unique brand with a highly valuable purpose. We are starting with a completely new branding that is built on our DNA, and we already have profiles of one million users, with whom we can immediately enter into dialogue. We need to improve significantly, especially in global marketing and sales where we have to present a more consistent face. The customers assume we are digitized and expect everything to run well with Ottobock. In terms of the aforementioned issues, designing processes, having an IT platform that functions consistently worldwide—we still have a number of things to do. That is a real task for us.

AG: But there are enough examples of how digital platforms have disruptively disempowered global players overnight. That is a warning that we take seriously.

HD: We are the best counterevidence for this. In the area of orthotics we have incredibly good products, but we are not the undisputed market leader. It is actually much harder to become the leading supplier on the global market than to buy this or that technology. For me global market presence is the key to success.

BG: For the decisive competitive advantage it is not enough to have a certain technology. You have to have superior technology and the relevant competence. By definition, core competences are very difficult to acquire. Our microprocessor knees (MPKs) are a good example. The combination of high-performance hydraulics, biomechanical knowledge, and control algorithms form a superior technology that is based on our core competences. Digitization alone or in combination with other new technologies promises new and better solutions for the problems of our customers and patients. As a company we accept the challenge to develop new core competences in exactly this technological field of tension. We are already on our way, and I am very optimistic that we will continue to considerably expand our competitive position.

Quantum Technologies

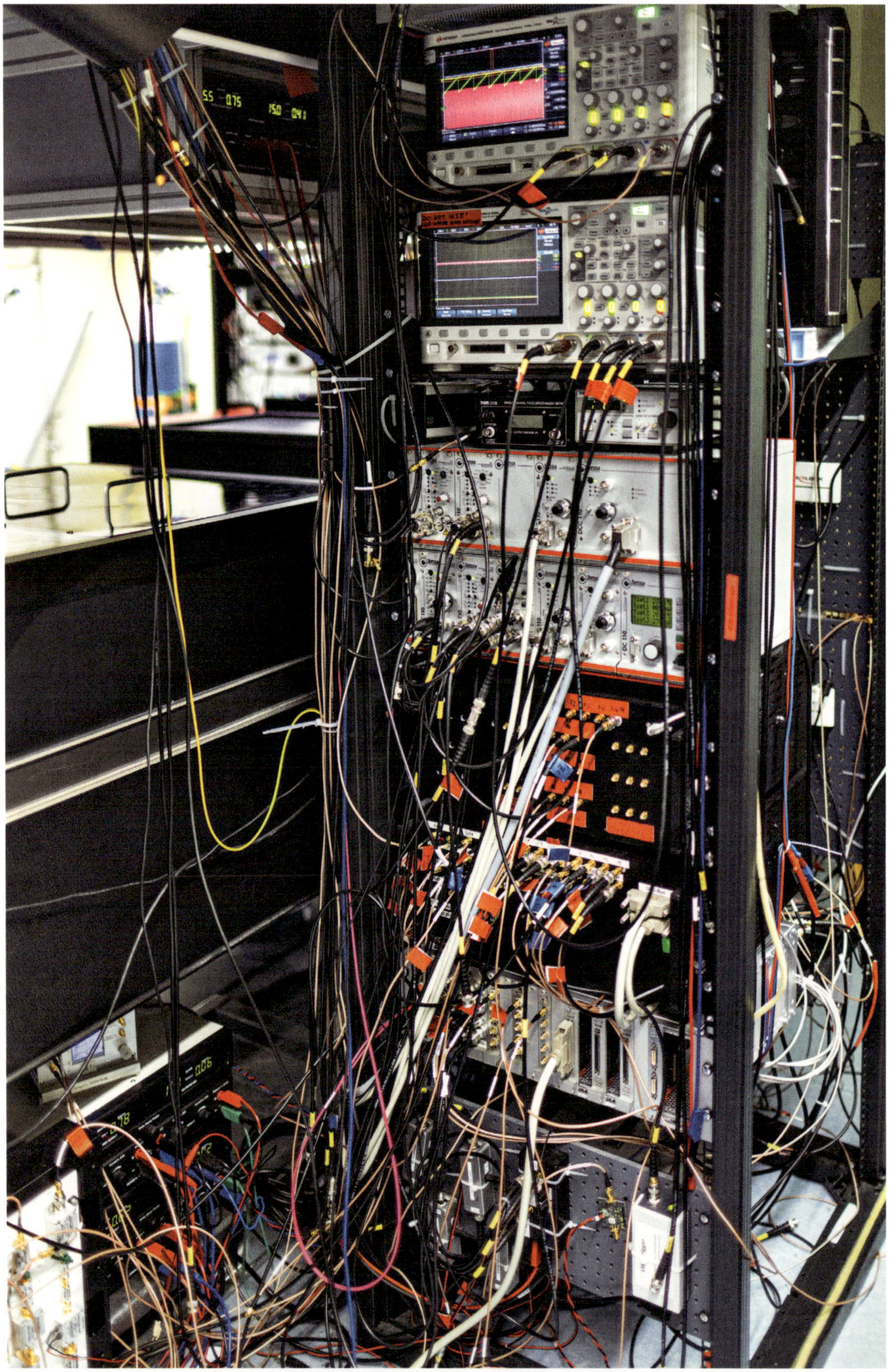

PIET OLIVER SCHMIDT AND CHRISTIAN OSPELKAUS

The Revolution of the Ticking Atoms

German civil servants, of all people, are now competing with major computer corporations and the geniuses of Silicon Valley to develop a quantum computer. These two experts at the excellence cluster QuantumFrontiers conduct research into the relationships between space and time at the atomic level. When they manage one day to control an atom's quanta more precisely, it will open up undreamed-of possibilities for science and industry.

Professor Ospelkaus and Professor Schmidt, quantum computers are considered one of the most promising technologies of the twenty-first century. How can it be that an agency such as the Physikalisch-Technische Bundesanstalt (PTB), which is part of the Quantum Valley Lower Saxony, has set a goal of developing a quantum computer that is superior to all current models by 2025? Is it true that your expertise in quantum mechanics is due to your decades of experience working with atomic clocks, which function on the same basis as the quantum leap?

Christian Ospelkaus: In the context of our research on the atomic clock, the PTB has conducted high-level investigations of ion traps and frequency standards for many years. To build a frequency standard you basically need an extremely well-controlled quantum system that you can precisely examine and manipulate. Many of the enabling technologies that are needed to control the ions and atoms have been developed in cooperation with Leibniz University Hanover and the Technical University Braunschweig. A range of initiatives exploring quantum mechanical control were financed by the Deutsche Forschungsgesellschaft (DFG). These included the former excellence cluster QUEST (Quantum Engineering for Space Time Research), which gave this institute its name, and other excellence clusters and special research centers such as QuantumFrontiers and DQmat.

These research programs are like layers of sediment on the ocean floor: one cut reveals what colleagues were doing twenty years ago. Is everything built on everything else? Did your knowledge cluster grow in an organic way?

Piet Oliver Schmidt: We are German leaders in concentrating and focusing know-how and expertise in the field of quantum metrology. As the national metrology institute in Germany, the PTB is the highest federal authority on this subject. It has been a part of our DNA since the discovery of quantum mechanics.

Did you say meteorology?

POS: [*Laughs.*] No, *metrology*—the art of precise measurement. There have always been fruitful research projects that, as you say, were built on one another. Ours also involved atomic interferometry, which employs the tiny wavelengths of atoms to achieve even higher resolution. For example, it can be used to accurately measure gravitational acceleration. A transportable system could make it possible to also conduct geological surveys, identifying resources with an unparalleled precision. These are all activities that were identified in basic research at the time.

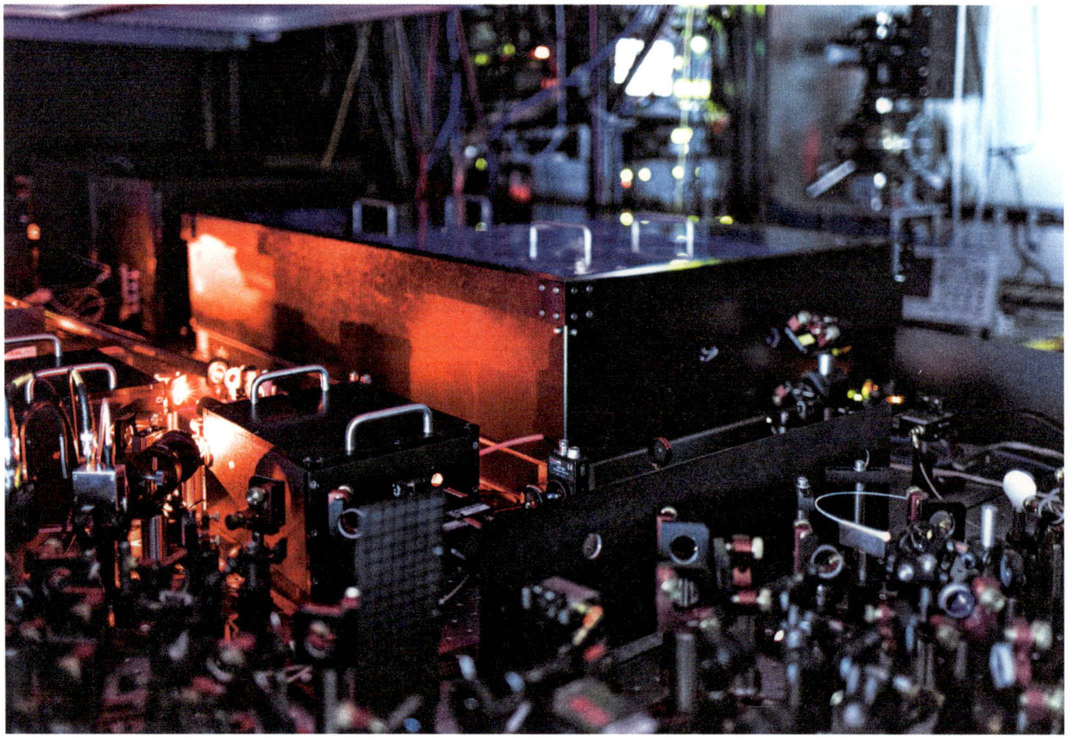

For which industrial application fields is quantum technology a key technology?

POS: We can basically go through the individual pillars of quantum technology, beginning with quantum communication to securely transmit sensitive information using quantum keys. Quantum simulation is important, as is, of course, quantum computing.

CO: For a long time, quantum computing was regarded with a great deal of skepticism, not only here in Germany. But then, in 2017, after a long preparatory phase, the EU's Quantum Flagship was established, to which we are also contributing a project, resulting in the systematic establishment of research programs in Germany.

What are the characteristics of a quantum computer? In conventional technology a microchip made from silicon processes information using minute transistors. The information is coded using zeros and ones. However, silicon is gradually reaching the limits of miniaturization. Eventually Moore's Law will come to an end with silicon chips, which use quantum logic gates. How exactly does they work? How is this zero-one information generated in your quantum logic gates?

CO: This can be nicely illustrated using a single bit or a quantum bit. The transition to a quantum computer is marked by the addition of a Q before the bit, turning the bit into a qubit. A bit can only be in two states: either zero or one—light on or light off. In contrast, a quantum bit is a two-level system, which can be described using all possible states on the surface of a sphere. Every qubit state corresponds to a point on the surface of this sphere, which is identified by longitude and latitude. The two eigenstates are located at the south pole and north pole, but in addition every state is possible on the surface of this sphere. Instead of the two discrete states there is a continuous state space that can be described by means of longitude and latitude, that is, by two numbers.

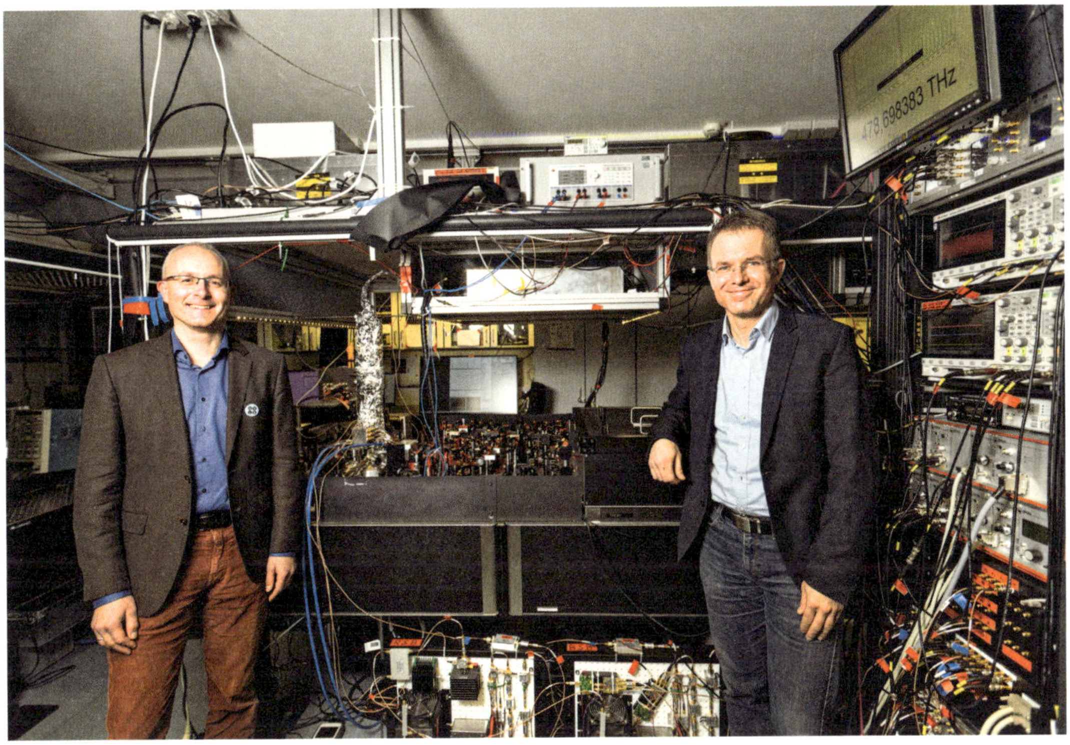

Prof. Dr. Piet O. Schmidt (left), cospeaker of the cluster of excellence QuantumFrontiers and head of the QUEST Institute at PTB and professor at Leibniz University Hannover (LUH). Prof. Dr. Christian Ospelkaus, Institute of Quantum Optics, LUH

So instead of zero and one states, the qubit has a possibility space with many states. How do you calculate this? After all, the atom is to provide results within a computing process. How is this possible when there is no zero or one?

CO: Computing operations are basically airplanes that fly around the sphere and connect the points to one another. They convert one state into another, that is, they convert one longitude and latitude into another longitude and latitude. A computing operation is, so to speak, described by a flight path around the sphere.

Is it three-dimensional?

POS: No, it is two-dimensional because it is the surface of a sphere.

How many number combinations are possible on the surface of this sphere? Is it 360 times 360?

CO: No, it's an infinite number. If you move an arbitrarily small amount on the surface of the earth, you have a different state. The state of a qubit can be described using two continuous numbers, while a classical bit is described using the discrete values zero and one. The problem is that the computer operations take place in the dark, and the light is always off. You can shift states around on the sphere using control fields, but you are not allowed to look where the qubit is currently positioned. When you start the quantum mechanical measuring process—

You destroy the sphere.

CO: Exactly. The moment you switch on the light, the qubit has to choose between the extreme points of the north pole or the south pole, in other words between zero or one. You only measure one or the other. You can't precisely determine where the system was on this sphere. All you know is that if it was directly at the north pole, I will find it at the north pole in one hundred percent of cases. If it was directly at the south pole, I will find it at the "south pole" in one hundred percent of cases. If it was at the equator, I will find it at the "north pole" or the "south pole" in fifty percent of all cases.

So the computational results are within a certain spectrum? Are the computational results precise down to the last decimal place?

CO: It basically provides probabilities. If I place a qubit at the equator, I expect either darkness or light with a probability of fifty percent. Here you can already see one of the essential aspects of quantum computers. You begin with a well-defined state such as the north pole. Then you have the whole sphere at your disposal as state space. However, at the end of the calculation you have to reduce the whole thing down to two measuring results.

So ultimately zero or one again?

CO: Yes. A sphere is very easy to understand and can be simulated on a classical computer. But if we look at more spheres, it gets complicated. Since COVID every child knows how exponential growth works. If you place more qubits next to each other, the number of possibilities increases. For two qubits the intuitive picture of the sphere no longer functions, as the space now has a lot more dimensions than we can represent with our language.

So that would be two to the power of what?

CO: Two to the power of n, according to the number of qubits. It becomes extremely complicated.

So it is a type of multitude?

CO: Exactly. In 1982 Nobel Prize laureate Richard Feynman established in his famous essay that simply to describe the behavior of 128 interacting spins on a grid, you would require more conventional bits than there are protons in the entire universe. That illustrates exponential growth, similar to the chessboard where you place a single grain of rice on the first square and then double it on every subsequent square. In his essay Feynman also demonstrated that only a new type of computing architecture that functions according to the principle of quantum mechanics would be able to solve such a problem in a finite period of time and with finite resources.

Could the quantum computer calculate this?

CO: Yes, that is its essence. The quantum computer functions precisely according to these laws. That is, in between you have a huge amount of information that can be stored in this system. But at the end you switch on the light. If you have a certain number of particles, N qubits, at the end you also have as your result N zeros and ones. If you want to design a quantum algorithm, the art is designing the control fields and the routes of the qubits in such a way that at the end you receive meaningful information for every qubit, which can help you solve the problem. You have to conceive the algorithms from the end, so to speak.

Are these informational possibility states or actually real information?

CO: It is not information until you have extracted it. They are states of the system that you manipulate. However, in principle, you only see what information you have extracted when you switch on the light.

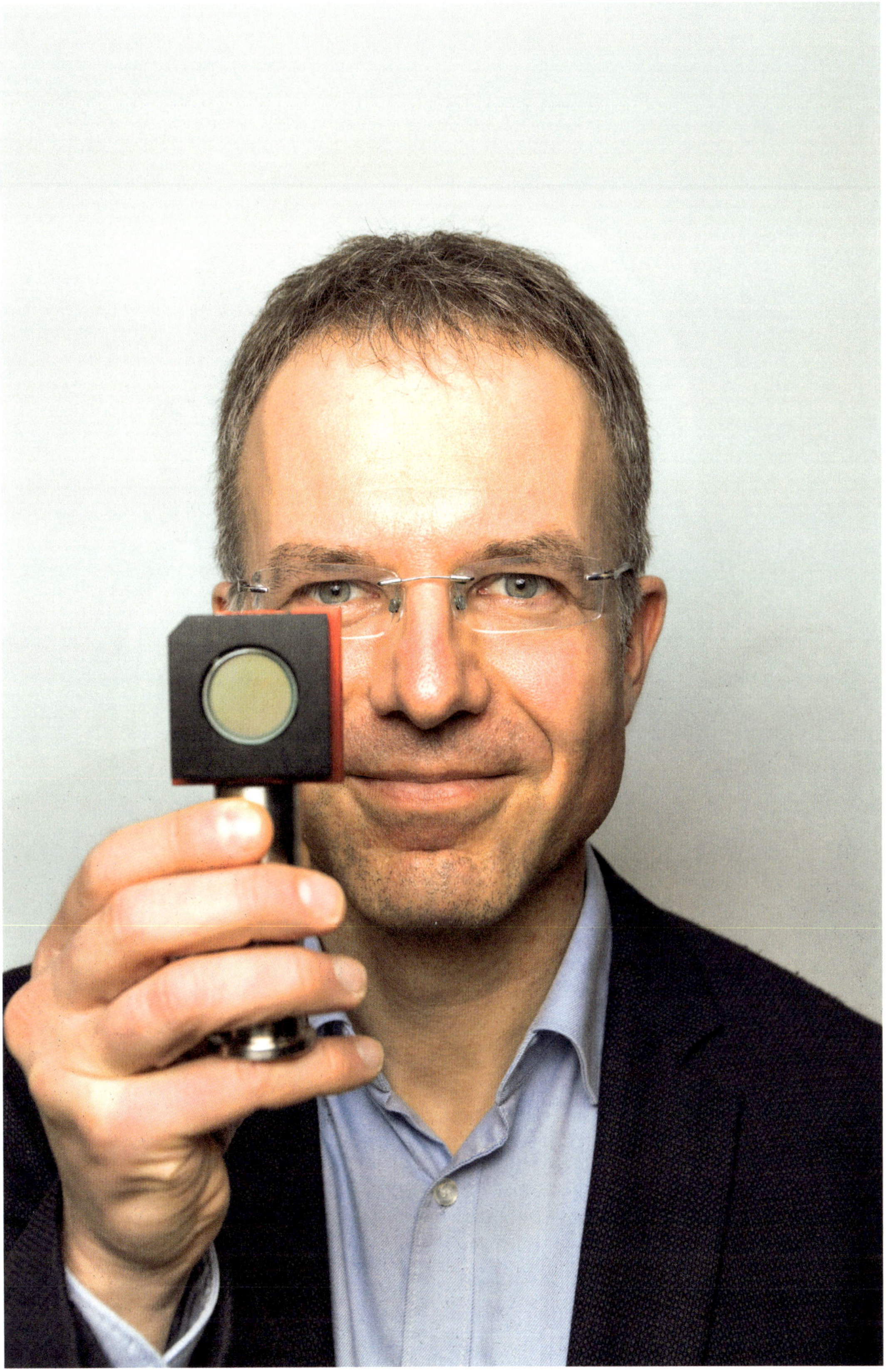

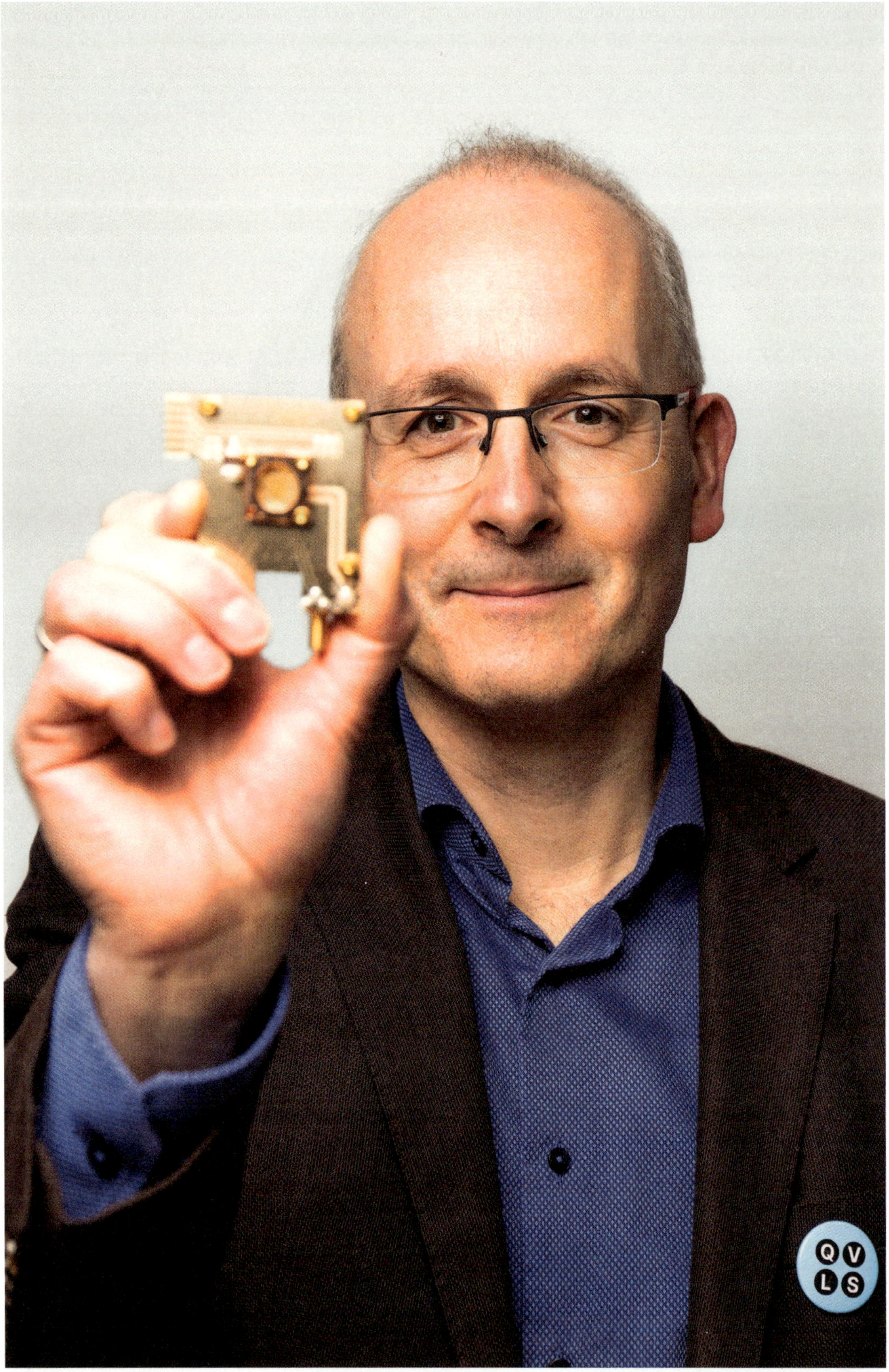
Q V
L S

POS: The big problem is that quantum algorithms are not intuitive. It can be better explained using quantum simulation, which is basically not an algorithm but an application. With quantum simulation you can also plausibly explain the usage. Bayer, Boehringer Ingelheim, and other pharmaceutical and chemical companies are already working in the field of quantum simulation. If you want to model or understand quantum mechanical systems or chemical reactions, conventional computers reach their limits for the reasons named by Feynman. There are a vast number of electrons, and you have to take into account all the spins and interactions. The matrices explode and the computers give up the ghost. However, in the case of quantum computers, the quantum mechanics is inherent. All we have to do is find an image of the molecule for a quantum computer, which works with qubits and spins. This is described as quantum simulation.

Only quanta can calculate quanta?

CO: That was the basic question that Richard Feynman asked in 1982: what can we learn from computer simulations, which were relatively new at the time, about nature? What happens when nature behaves like quantum mechanics? His answer was: the workload explodes exponentially with the number of degrees of freedom. This task can only be realistically solved by a computer that also calculates according to the laws of quantum mechanics. And so the term *quantum computer* was born. However, the difficulty is in explaining why they are so much more efficient. You can't say: the quantum computer has a higher clock rate or has more RAMs than a conventional computer. The reason why a quantum computer calculates faster cannot be simply reduced to conventional terms.

So you are building a quantum computer, but you can't explain why this computer can be faster than a conventional computer by an unbelievable order of magnitudes?

CO: You can explain this for quantum simulation because the quantum computer behaves according to the same laws. The possibilities available to it for the intermediate storage of information grow just as exponentially as the quantum systems that it simulates. An almost one-to-one representation of the system is generated. In the case of algorithms that we know from computer science, it is far harder to make unambiguous statements.

Do circuits also play a role here? Does the quantum computer contain a chip, a circuit board, or even small diamonds?

POS: No, captured, charged atoms are employed here.

So the ion trap is again involved.

POS: It is really quite impressive. There are also fluorescent barium ions that you can see with the naked eye.

CO: For me the key moment was in the initial period, when you no longer controlled the ions with plugged-in equipment, but instead could control them with a kind of computer language. It felt as if you could speak to a single atom via the keyboard. I was extremely fascinated by this.

Is the atom manipulated with a magnetic field or a laser?

POS: With both, as a matter of fact. You always use the one that is most suitable. We still always use lasers for cooling the atoms and afterwards for selection.

But doesn't a gray area always remain if the sphere has an infinite number of possibility states?

POS: No. As soon as I illuminate it with a laser, it decides. It is either dark or light—zero or one.

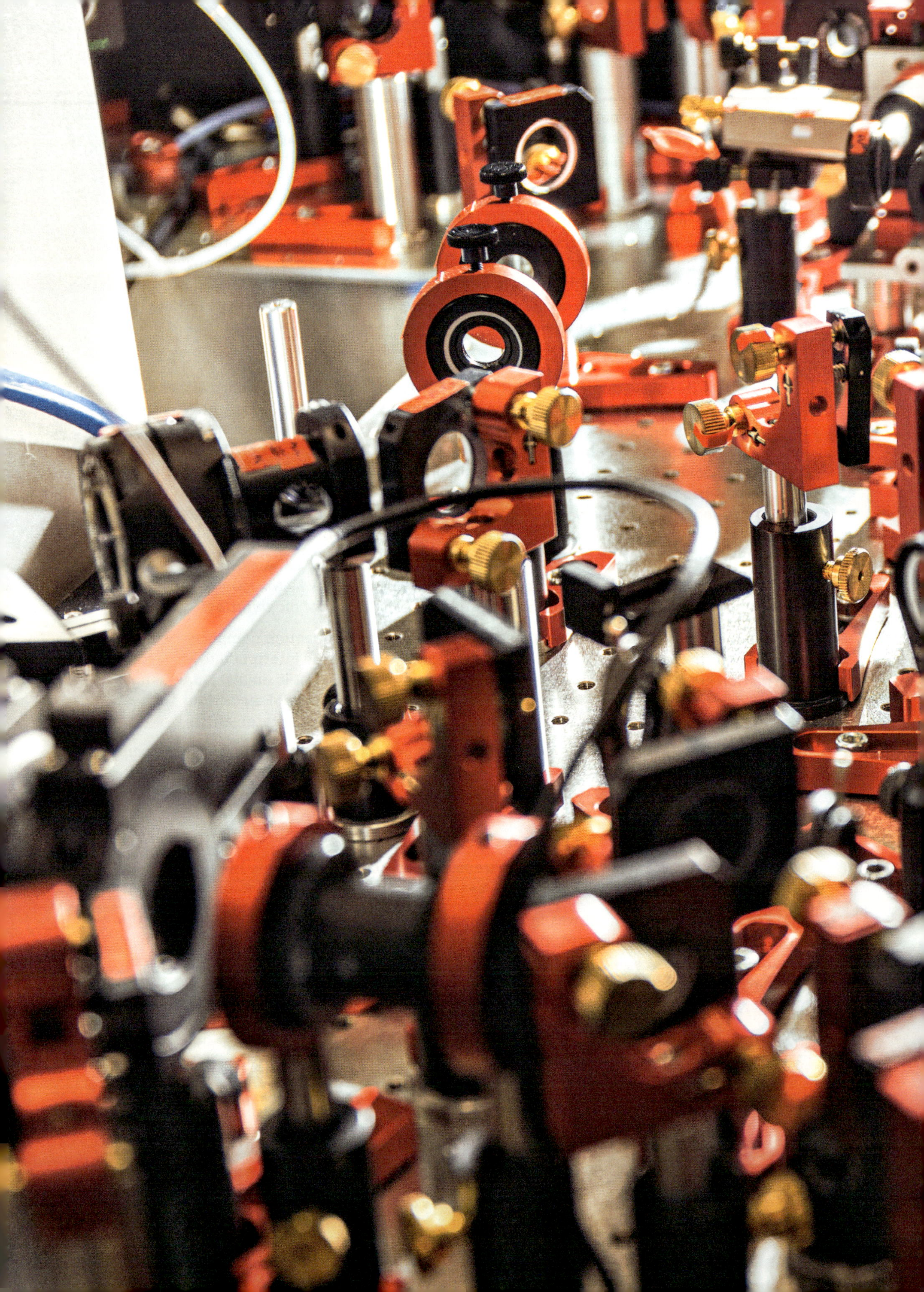

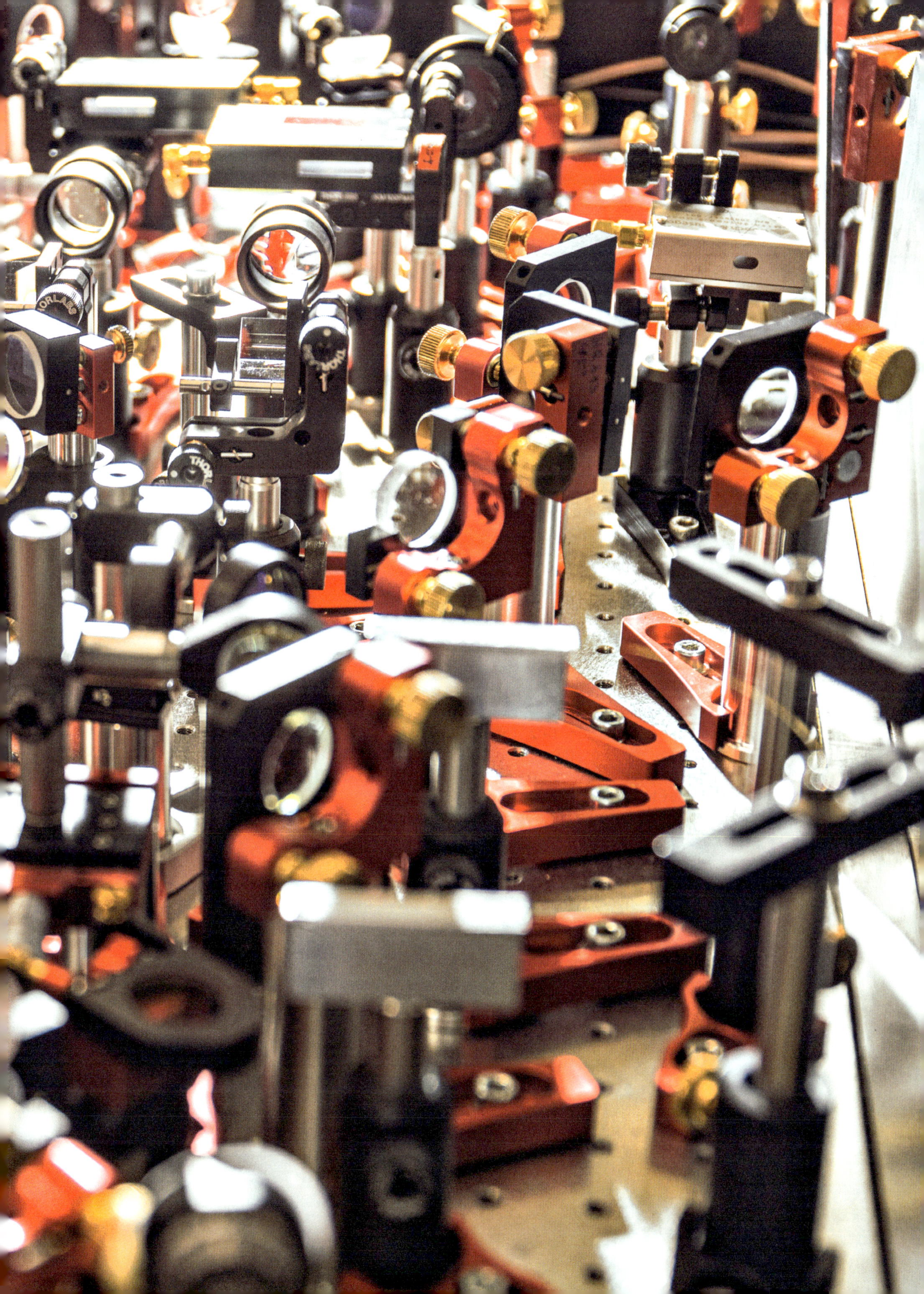

CO: The moment that you determine the state, you also destroy information.

And along with it the other possibilities, so to speak.

CO: You have to manipulate these states as well as possible, without anything spontaneous happening. For one, you need an irreversible preparation operation. You prepare either zero or one. And secondly you need reversible operations, which you have to be good at, and at the end you need a measurement process, in order to achieve good projections.

But that is no longer reproducible. That's doesn't sound at all clear-cut, because the atom has to decide. According to which criteria does the atom decide to be zero or one now?

POS: [*Laughs.*] Einstein also wondered about that, and he never found a satisfactory answer.

Einstein discovered that light consists of particles and received the Nobel Prize for it in 1922. Today these light quanta are called photons. Nevertheless, he criticized quantum physics from a philosophical perspective because the new role assigned to probability made him uneasy.

POS: Since then many experiments have disproved Einstein on this point. The measurement result can appear random because, for example, the state on the equator can appear either light or dark with a probability of fifty percent. But if you repeat it often enough, and you always get exactly fifty percent, then you know that the state was on the equator. But you still don't know where on the equator.

So you still don't know the longitude.

POS: Exactly. Then you have to take another measurement. And that is why you have to design the algorithm or the calculation so that the answers are always unequivocal. You have to have asked the right question with your algorithm, so that at the end zero and one is a meaningful answer.

CO: You have to orchestrate these coherent rotations on the sphere in such a way that at the end of the measuring process you get a meaningful result. Conventional computers continually destroy information. Bits, for example, are coded in voltages. Everything from 0 to 0.6 volt is zero, and everything from 2.4 to 5 volts is one. At the end it is projected to zero or one. In contrast a quantum computer always has to calculate these precise rotations on the sphere. And this results in errors that you have to get under control, as the quality of these operations is decisive for how you evaluate such a machine.

The better it is controlled, the more precise the result. But isn't that a manipulation? Aren't you influencing the result by the way you treat the atom?

CO: No, that is, so to speak, the algorithm. Put simply: if I have a tube through which I can observe the result on the other side, the algorithm tells me how I have to position my head so I can see the correct result through the tube. But errors occur with these rotations. And the question is: how do you cope with them? And when you have a lot of qubits and make a lot of these rotations, naturally the errors accumulate.

They don't cancel each other out?

CO: No, because you are not allowed to destroy any information. But there are smart procedures for correcting quantum errors, although they only function when the error per operation is smaller than the error tolerance threshold. If you reach this, you have found the holy grail. If you are below this threshold value, the number of additional qubits skyrockets to correct your errors. That is why the decisive question with a quantum computer is not just the number of qubits but the quality of the calculations.

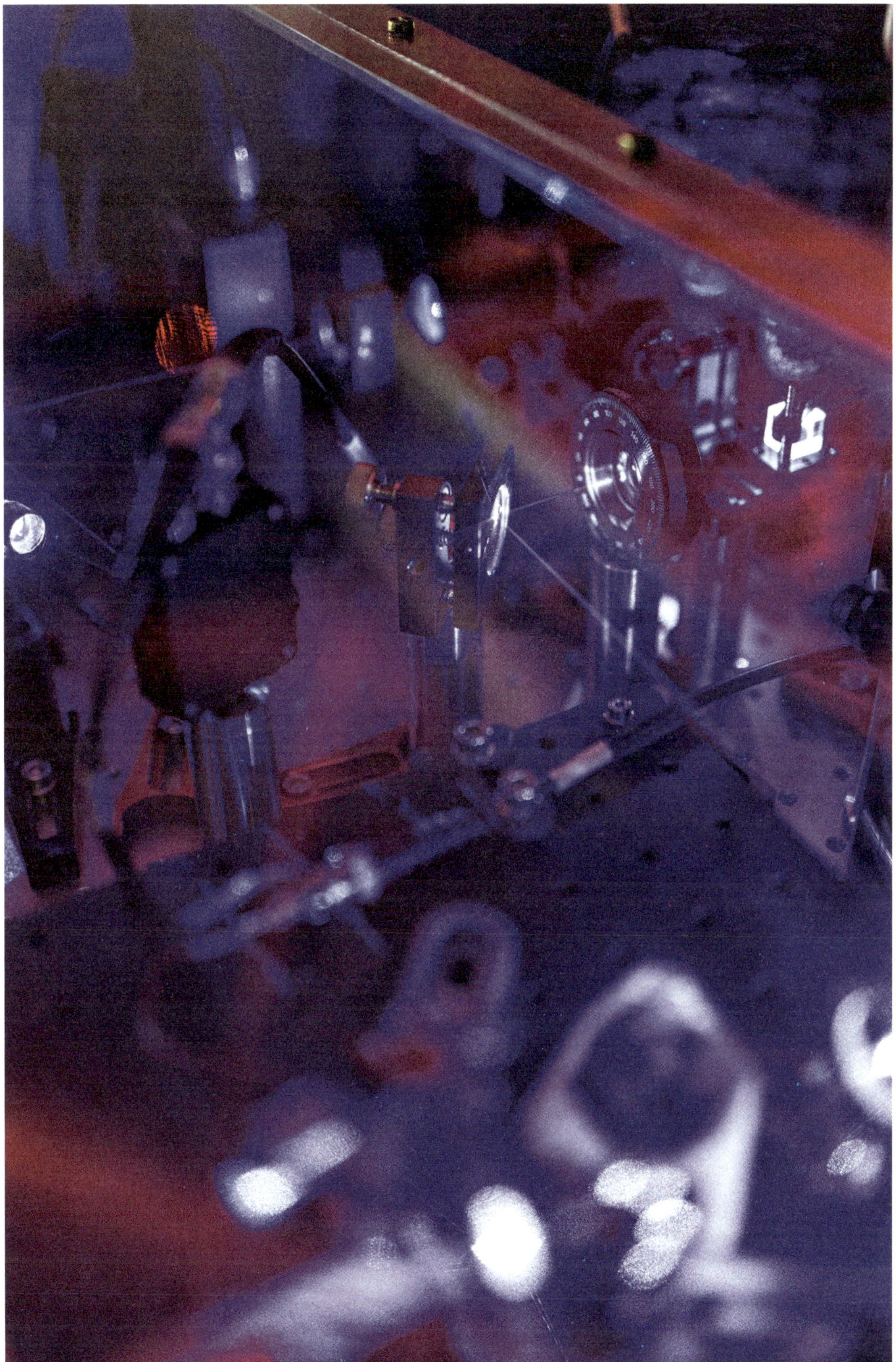

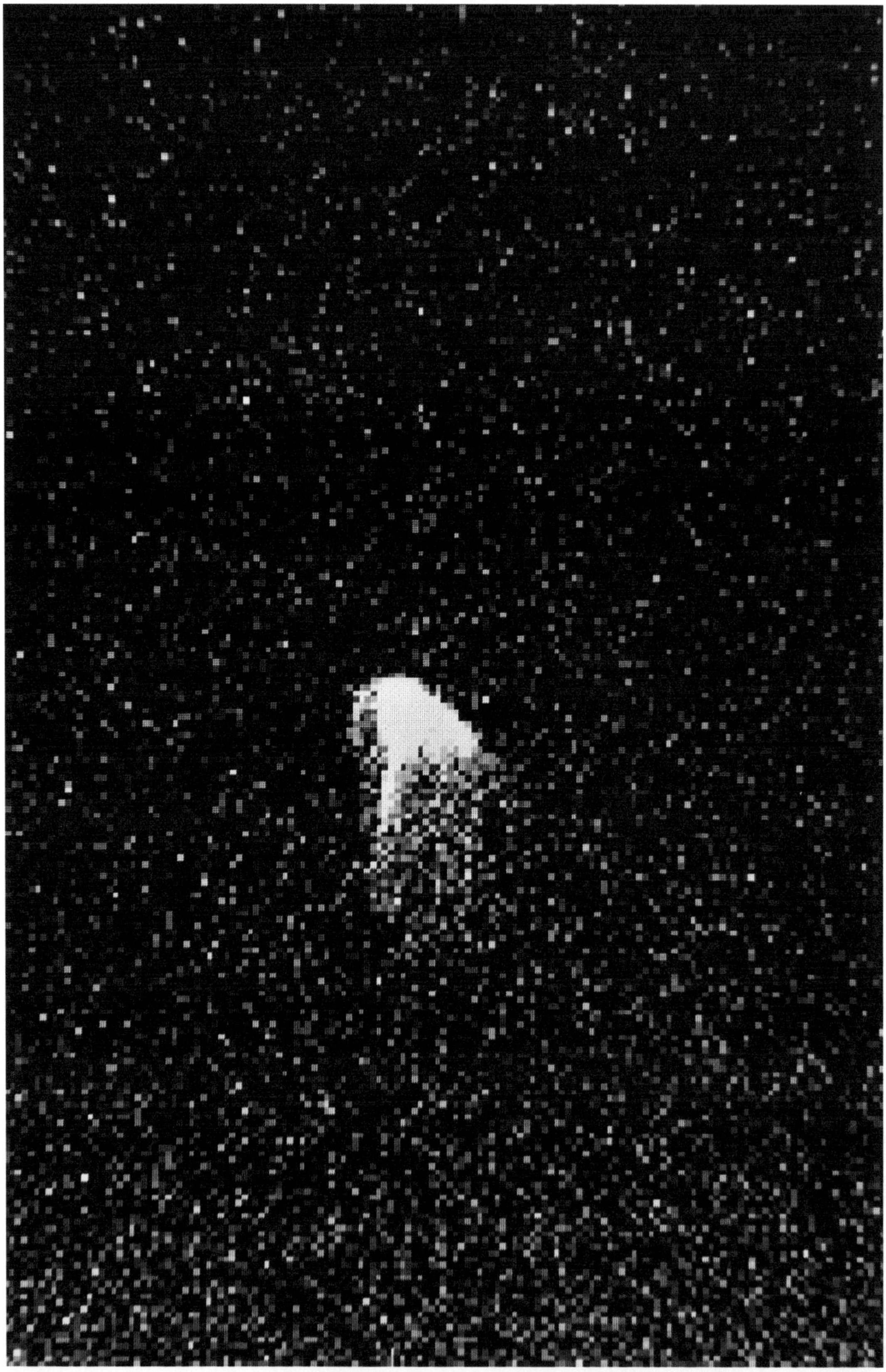

 A single beryllium ion imaged with an optical lens system.

What application fields is the quantum computer suitable for?

CO: If you construct this quantum computer with one thousand qubits and if it computes to a sufficient degree of accuracy, this will open up a field of untold possibilities, for example, in finding new materials for building better prostheses or developing molecules for new active agents.

Or folding molecules. Calculating active agents from endless combinations of molecules and assessing their effect is a very large growth area. It is a possibility calculation, which a quantum computer can do very well.

CO: Exactly. The quantum computer opens up huge possibilities for problems where conventional computers have completely failed in the past. Quantum computing enables you to tackle a completely different class of problems.

POS: For example, if you could improve the Haber-Bosch process, a chemical reaction for producing fertilizer, the global CO_2 balance would already look a lot better. A lot of energy goes into manufacturing this fertilizer. Battery technology is another example. Volkswagen is pursuing quantum computing to improve traffic flow, but eventually it will be used to simulate battery chemistry, so that you can store more energy per volume.

CO: Or the processes in fuel cells.

POS: These are all quantum mechanical processes, and with a quantum computer you have the appropriate resources to compute them.

So for very complex chemical-physical processes …

CO: Anything that involves quantum mechanics.

POS: Naturally they have to be interesting problems in need of a solution.

How do you find these problems?

POS: First the quantum computer has to be running and the error tolerance reached. But our theoretical physicists at Leibniz University Hannover are already collaborating with companies on selecting which algorithms and simulations could run on it. A type of electronic digital twin for the quantum computer is being established, so that you can already predict this for smaller systems.

But which computer computes the digital twin?

POS: That is a conventional computer that can simulate the quantum computer with a reduced system size. You can then extrapolate using this as the basis for your calculations.

CO: Building such a machine is an interdisciplinary challenge. For this you need theoretical physicists, mathematicians, computer scientists as well as high frequency technology, laser engineers, and atomic and molecular physics. You also need mechanical engineers for drawing the construction plans. This works very well for us as we have great people who all work together in all areas.

This is just the beginning of the possibilities that are opened up by this technology.

POS: I am quite certain that once they are really up and running, quantum computers will be very useful in a lot of areas. Quantum sensor technology will also definitely play a major role; in addition to atomic sensors there are also, for example, nitrogen-vacancy (NV) centers in diamonds that can be made very small. It also opens up new possibilities for measuring temperature and magnetic fields. But in the field of material research it will also be possible to develop more stable, flexible materials, or those with different qualities, and as soon as, for example, optimization problems in production sequences are effectively resolved with a quantum computer, there will be a multitude of new application possibilities.

Such a quantum computer could prove to be quite important in Mark Zuckerberg's metaverse, the simulated world where augmentation is calculated from a multitude of possible forms.

POS: Of course, expectations are also very high for the research discipline of quantum machine learning, although that is very much in its infancy.

But, to return to the beginning of our conversation, what does the development of the quantum computer have to do with the atomic clock?

POS: The technology employed for the construction of these ultraprecise atomic clocks is basically very similar. In the case of the atomic clock we need a single atom that is caught in an ion trap, for example, and isolated from the surroundings as well as possible so that it doesn't start ticking incorrectly. In the quantum computer this incorrect ticking translates into incorrect calculation. It is exactly the same principle. And that is why we can take the technology developed here in the field of atomic clocks and employ it directly in the development of a quantum computer.

CO: Allow me to explain in simple terms what an atomic clock is. An atomic clock has two different states, two quantum mechanical states, which are separated by a specific energy distance. This energy distance corresponds to the frequency needed to trigger the transition. When the distance between these states changes, the frequency changes—as does the time. Now if you have one qubit, it also has two states. If you change the distance between these states, in a certain sense the information that is stored in the qubit is destroyed. And that is precisely the same challenge that you have with the quantum computer. You have to be able to manipulate the two energy levels as precisely as possible.

POS: So, bluntly formulated, an optical clock is nothing other than a one-qubit quantum computer.

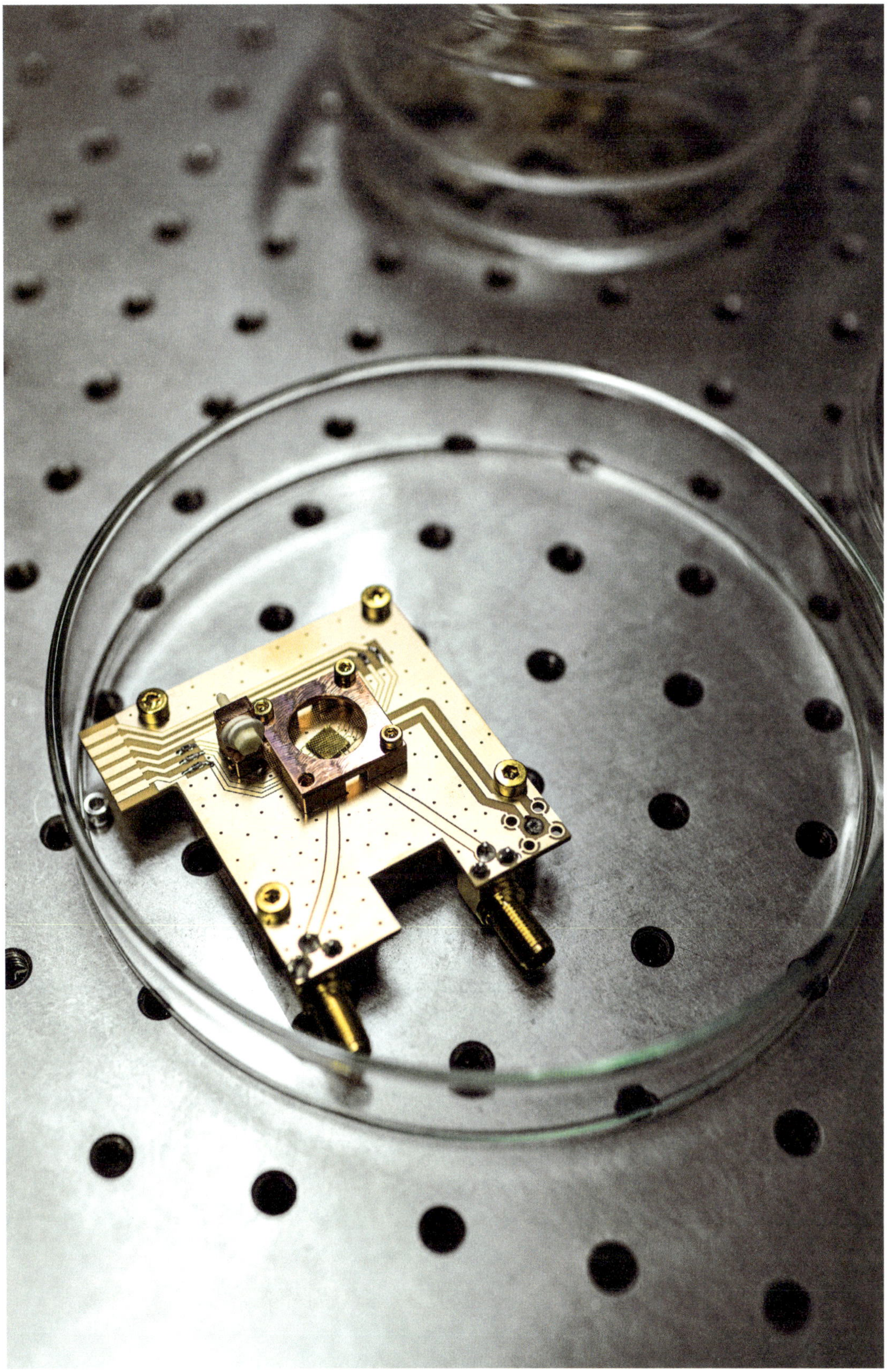

Ion trap chip with socket

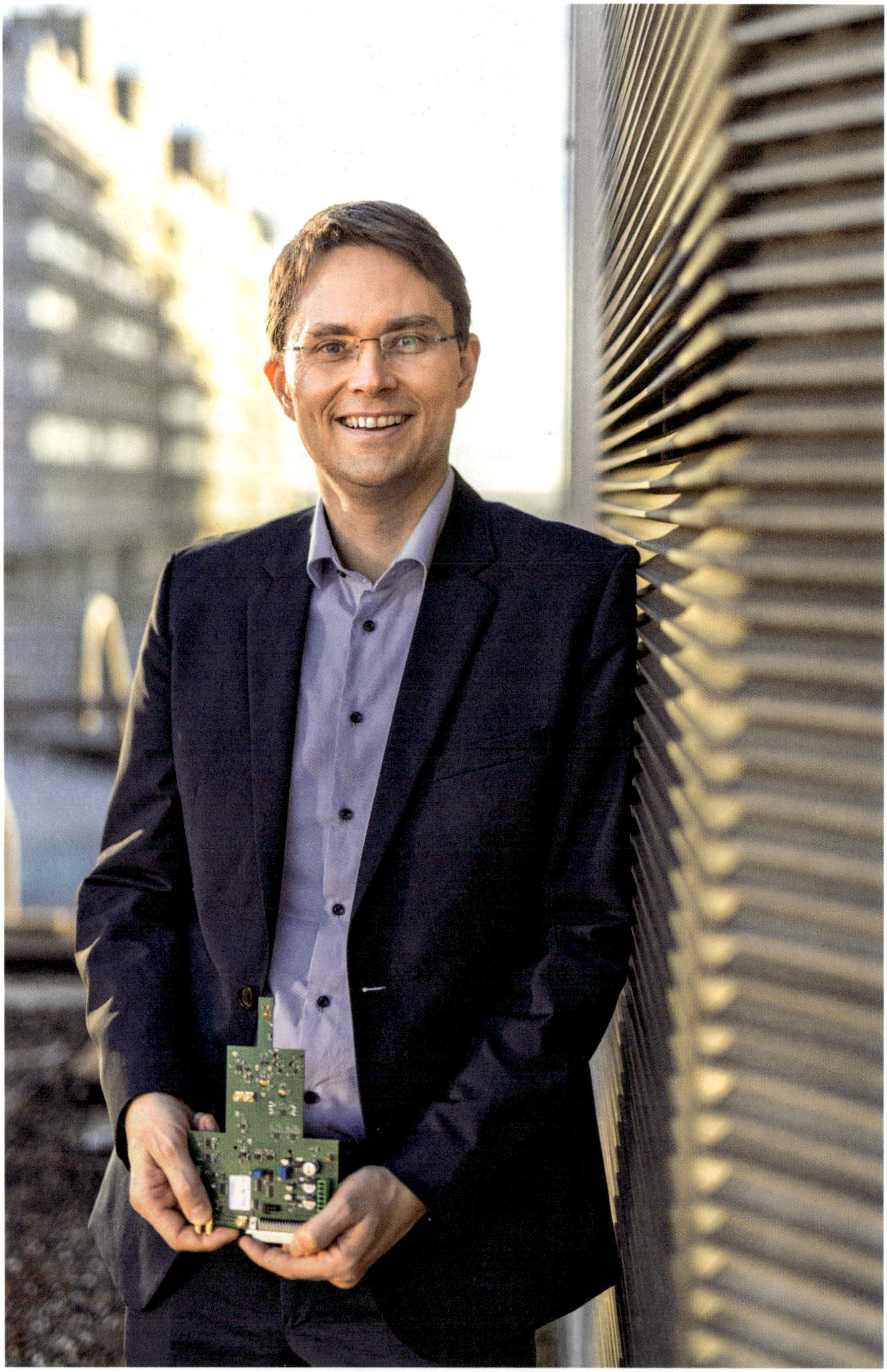

JENS ANDERS

The Quantum Leap of the Frog's Leg

Back in 1791 an experiment proved the presence of magnetic fields in animal and human bodies. Today the director of the Institute of Smart Sensors at the University of Stuttgart measures the tiny magnetic field in nerves and muscles with the help of minute diamonds that function as quantum sensors. His vision of the future opens up a broad spectrum of applications within the human body: from portable measuring devices to implantable sensors that, with the aid of quantum technology, can detect illnesses at an early stage and enable the continuous monitoring of a person's state of health.

Professor Anders, you are the spokesperson for the Cluster4Future QSens, which is engaged in developing quantum sensors. Over the last ten years quantum technology has become a major trend, and the state is investing a lot of money in research. What is behind the quantum hype?

Quantum technologies comprise three areas: quantum computing, quantum sensor technology, and quantum communication, the last of which, for example, makes possible the intrinsically bug-proof transmission of information. Quantum computing brings huge value propositions with it. When a quantum computer with one thousand logical qubits is released—requiring one million physical qubits—it will really change the world. But this will definitely take another ten to fifteen years. Along the way there will be "quantum leaps," for example, through quantum simulators with one hundred qubits, which can compute the characteristic of new molecules with great precision and with an efficient use of time, instead of having to arduously test them in the laboratory. However, our field of quantum sensor technology is so exciting that even one qubit can make a difference. You don't need one hundred, one thousand, or one million qubits—you only need one. This one qubit measures a magnetic field, an electrical field, or the composition of a substance far more accurately than is possible with a classical sensor. That is why quantum sensor technology is far closer to market entry. Today start-ups are already offering quantum sensor technology products. For example, an American company markets a quantum magnetic field sensor for measuring brain activity. This magnetoencephalography (MEG) unit looks like an oversized bathing cap with up to 512 sensors and enables precise measurements of the neuronal currents in the brain.

But these bathing caps with suction cups have been around for ages ...

There are a number of clinics in Germany that offer MEGs. But as these are based on a different measuring principle—what are known as *SQUIDs*, a first generation quantum sensor—such devices are very complex and expensive. They only operate at very low temperatures, which means they have to be continually cooled, and if the patient moves during the measurement the image will be blurred and unusable. In contrast, quantum sensors of the second generation frequently operate at room temperature and are small enough to be attached directly to the patient, moving with the patient. That is a quantum leap. In the next step, we want to launch less expensive and even smaller quantum sensors that can also cope with harsh environments. For example, cars are known to have high temperatures in the engine compartment. In the future our quantum sensors must also be able to function in such environments. This is a very important aspect of our cluster QSens, as electromobility will grow strongly over the coming years. In the case of the battery, the charging cycle has to be optimized so that it can be charged in ten minutes without damage. With a classic sensor, it is not possible at the moment to accurately and reliably measure the charge, health status, or remaining lifetime of a battery. A quantum sensor can do this a lot better if we succeed in making the sensor small enough to integrate it directly into the battery.

What other applications are there in your cluster QSens?

For example, in biomedicine, we measure biomolecules using nuclear magnetic resonance (NMR) or free radicals using electron spin resonance (ESR). We are familiar with free radicals from the saying "an apple a day keeps the doctor away." In the body, there is an interplay between free radicals and radical scavengers, antioxidants. Free radicals have a free electron spin that one can measure in a highly targeted fashion in the body using ESR, for example, as markers for cancer. If you look further into the future, it is conceivable that you will be able to stick a smart patch onto the skin and predict skin cancer with a noninvasive sensor. In the cluster, we have a project that—simply speaking—aims to monitor patients who have recovered from cancer with a small device to detect at an early stage whether the cancer has returned.

Fantastic. Such noninvasive measuring procedures could then be integrated into the Apple Watch, for example. But Apple and Google have both tried this, and they both failed.

It will definitely not be easy, but with ESR, we have the right method to hand, and with the quantum sensors, the right detectors too. Now it is about giving our best in our research cluster QSens. Over the first three years, our cluster is very generously financed by the federal government to the tune of over twenty million euros. In order for a company to be successful with quantum technologies, we need to have a business model on the market within three to five years, with which you can earn money. That is why in our cluster all the projects are conceived with a view to the market, and together with our industry partners we have identified the low-hanging fruit.

What are you doing in your cluster with Ottobock?

Ottobock is a research partner in our cluster and is participating in the project "QHMI—Quantum-Human-Machine-Interface Based on Quantum Sensors," which is scheduled to last for six to nine years. The long-term vision is to measure muscular activation with magnetic field sensors in order to intelligently control prostheses in the future. Today one can already measure muscular activity electrically, using electrodes that are attached to the ends of the muscles to detect electric potential signals. However, the big disadvantage is that the attached electrodes degrade over time so that the long-term stability required of a prosthesis cannot be achieved. In contrast, magnetic field measurement is noninvasive and contactless and, therefore, just the right thing for a prosthesis that has to function for years. One can now deploy different sensors. Conventional sensors have a sensitivity problem, but with quantum sensors, it will be possible to reliably measure the relevant small muscular magnetic fields in the range of just a few picoteslas.

What does picotesla *mean?*

Tesla is the unit of the magnetic flux density, and *pico* means 10^{-12}—these are very small magnetic fields. In comparison: the Earth's magnetic field in Germany is around fifty microteslas, in other words, about one million times greater than the muscular signals to

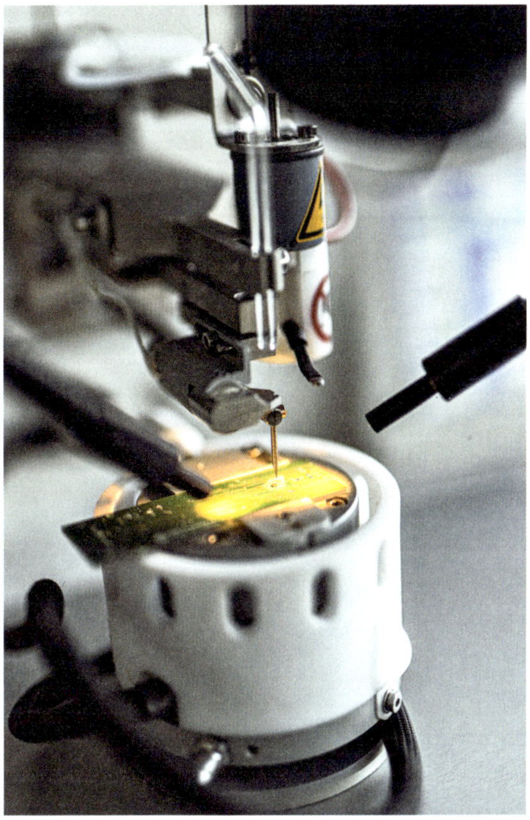

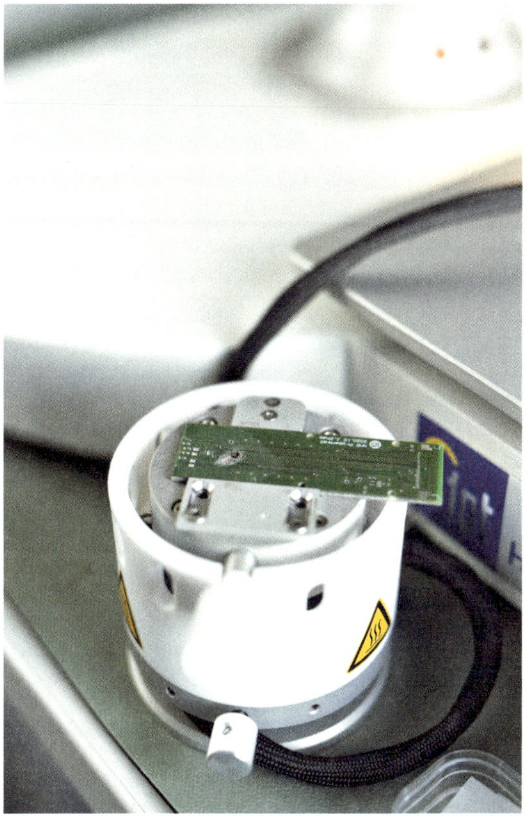

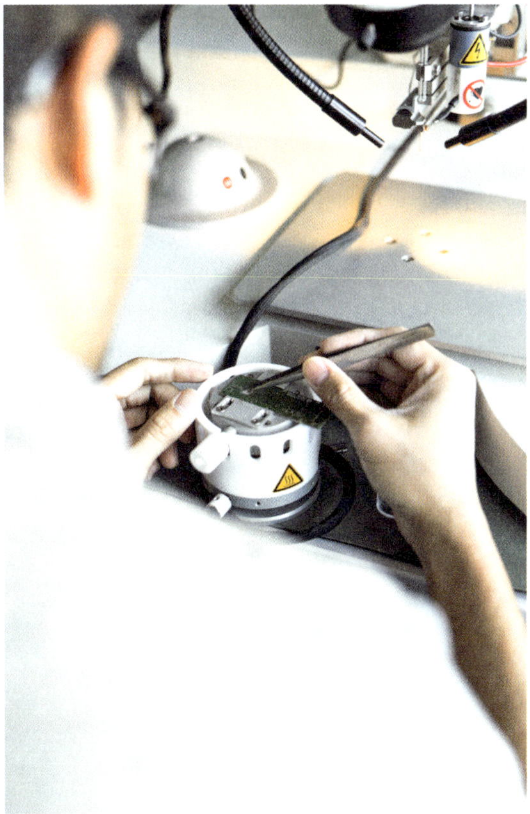

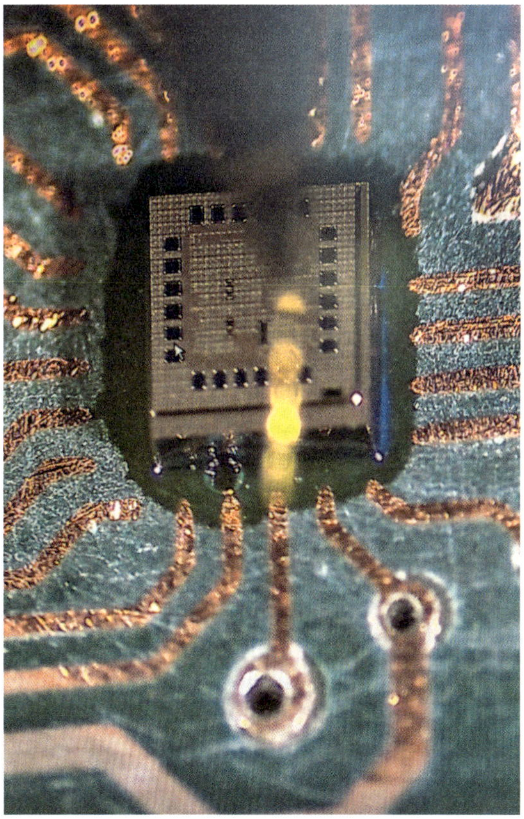

be measured. Our magnetic field sensors are based on nitrogen-vacancy (NV) centers in a diamond and can be even more sensitive.

What does nitrogen vacancy *mean?*

Nitrogen vacancy—as in nitrogen-vacancy center—is a nitrogen defect. There is a vacancy in the diamond's crystal lattice; a carbon atom is missing, and a nitrogen atom sits next to this vacancy.

Like on a microchip in semiconductor electronics?

Yes, exactly. In semiconductor technology, we dope semiconductors—typically silicon—with certain dopants such as boron, and here we dope the diamond crystal with NV centers. This nitrogen-vacancy center has fantastic characteristics. The most important feature for us is that it has a spin. A spin is like a tiny compass needle that aligns itself within the magnetic field. I can control the spin from the outside. The chips that we are researching at our institute lie beneath the diamonds and control these spins, which react sensitively to the external magnetic field. By stimulating the spins and reading their response, one can precisely measure the magnetic field. That is the principle.

Who discovered that the body itself generates magnetic fields?

That goes back to people like Luigi Galvani. In 1791 Galvani experimented with the leg of a dead frog, which he placed on the wire grating of his balcony on a thundery day. It started twitching during the thunderstorm.

And afterward he enjoyed the grilled frog's leg with tartar sauce?

[*Laughs*.] I wouldn't rule it out. But that was the start. The research on NV sensors goes back to people including Professor Jörg Wrachtrup, a scientist from Stuttgart and Leibniz Prize winner who is the cospokesperson of our cluster.

Is he the spiritus rector?

He is the physicist in our team who is developing the methods in quantum physics. I am the electrical engineer who ensures that his sensors can be miniaturized and also function in a normal environment. That means they have to be small, resistant to interference, and affordable. In all the projects in our cluster, we aim to produce quantum sensors scalably and in large quantities so that they can find their way into everyday use. One application is the project with Ottobock, where it is about producing corresponding magnetic field sensors for prosthetics.

So the diamond sensor in the prosthesis generates a very small magnetic field in order to measure nerve signals?

Put simply, what happens in the body is as follows: The brain controls our muscular activity via electrical impulses that are sent to the nerves. The nerve impulse is a type of voltage pulse that passes through the body. The electrodes measure this voltage pulse. In order for this voltage pulse to propagate, membranes have to be polarized. To this end, electrical charges must flow, which correspond to an electric current. Currents generate magnetic fields, and the NV sensor measures these magnetic fields.

The muscle fiber itself has a minimal magnetic field?

Yes, a minute electric current passes through the muscle fiber, which generates an even smaller magnetic field. It is extremely fascinating.

And the sensor can measure this change?

Exactly. And the idea is that it is not necessary to measure invasively at all. Instead, the sensor is attached to the skin. That means the sensor is contactless, noninvasive, and can still measure the magnetic field over a distance.

These noninvasive methods make operations—which can also be very stressful psychologically—superfluous.

Absolutely. The magnetic field sensors, which are based on diamonds, are so sensitive that we can even take another step. In the project with Ottobock, the first step after six years is the detection of the nerve signals at the ends. If the nerve conduction from the brain still works, then one can make use of this. But what if the brain functions and this connection is interrupted? In principle, the magnetic field sensors are sensitive enough to measure the signals directly on the cranium. Here, depending on the thickness of the skull, there is a relatively large distance to the neurons. However, the sensor can still reliably measure the magnetic field on the cranium. Here we are in the femtotesla range, in other words, 10^{-15} tesla, another three orders of magnitude smaller. This type of detection could be used by paraplegic patients, for example, in order to tap into the signals directly at the brain.

But aren't these signals totally mixed up? Professor Aszmann told us that the quality of this information, which one taps at the brain, is not sufficient as the actual quality of the information is only generated in combination with the responsible muscle cells. Many movements take place routinely and redundantly in the limbic system without the participation of the cerebral cortex. So there is not a lot to measure there.

Exactly. The muscle is the better choice for controlling the prosthesis with small movements, as a lot of information has already been filtered out there.

The muscle knows what it has to do.

Yes, it is already very specific. However, if you have a paraplegic patient, the signals never get to where they are meant to go. That means I have to go directly to the brain and solve the problem of the what is known as *source localization*. You don't just use one sensor, but an array of sensors that are distributed across the head, and then, on the basis of the large number of points at which you measure the magnetic field, you have to draw conclusions about the sources of these magnetic fields.

It would be better if you measured the person before they suffered the paralysis, so that you know which neuron area is responsible for which movement.

Yes, you are absolutely right. But as individuals differ, you always have to calculate back.

But doesn't it always take place in the same sector?

That is true. You can map the brain and know which activities are controlled by which brain areas. In order to compile this map for an individual, you can make an MRI scan of the head. But then it is about precisely locating the sources on this map and also providing a temporal resolution in order to differentiate the brain commands. Naturally, for Ottobock the control of the prosthesis is of relevance in this first step. But the second step has many implications, such as the provision of medical care for paraplegic patients. But beyond this is also the field of the human-machine interface. Everyone knows Alexa, who can be told to turn the radio on or off. But we want to go a step further in our project and not just issue simple commands such as turning on and turning off via speech, but via thought. We want to proceed from thoughts to the sensor via signal data processing to the execution of the command. And you are completely right: the signal and data processing is decisive here. The sensor captures a great deal of unspecific information and magnetic field intensities, which one first has to calculate back to what the person actually wanted to do.

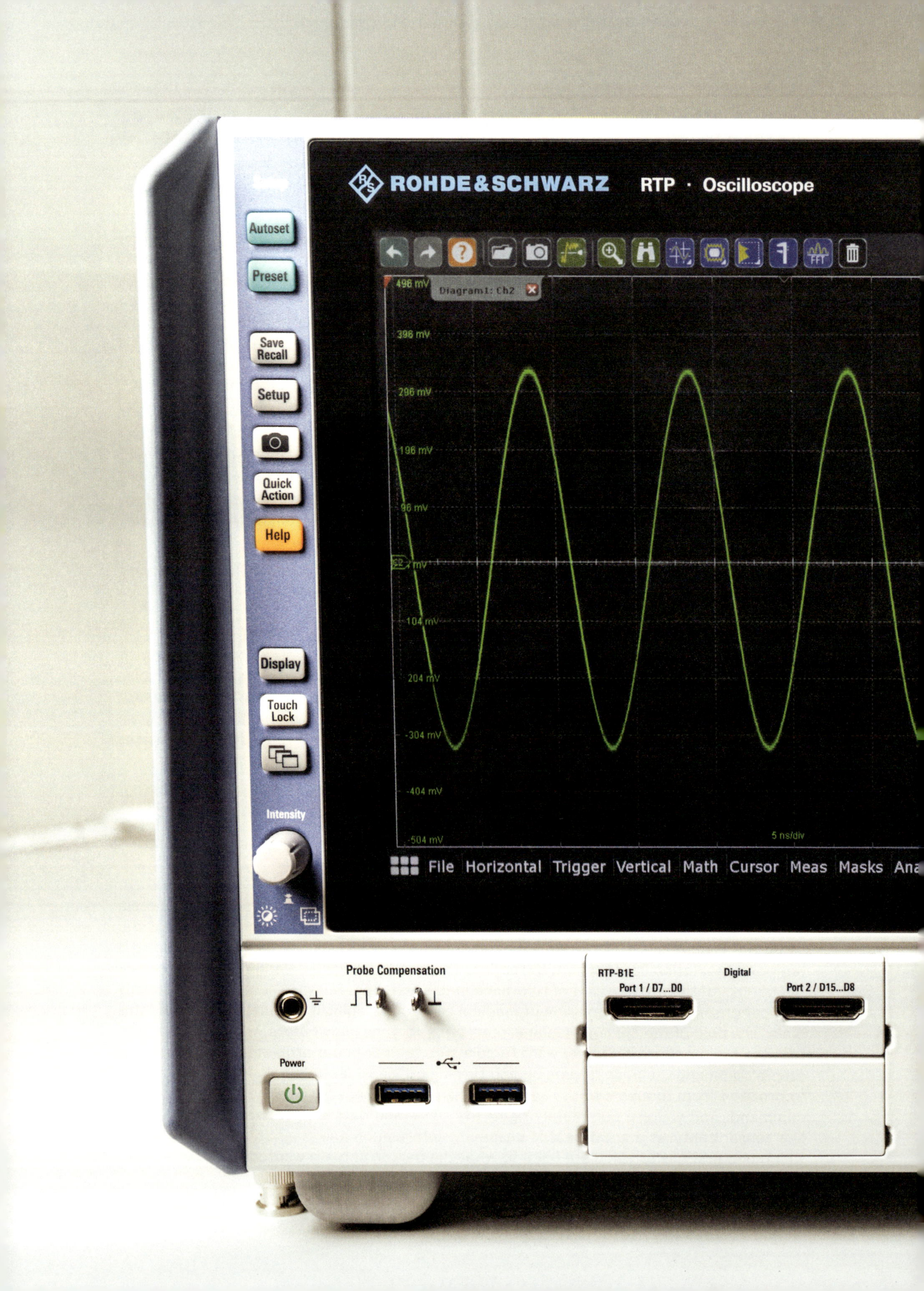
ROHDE&SCHWARZ
RTP · Oscilloscope
Autoset
Preset
Save Recall
Setup
Quick Action
Help
Display
Touch Lock
Intensity
File Horizontal Trigger Vertical Math Cursor Meas Masks
5 ns/div
Probe Compensation
Power
RTP-B1E
Digital
Port 1 / D7...D0
Port 2 / D15...D8

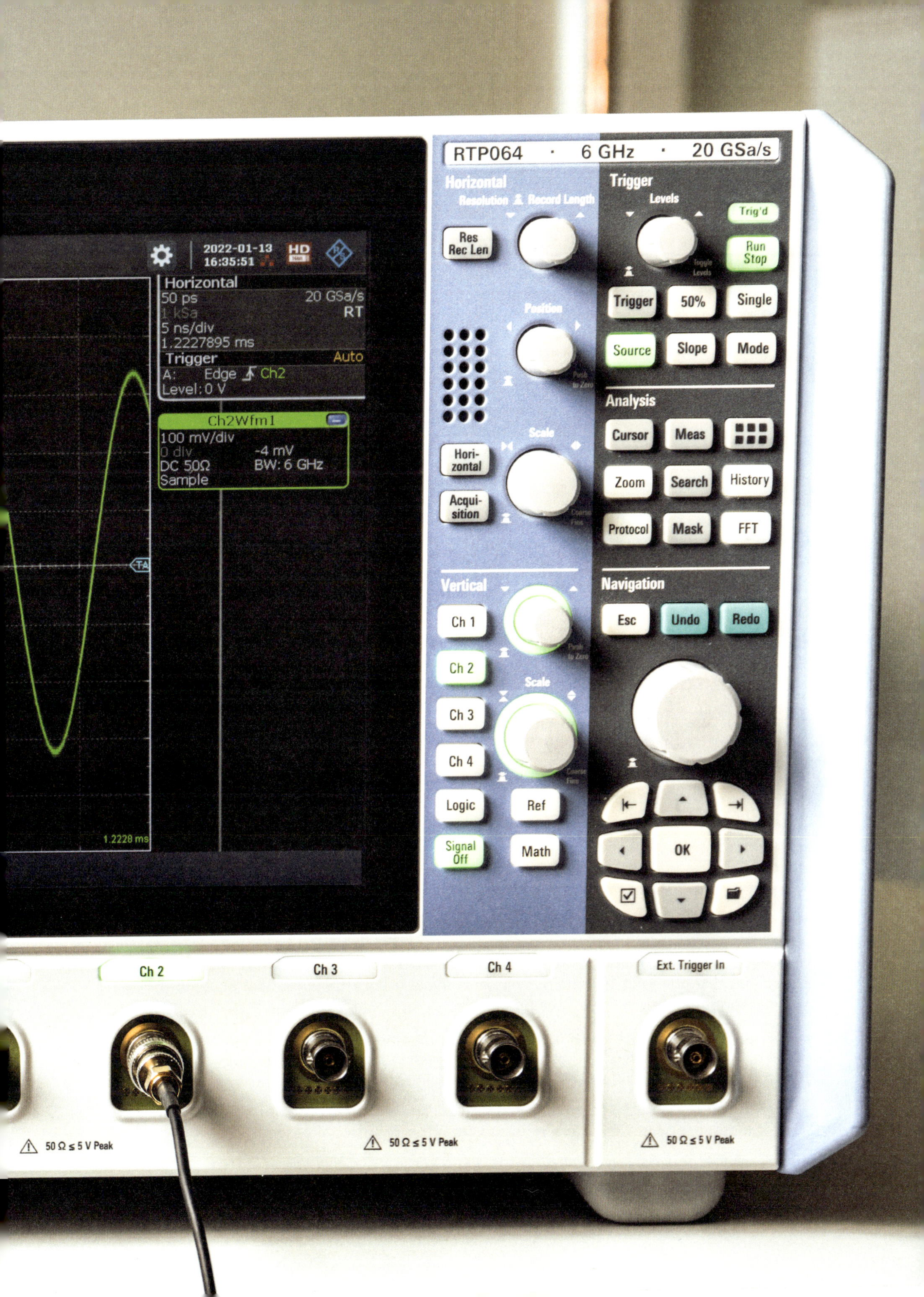
RTP064 · 6 GHz · 20 GSa/s
Horizontal
Resolution
Record Length
Res Rec Len
Position
Scale
Hori-zontal
Acqui-sition
Trigger
Levels
Trig'd
Run Stop
Trigger
50%
Single
Source
Slope
Mode
Analysis
Cursor
Meas
Zoom
Search
History
Protocol
Mask
FFT
Navigation
Esc
Undo
Redo
OK
Vertical
Ch 1
Ch 2
Ch 3
Ch 4
Logic
Ref
Signal Off
Math
2022-01-13 16:35:51
HD
Horizontal
50 ps
20 GSa/s
1 kSa
RT
5 ns/div
1.2227895 ms
Trigger
Auto
A: Edge Ch2
Level: 0 V
Ch2Wfm1
100 mV/div
0 div
-4 mV
DC 50Ω
BW: 6 GHz
Sample
1.2228 ms
Ch 2
Ch 3
Ch 4
Ext. Trigger In
50 Ω ≤ 5 V Peak
50 Ω ≤ 5 V Peak
50 Ω ≤ 5 V Peak

How many diamonds will you have to shoot into the head in order to read thoughts? A few hundred at least?

[*Laughs.*] If we knew that, we would already be a good step further, and luckily all the sensors sit on the head. We already have methods such as magnetoencephalography (MEG) that can be used to measure the brain's magnetic field.

It is not possible with magnetic resonance imaging (MRI)?

Conventional MRI doesn't supply a picture of brain activity; instead it displays its structure. We don't just want to record static structures but also dynamic structures. With functional MRI (fMRI) it is possible to measure the activity, but the signal employed there is not very specific. However, above all, we want to train neuronal networks through machine learning with a finite number of sensors, for example sixteen or thirty-two, so that in future paralyzed functions can be carried out again.

But the brain doesn't need any training because it doesn't forget the routines. If the patient lost his arm ten years ago and then an advanced sensory technological system is attached, does the brain send exactly the same signals as before?

Even when the brain doesn't necessarily need any training, the neuronal network definitely needs to be trained through machine learning. That means the patient or the Alexa user has to calibrate their system. They must think, "Please make it louder," so that the sensor system knows that a command is now coming: "Increase volume." You have to train the neuronal network to recognize these simple commands. In addition to the sensor, which detects the magnetic field intensity, the signal and data processing with neuronal networks has to be trained, in order to filter out what is relevant from the mass of information.

So the point is not complex things such as emotions, feelings, or memories, but redundant routines that are extremely trivial from the perspective of the complex system of the "brain." These routines can be generalized: the brain always generates the same pattern in reaction to the command "lift leg." In many points, artificial intelligence does not reach the level of human intelligence, but when it comes to pattern recognition, it is very good and actually better. In principle it is higher mathematics, statistical procedures, but not human intelligence.

Absolutely correct.

I picture it as follows: the sequences for walking are relatively similar in all people, as they are located in the same brain region and generate the same electrical patterns. So if one hundred thousand people were to feed their walking processes into a cloud, it would be relatively easily to feed these back into an amputee or a paraplegic patient.

Yes, definitely. Collecting data from many patients is a good method for obtaining a good initial value. But for optimum performance, the prosthesis has to be trained individually. In contrast, with applications such as Alexa, the data from the cloud may already be sufficient because the fault tolerance is higher if Alexa makes the music louder, even though you wanted it quieter.

However, the electrical impulse is not the only thing that arrives via the nerves and activates the synapses. This is also done with biochemical processes in the cells. How do you measure them? You are merely the electrical engineer in this game.

[*Laughs.*] You're right. We do not worry about all these complex sequences. It is irrelevant to us if messenger substances are released from the synapses via exocytosis—

That is what I thought: you are merely the electrician.

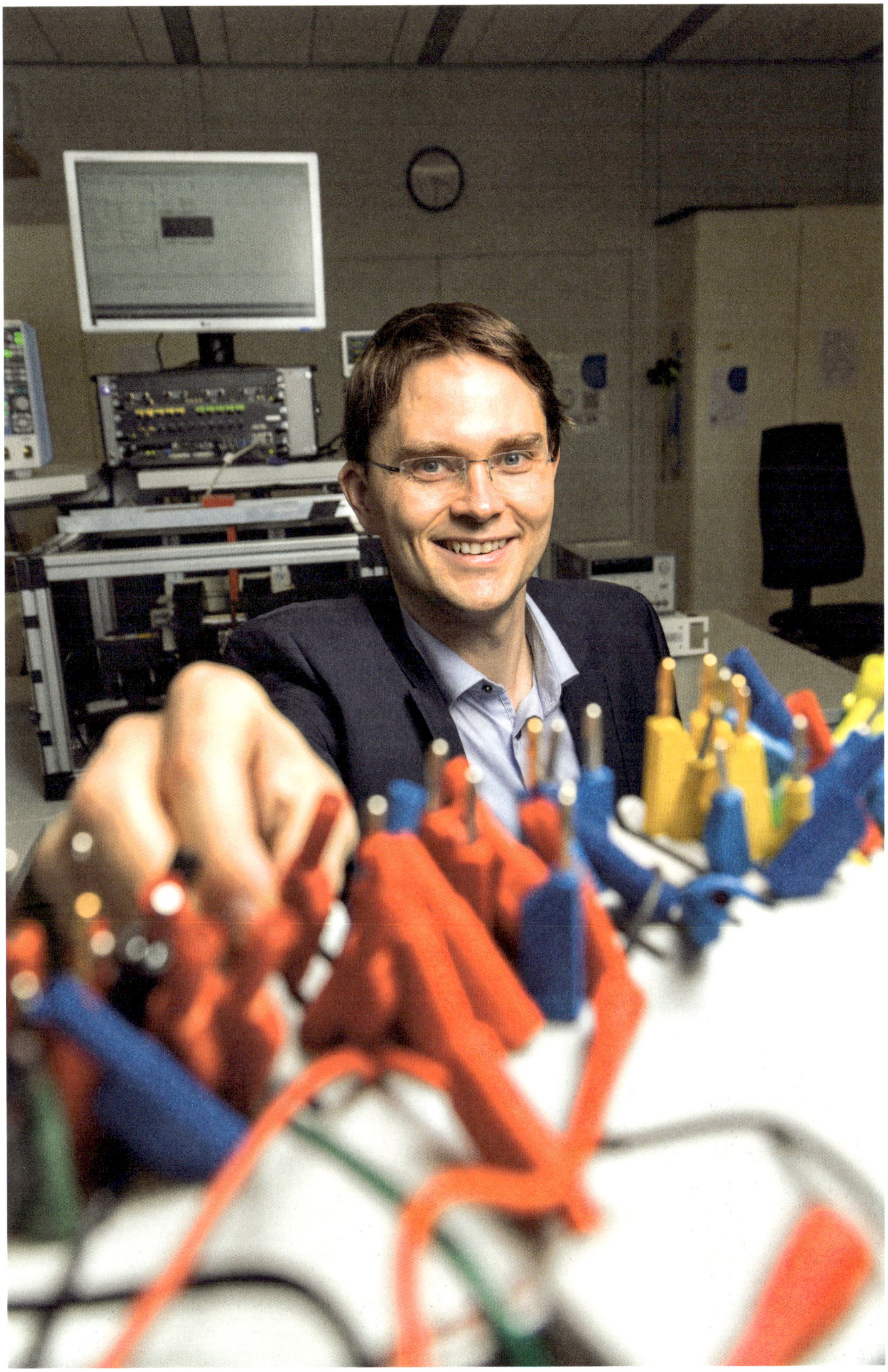

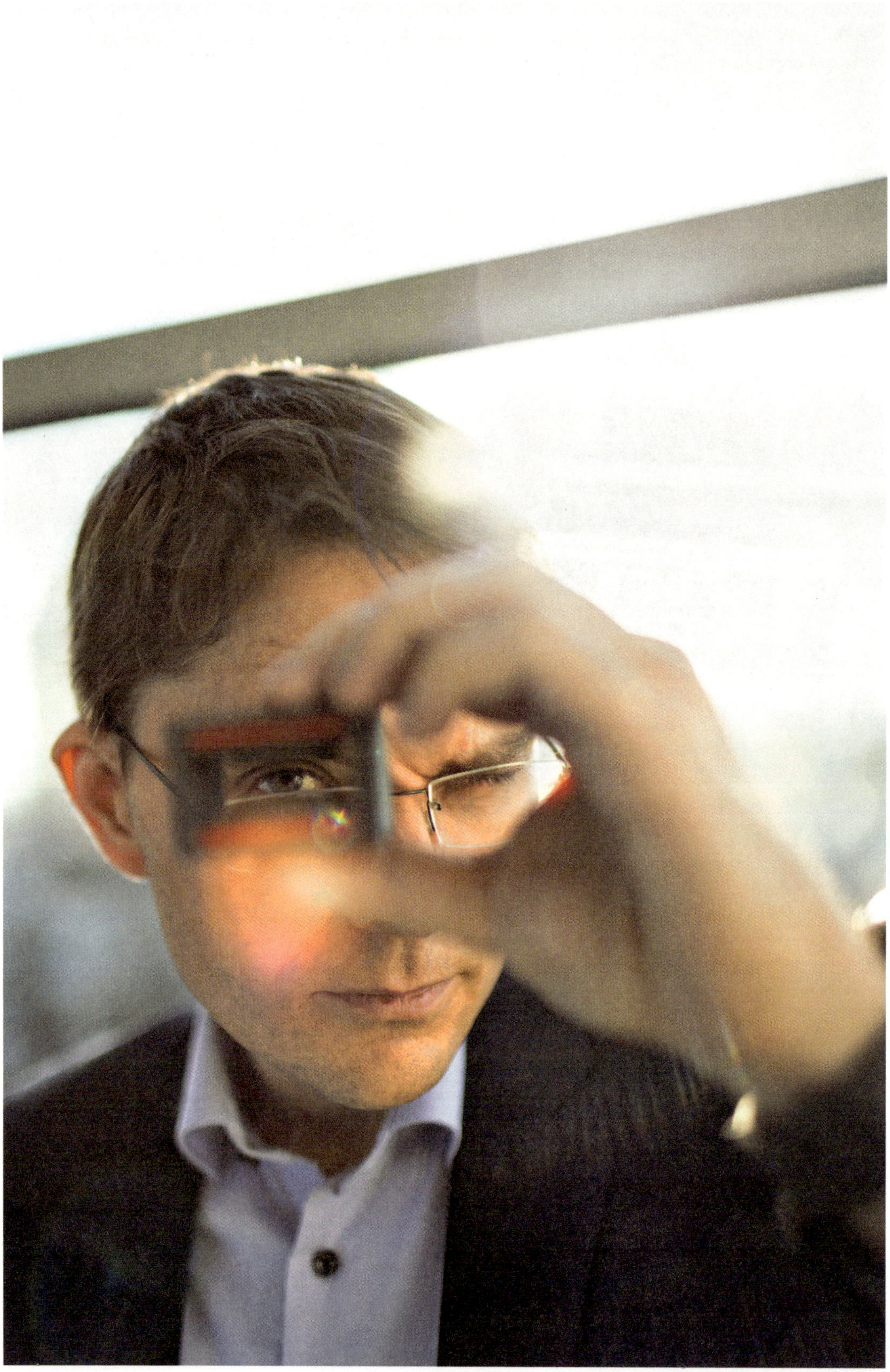

—as long as a current is flowing. And after the synapses, there is always a nerve fiber. Then there is a magnetic field that I can measure and identify as a temporal waveform. Action potentials are sent, which we can recognize—the key idea is pattern recognition—and we measure the magnetic signature of the action potential. That means we are not interested in the biochemistry behind it.

Does the biochemistry have anything to do with the magnetic field at all? Is it just an exchange of electrons that generates this magnetic field?

That depends where the signal is intercepted. In the case of a signal path along nerve or muscle fibers, the signal can be modeled as purely electrical signals, which can be described by means of relatively simply equivalent circuit models, although the underlying processes are still biochemical in nature. During the generation of an action potential in a neuron, the biochemical nature of the underlying processes that ensure that certain ions in a cell are pumped in or out have to be taken into account. Either way, at the end a current flows whose magnetic field we can measure.

If we take a closer look at a synapse: an electrical signal arrives via a nerve, which is to be passed on unchanged to the next neuron. In order to do this, the electrical signal is biochemically translated into messenger substances, which then trigger the same electrical impulse at the opposite synapse. How high is the voltage running through our nerves?

Individual action potentials generated in the brain have peak to peak potentials of up to 120 megavolts.

There are thousands of kilometers of nerves before one reaches the little toe—

[*Laughs.*]

—so the current would immediately dissipate if these transformers didn't continually amplify the effect. Or am I on the completely wrong path here?

Not at all. The principle behind it is the regeneration of the signal. You are absolutely right, if I were to simply send out the signal it would get smaller and smaller and eventually disappear. It is the same principle in a neuronal network, in which commands come from different directions. The signal is continually regenerated along the signal chain and thus can overcome distances. The signals from a number of actions arrive at the synapse at the other end, if they are sufficiently strong. If the neuron's threshold is exceeded, it continues to fire. In fact, even the process of transmitting the signals along nerve and muscle fibers typically involves signal regeneration.

Can this magnetic field measurement also be used to explore the body and detect illnesses? For example, would it be possible to diagnose Alzheimer's earlier through a certain pattern of magnetic fields?

Yes. Absolutely correct. There are a variety of research paths that are employing these technologies. But until we have reached this point, a lot of research work will be necessary, and we will have to orient our team more strongly toward medicine.

Can your sensors also be used for the noninvasive measurement of things such as blood sugar, lactate, cancer messengers, and inflammatory markers? Are you also pursuing this direction? That would be a multibillion market.

Yes. For example, we have a project "QMED—Quantum Sensors for Biomedical Diagnostics" with Boehringer Ingelheim and Rentschler Biopharma. In this project, we want to replace classical nuclear magnetic resonance spectroscopy (NMR spectroscopy) with second-generation quantum technology. A normal nuclear magnetic resonance spectroscope costs between one and fifteen million euros and requires a huge magnetic field in order to align the spins and read them off. The great thing about the NV center magnetic field sensor is that instead of a ten-tesla base magnetic field, we only need 10^{-3} tesla—four orders of magnitude less. This enables the NMR measurements to be carried out a lot more economically, and the results are still highly specific. This means that one can still differentiate different chemical substances and, in some situations, is even more sensitive.

What does the project with Boehringer Ingelheim measure?

They want to measure the chemical composition of the proteins and, in the medium term, the folding of protein molecules in drugs.

Protein folding is a real sensation, which is just one or two years old. This enables a wide range of biologically generated medicines to be produced. Sartorius is also working on protein folding.

Naturally they are all working on it. And the good thing about the sensor is that it doesn't need a space-consuming magnet but only a device the size of a shoebox. If this approach is pursued, maybe at some point the NV-based sensor will be so small that it can be inserted into the body invasively, for continuous monitoring, instead of taking a drop of blood from the body.

But it would be enough if I were to apply the blood to a test strip—as in a blood glucose measuring device, which is merely a banal microchip—and then stick it in your little shoebox in order to identify 130 different markers.

That is an approach that one could take to measure other biomarkers in addition to the glucose. If you want to take a sample once a week or month, the needle prick is not a problem. But with diabetics, it is about avoiding the needle prick as they have to take blood numerous times during the day, and in time the fingers become pricked all over. There are already devices on the market that enable a sensor-based measurement for up to fourteen days. But we want to make it even easier. For example, today, placing a stent in the bloodstream is no longer a major operation. It would be possible to place a stent in the bloodstream every five to ten years, which one could read off inductively. With the stent, we have a kind of coil to which the spins are coupled. When the magnetic field in the coil changes, a potential is induced, and current flows through this stent. I can place another coil outside and read off the magnetic field via inductive coupling.

Or you introduce small robots into the bloodstream. This is just a question of miniaturization. Your diamond is almost small enough; it is just the chip that needs to be smaller.

Small robots in the bloodstream that can simultaneously analyze and repair is a very nice vision. But we are still a long way off from this. It would be a great step forward for us if we could read off a stent from the outside with a credit-card-sized device. That would be very convenient. My personal vision is to create a digital touchpoint by this means, where you can read off the person's complete state of health.

I have one final question: why are there so many Chinese students in your laboratories? The state is pumping a lot of money into this, and all we are doing is training students from Southeast Asia. Something is not right here.

Yes, that is a completely justified question. And here, two hearts beat in my chest. On the one hand, my heart sometimes bleeds when I see how few domestic students choose the disciplines of physics, electrical engineering, and mechanical engineering. And you have to remember Germany is a country poor in raw materials and our capital only comes from our ideas. The companies also need far more engineers in the fields of mechanical engineering, electrical engineering, physics, et cetera. But studying engineering is no longer popular today. We have just as many intelligent people compared to twenty or fifty years ago. But they prefer to study business management in order to earn millions at a consulting company or as an investment banker, instead of eighty thousand as an engineer in a mechanical engineering company. I currently have six vacant postgraduate positions that I would be happy to give to German students—and would greatly help me in teaching. I have several lectures in the bachelor's courses taught in German, and I need German-speaking tutors to supervise the associated exercises. However, it is very difficult to find top-qualified graduates in the required numbers. On the other hand, I am delighted at the diversity in our international group, which is highly qualified and motivated. I also studied abroad. I think it is a fantastic experience when the students go out into the world. This being said, if I had the choice, I would like to see more students from Germany and Europe in our department. Our department profits enormously from our international students, but when I think about Germany as a business location, then it is time to change course and, once again, generate enthusiasm among young people for the natural and engineering sciences.

MARTIN BÖHM, HANS DIETL, ANDREAS GOPPELT,
BERNHARD GRAIMANN, AND GÜNGÖR KARA

Reflections on Quantum Technology

The five experts from Ottobock discuss how relevant quantum technologies could become for the company's future products.

Bernhard Graimann: People are probably wondering what Ottobock actually has to do with quantum technologies. Research at Ottobock is applied, industrial research. The maturity level of the respective technology that we invest in is normally relatively well advanced—and we develop prototypes that we then pass on to product development. But we also have the task of looking behind the scenes at what is happening in the academic world in terms of emerging technologies. We want to understand this better. Naturally we could just read the papers, attend conferences, and discuss with experts—but it is far better to work directly with the specialists on concrete research projects as partners. In the process we generally define requirements for a possible product, and over the course of this project have the opportunity to evaluate prototypes. This gives us a deep understanding of all that is possible with this technology. We are also participating in the project "Quantum-Human-Machine-Interface Based on Quantum Sensors" as an observing research partner in the research cluster QSens. Here we are interested in the intelligent control of prostheses and orthoses. Until now, in prosthetics we have captured myoelectric signals on the surface of the skin to control a prosthesis with this information. However, there are a couple of challenges here. The number of degrees of freedom—in other words, how many joints can be controlled simultaneously—is limited, just like the resolution. That is why we are looking for better possibilities. One possibility is to become invasive and measure the myoelectric activity inside the body—or you enter the peripheral nervous system directly. We are not interested in brain-computer interfaces because the information in the peripheral nervous system has already been prepared accordingly and is better for our use. We are currently concentrating on noninvasive systems, but we are also observing what is happening in the invasive world. And now quantum sensors are coming into the game, promising to make the quality of invasive measurements noninvasive in the future, while at the same time enabling much better and more precise control of prostheses or orthoses. But this is still a long way off, and there are still a lot of questions. For example, a very important question is how to deal with other background magnetic fields that could interfere with the measurement. I can measure the minute magnetic field that is generated when an electrical signal flows through the nerves with a quantum sensor. But in the surrounding area there are a lot of other magnetic fields that could interfere with the reading. Solving this problem will be an important step in very efficient noninvasive man-machine interfaces. As always, it is important for us to be familiar with alternative approaches to solutions and to compare them. We are particularly interested in approaches that are minimally invasive, such as what Hugh Herr is doing at MIT, who is injecting small magnetic capsules into the muscle so that muscle movement can be measured quite easily from the outside. It hasn't been decided which approach will catch on, but for us it is important to understand such technologies at an early stage and to monitor developments.

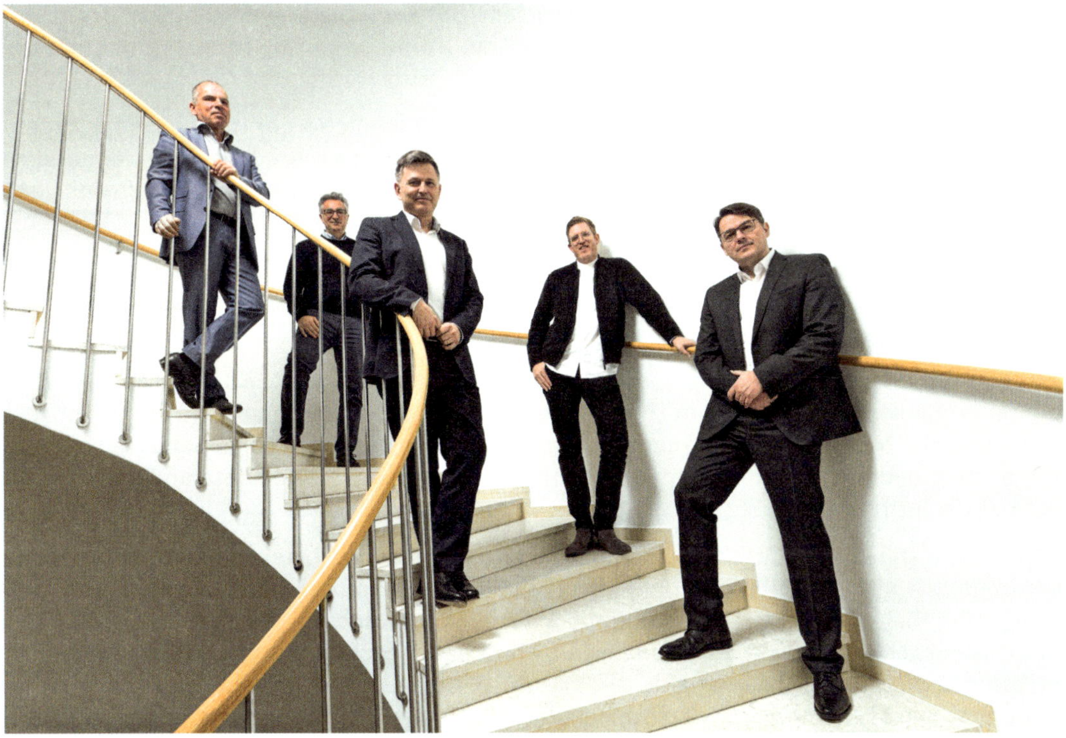

Hans Dietl, Güngör Kara, Andreas Goppelt, Martin Böhm, and Bernhard Graimann (from left to right)

Andreas Goppelt: I am excited about another dimension of Professor Anders's research: SQUIDS, the quantum sensors of the first generation, only function at very low temperatures and need to be continually cooled. Quantum sensors of the second generation are stable at room temperature and could be decisive for electromobility as the batteries can be charged much more quickly. If this works, it will have a huge effect on Ottobock's technologies and could be extremely relevant for us.

BG: If we take a look at quantum technology as a whole, a lot of application scenarios are still in the distant future. Nevertheless, it is interesting for us to understand quantum computing. Over recent years there has been a lot of hype around it. Computer structures are generated with more and more qubits, opening up completely new possibilities. When this actually becomes feasible, it will start a revolution. For example, modeling and simulation are very important for us in many areas for optimizing technical solutions without actually having to build the things or calculate the effects of medical aids on the human body. Even though the computing power is considerably better than it was ten years ago and continues to improve all the time, the real revolution will come with the quantum computer.

Hans Dietl: The cryptography and encryption will also open up a new world.

BG: Yes, this will turn the whole world upside down, especially in the financial industry. That is not directly our issue. What is important for us is that we can model highly complex interconnections. Quantum computing is very promising here. Then completely different algorithms will play a role, compared to those we have understood to date. That is a huge research field.

HD: There is an interesting start-up that writes operating systems in emulation of quantum computers. There is a lot happening. It is important that these things develop in parallel.

BG: A major issue is how to establish the interface between the standard binary world and quantum computing. In other words, the interface from computing to quantum computing—and back again.

AG: I see it slightly differently. With quantum technology science frees itself from the determinist perspective of Newtonian physics and attempts to gain a more holistic understanding of the world, which still cannot really be grasped intuitively. This brings us to Schrödinger's cat and similar thought experiments, which can still be reconciled with the laws of physics.

Güngör Kara: To establish the link to digitization: these new technologies open up new possibilities for us with respect to our core competences and our value creation chain that we can't even imagine at the moment.

Martin Böhm: The exciting thing about Ottobock is that we deal intensively with almost all digital technologies presented in the Gartner Hype Cycle, apply them, and try to convert them into the business model. Quantum technologies are relevant to us because we need the corresponding computing capacities for ever-larger volumes of data and increasingly complex algorithms.

AG: What distinguishes Ottobock is simultaneously a huge challenge. If we don't just think about people with special bodies, but also about empowerment, human augmentation, and everything that goes with this, it requires a deep understanding of an enormous number of different technologies. This distinguishes orthopedic technology from dental medicine or the manufacture of hearing aids. Ultimately, it is this breadth of technology at Ottobock that makes the company so fascinating. This means that we have to be very smart in selecting which technologies we will codevelop ourselves as a company and which technologies we will engage in active listening and remain in tune with the times through research collaborations in the academic sphere or with start-ups and industry players. We have to understand what is happening with the technologies, and then recognize at a very early stage what impact this has on our business models. Doing this correctly is enormously valuable and a high art. Quantum technology is a research area where we have decided to actively listen.

Advanced Materials / 3D Printing

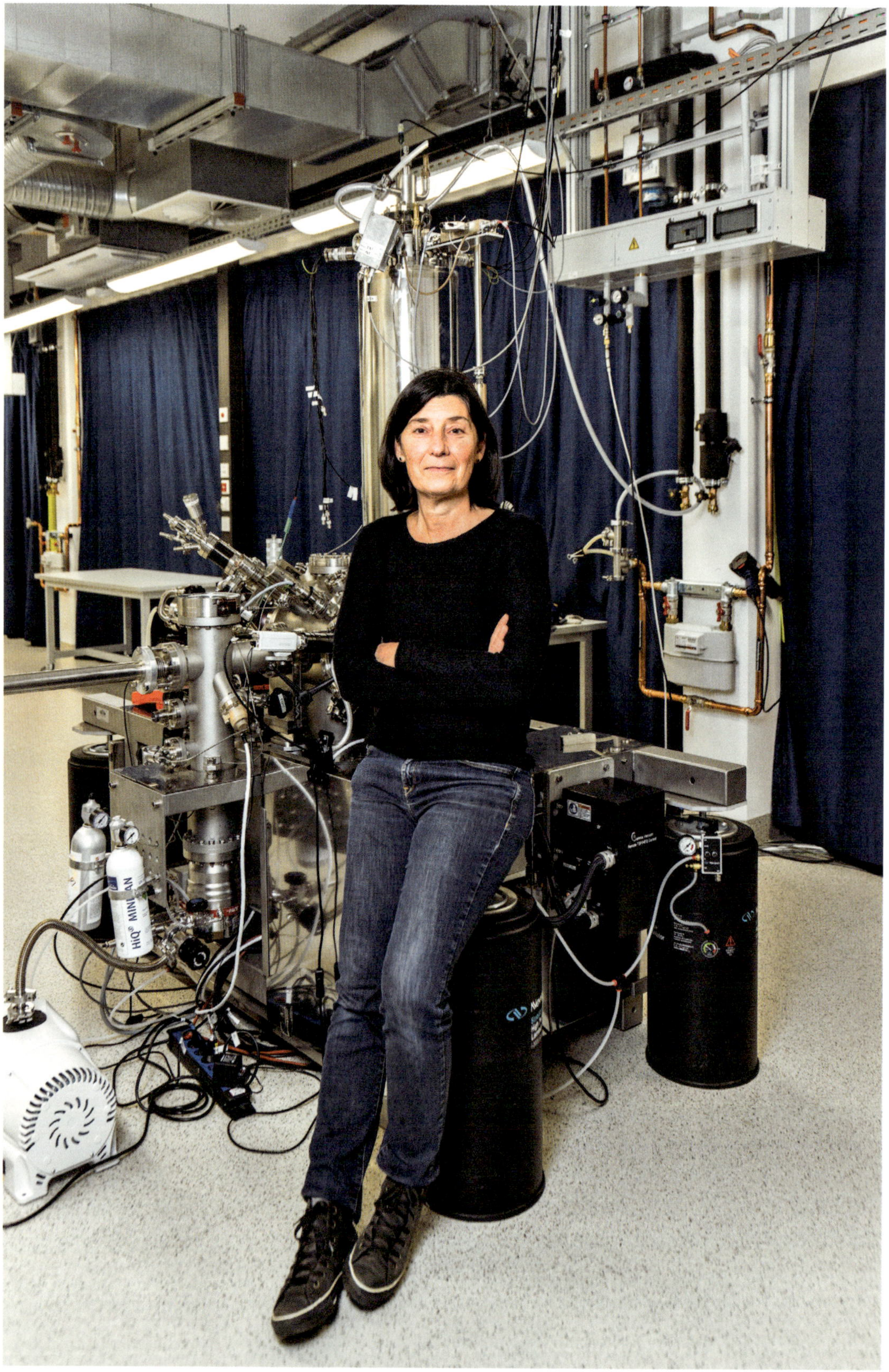

FRANCESCA MORESCO

Competing for Every Nanometer

A nanometer is a millionth of a millimeter. In this range, things are so small that they are invisible to both the naked eye and optical microscopes. Nevertheless, this microcosm could be decisive for the future of chemistry and medicine. Molecular machines, which are currently in the starting blocks, could one day build completely new molecules or distribute active agents in the human bloodstream and repair cells. The worldwide community of molecular machine engineers is very small and competitive, and meets regularly for the Nano-car Race, in which miniscule cars made from the tiniest molecules compete against one another, reaching top speeds of 16.66 nanometers per hour.

Dr. Moresco, you are the head of the research group Single Molecular Machines at the University of Dresden's Center for Advancing Electronics. You conduct research into nano-materials at the molecular level and their electronic characteristics as an alternative to silicon technology. What does this involve?

Within the field of nanotechnology we are concerned with molecular electronics and molecular machines, which are closely connected. Within semiconductor technology the structures of the electronic components are currently getting smaller through the advances in miniaturization. Several billion transistors fit onto one processor, but silicon technology is increasingly reaching physical limits, as the electronic properties of the extremely small crystals can no longer be adequately controlled.

So Moore's Law, which states that the integration density of microprocessors doubles every eighteen months, is coming to an end? Silicon is becoming increasingly expensive and although it should be as freely available as sand on the beach, a lot of people don't know that desert sand is not suitable for the extraction of silicon. How can you achieve smaller and better structures?

Molecules are the smallest stable structures that can be used as electronic components. We work with molecules in the nanometer range (10^{-9} m) that behave according to the laws of quantum mechanics, not classical physics. Nanomaterials are quantum mechanical objects. My colleagues in organic electronics are developing microelectronic components for solar cells, transistors, or coating for screens from organic molecules and polymers. That functions very well, because molecules behave in a similar way to semiconductors. The gap energy is about the same as that of silicon. With the right organic molecules we can synthesize as many materials as we need.

You can reproduce the properties of silicon with organic molecules. For example, solar cells are composed of polyamorphous silicon. Is it possible that one day solar trees with a greater efficiency than silicon will be grown from organic molecules?

Solar cells or transistors will not really be any better until we have understood the mutual interaction between electrons and light at the level of the single molecule. The molecules are not really like silicon, as they are very small and imperfect crystals. In electronics—as the name implies—everything has to do with electrons. If one applies a voltage to a crystal, electrons flow through it, encountering atoms of a uniform lattice structure that diffract them as if they were light. The electron is simultaneously a particle and a wave. Its wavelength is almost as long as an entire molecule. This can't be compared with the properties of silicon. That is why we take a bottom-up approach in molecular electronics and investigate the electronic and mechanical properties of individual molecules. We want to understand the movement and the intramolecular mechanics of a molecule on a surface. That is the first step toward developing controllable molecular machines at the atomic level. We can already build a small transistor or a logic gate from just one molecule. And here quantum mechanics comes into play again.

Can I picture the chip of the future as a small gold plate with lots of individual molecules as transistors?

A computer can be built entirely from molecules. There are a variety of approaches. For example, one can process quantum mechanical signals in the molecule itself, because the molecule, as a quantum mechanical object, can function like a logic gate that translates binary input signals into binary output signals. In just one molecule! A central element combines numerous input and output elements and thus needs fewer connections. But naturally there are many challenges and questions: How do I connect a single molecule to the rest of the world? Can I also use additional molecules as wires and thus design a completely molecular-based structure comparable to a circuit board? What would a hybrid structure composed of metals, semiconductors, and molecules look like? How can a transistor or a switch be configured?

As an on-off switch, a transistor is an essential component of electronic circuits and always zero and one. How does a molecule become a transistor?

Molecular transistors function according to the resonance principle. In contrast to silicon, the molecule does not have any energy bands with numerous electronic states and band gaps, but individual energy states. During resonance, that means when the applied voltage reaches a molecular energy state, more current flows through the molecule. The problem is that everything is so incredibly small.

That is why you use a scanning tunneling microscope to test the organic molecules on surfaces such as gold, silicon, or semiconductor surfaces. What kind of experiments are these?

The functional principle of the scanning tunneling microscope (STM) is based on the quantum mechanical tunnel effect and is capable of imaging the topography and electronic properties of structures on surfaces at minus 268 degrees Celsius. This temperature is not far from absolute zero. However, with STM we don't just observe the behavior of organic molecules on metal and semiconductor surfaces; we can also precisely measure the electronic structure and the electron density of states and build nanostructures, atom by atom.

In the molecule?

In the molecule. The STM tip provides the necessary drive force to drive the molecule with electrons. The molecular conformation—in other words the spatial arrangement of the atoms in a molecule—is precisely controlled, while the STM tip can induce movements and conformation changes in the molecule. A molecule is not just a series of atoms, but also electronic states. If I change something in the molecule, this state also changes.

The bonding forces between atoms are electrical.

Exactly. With the right energy I can achieve a free and an occupied electron state in the molecule and increase the probability that electrons will flow. This increases the electric

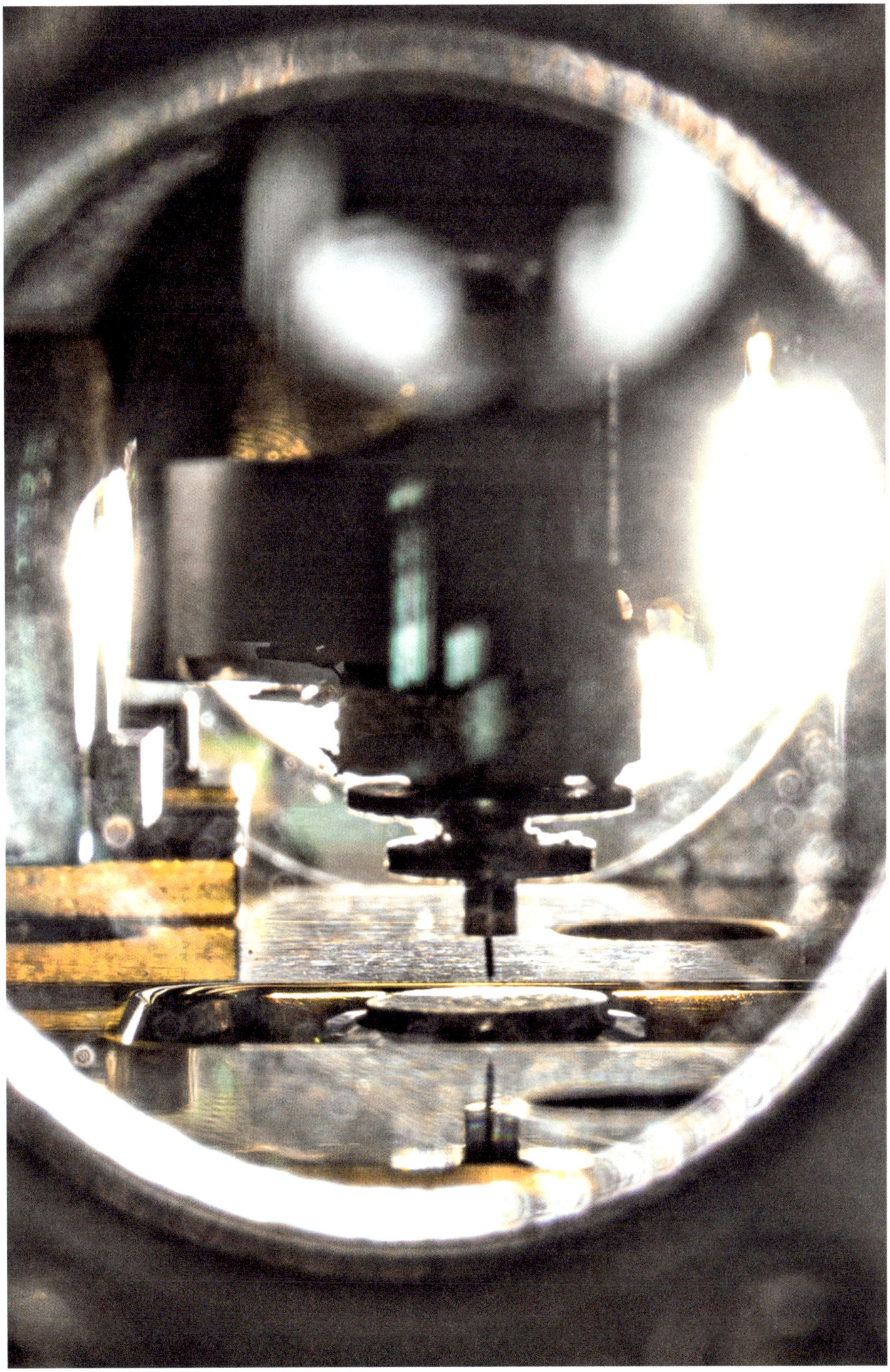

Tip and scanner of a scanning tunneling microscope

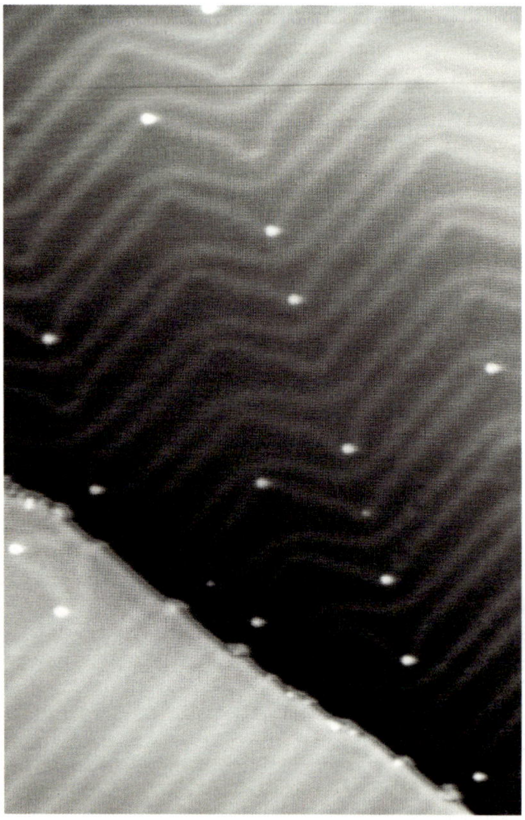

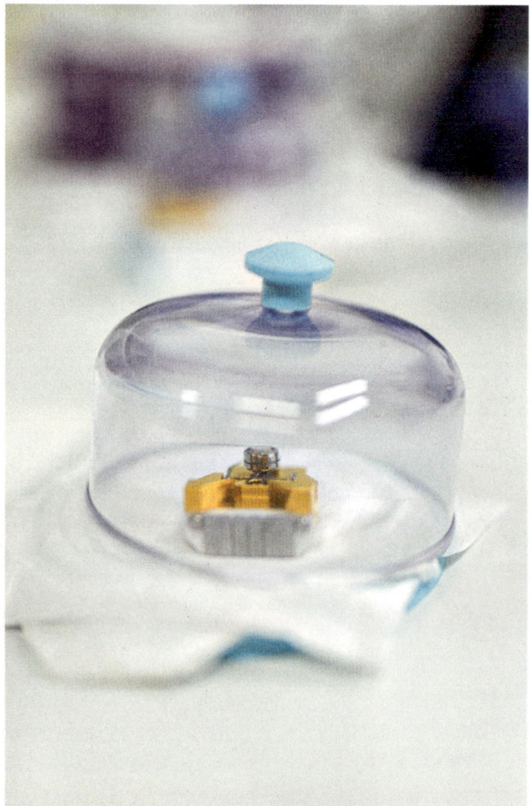

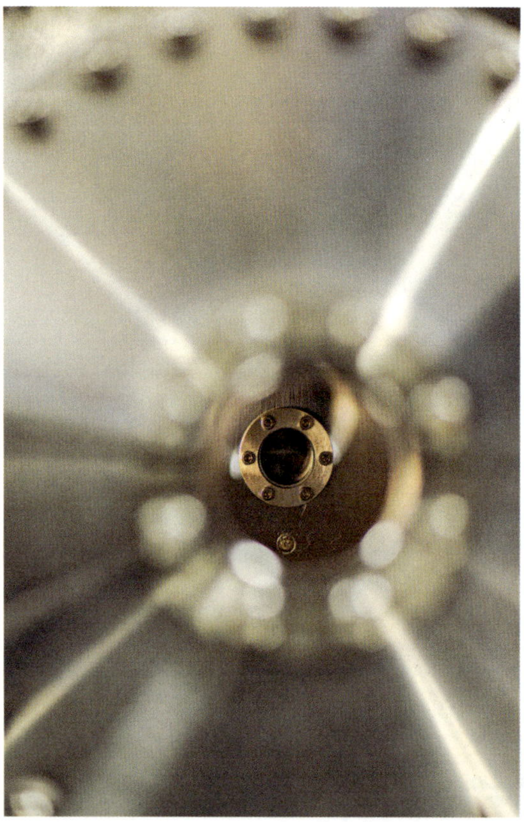

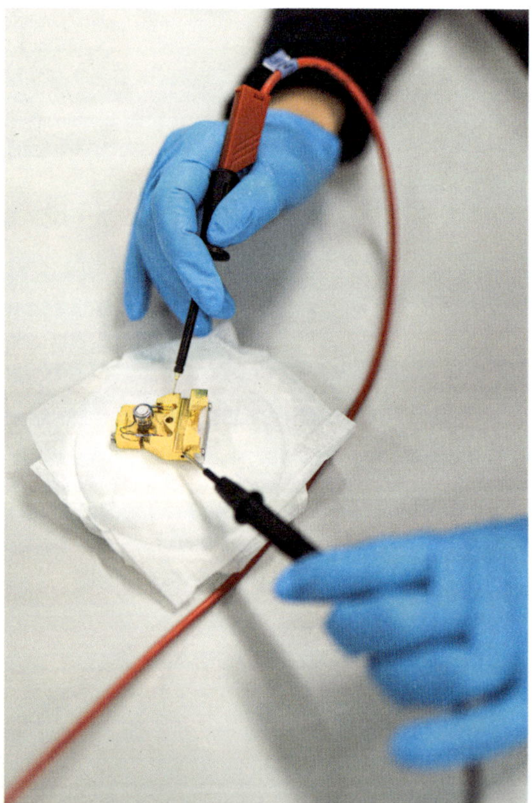

current, which I can measure with the STM. The STM indicates this state as a peak. This enables me to measure the molecules occupied or unoccupied states.

If you allow the occupied and unoccupied states in the molecule to interact with one another, the molecule computes, so to speak.

The simplest variants are molecular switches, where we change the form of the molecule by means of light or electrons, and thus automatically produce a different electronic state. Or we combine the molecule with other atoms and produce a logic gate.

A carbon molecule can be used for computer operations by changing the electronic states by means of quantum mechanics. This enables you to exactly reproduce the silicon gap?

You always have the gap. But with small structures the electrons can sometimes flow through unobstructed. This is referred to as *ballistic transport*, which is very exciting, but also difficult to control. All you need is a small defect and the whole electron flow is destroyed. However, with the molecules the structures are so small that it is the tunnel effect, not ballistic transport, which is the main transport mode. That is why it is so interesting for both electronics and chemistry, because one can also use modern, nonpoisonous, and sustainable molecules.

Can you also build new molecules? New organic molecules for the chemical or food industry?

Yes, normally we work with the chemists who synthesize the molecules that we want to investigate. One can also induce chemical reactions on a surface with the STM tip and create new molecules that no one else is able to synthesize. That is very interesting.

So these are completely new molecules that don't exist in nature?

Yes. For example, the carbon nanoribbons are very exciting. Maybe you are familiar with the carbon nanotubes. These are very small, tube-shaped formations composed of carbon with a honeycomb structure. The walls are composed of individual layers of graphite. These nanotubes are a very interesting material for the electronics industry. They have interesting physical, electronic, thermal, and optical characteristics as a result of the interplay between the atomic and electronic structure. The precisely cut, narrow strips of graphite, known as *nanoribbons* are also very interesting. Until now nanoribbons only existed in theory, but now with the aid of STMs they can be precisely synthesized on a surface at the atomic scale.

In what areas can your research be applied?

Molecular electronics is naturally relevant to developing model systems for applications in electronics and computer technology. In a molecule electronics and mechanics are combined. A molecular switch regulates the flow of electrons and moves the molecule between two positions. An electronic stimulus can be transformed into a mechanical movement that is controlled at the molecular level.

Is this also conceivable within the human body?

The classical nanomachines, which won the Nobel Prize a few years ago, function in a solution and are activated by chemical reactions, like a molecular robot that can pick up amino acids and combine them. It is possible to imagine a future in which small robots move around the body, distributing medicine and repairing cells. But that is still in the distant future.

What exactly is a nanomachine?

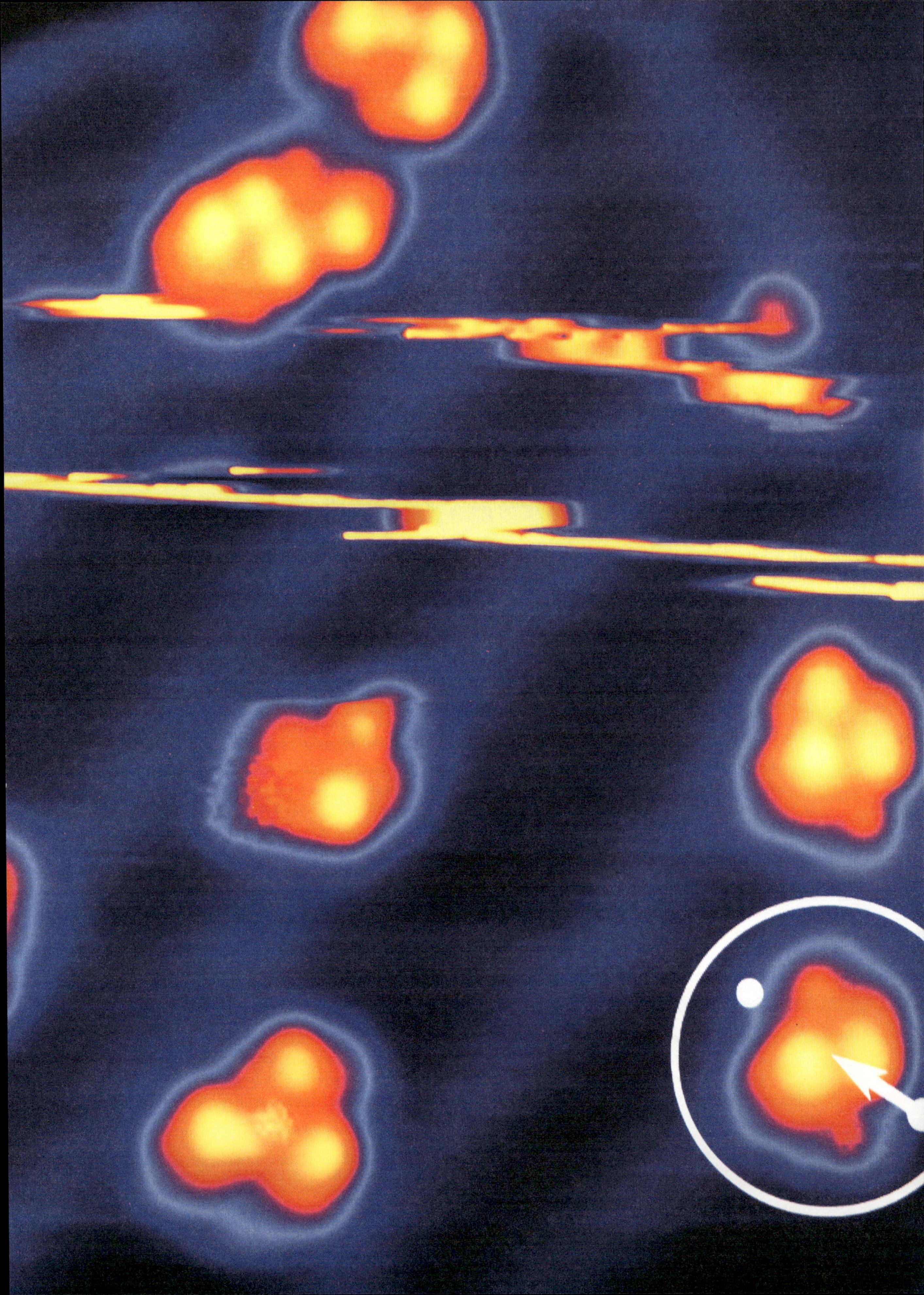

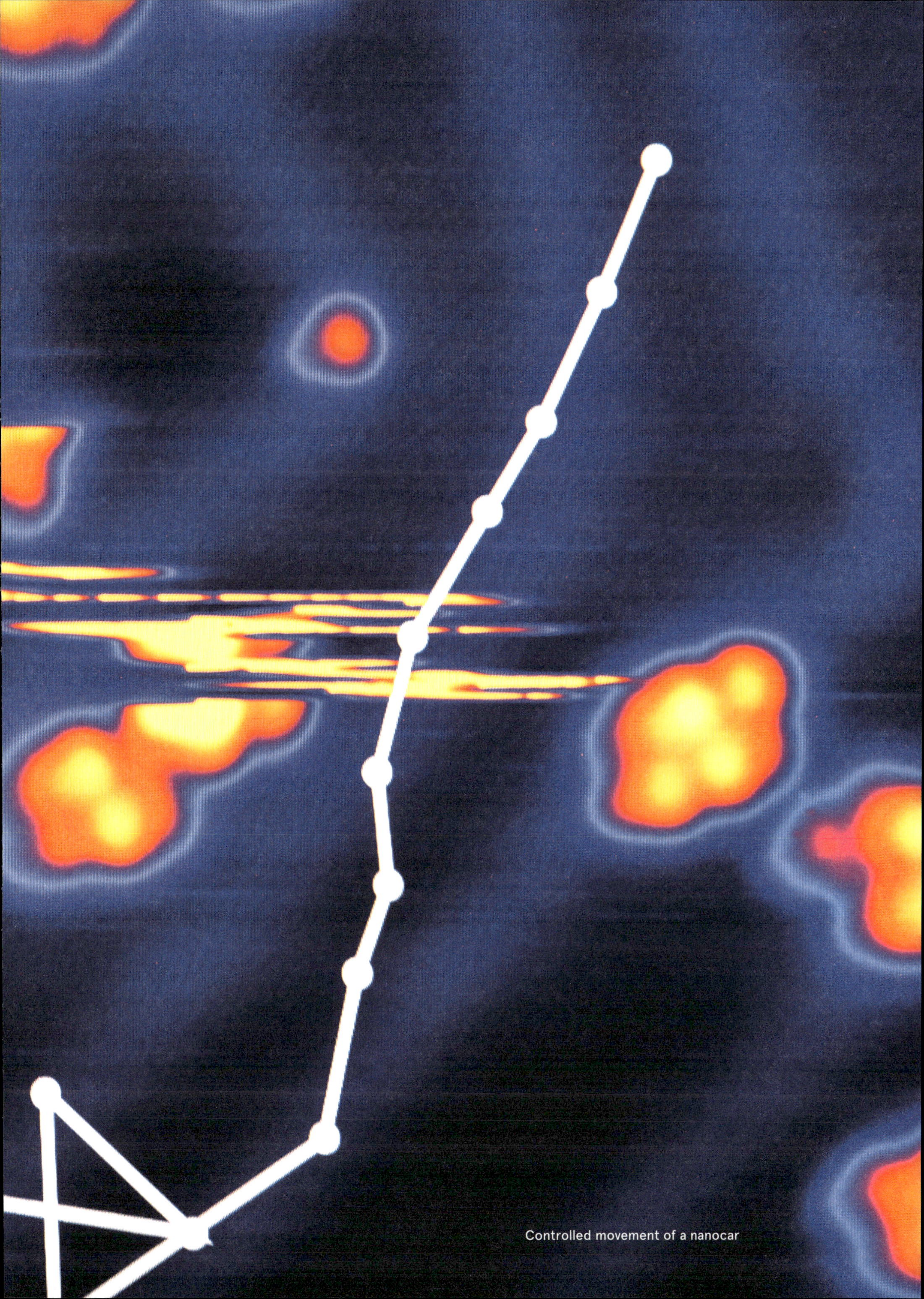

Controlled movement of a nanocar

A nanomachine or molecular machine is a molecule that can execute mechanical movements.

So the molecule activates something?

The molecule is activated and can then activate something. Just like a classical machine, energy has to be fed in from outside. The first molecular machines were developed by Jean-Pierre Sauvage, Sir J. Fraser Stoddart, and Bernard L. Feringa, who also received the Nobel Prize for Chemistry in 2016 for the design and synthesis of molecular machines. They have developed molecules whose movements can be controlled and that complete a task when supplied with the requisite energy. As early as 1983 Sauvage connected two ring-shaped molecules in such a way that they could still move with respect to one another, like two links in a chain. In 1991 Stoddart attached a molecular ring to an axel that was also constructed of molecules. Later two structures were added to the axel, and a molecular wheel was formed that could be pushed between the two structures through the application of heat. In 1999 Feringa developed a molecular motor powered by UV light. In 2011 the researchers presented a four-wheel-drive molecular car.

What molecular machines have you developed?

For example, a rotor that is operated electrically, not with light. Through voltage pulses the molecule always rotates in one direction. In addition we have developed molecular gears that can transmit the movement from molecule to molecule or that function as a rack-and-pinion drive.

And with the tip of the STM you fire electrons at the molecule and the molecule starts turning. What kind of connector is used to anchor the molecule to the gold plate?

In our case it is oxygen. You could also use sulfur, which would make an even stronger connection. This also generates many additional, exciting questions: will the rotor still turn on other surfaces? For example, it is difficult to get a molecule to roll reliably on a metal surface for a number of reasons. On the one side, a relatively strong interaction between the molecule and the surface is required to generate the rotation, so that the molecule doesn't simply glide over the surface. On the other side, a strong interaction with the surface makes it harder for the wheel to turn completely on its axis. A further question is whether one can move the molecule with the same energy when it is anchored differently, or why the rotation only occurs in one direction. If it were only thermally conditioned, then we would have the so-called Maxwell's Demon, which violates the second law of thermodynamics. Then I would have found a molecule that rotates on its own without electrons.

This doesn't occur in nature?

No. You always have to input energy, so that something can move out of its ground state.

So it is not a perpetuum mobile?

No. It would be a type of perpetuum mobile if it always rotated in the same direction just with thermal energy. So there are a lot of questions, and we want to gain a better understanding of the mode of action of the individual molecules. And here the scanning tunneling microscope is of help again, which has developed from an analytical technology into a tool, allowing nanomachines to be driven with atomic precision in order to restructure metallic and nonmetallic surfaces. Furthermore, this manipulation is also suited to the investigation of fundamental physical and chemical properties at the level of individual atoms or molecules, including quantum phenomena or magnetic properties.

So you are conducting real basic research.

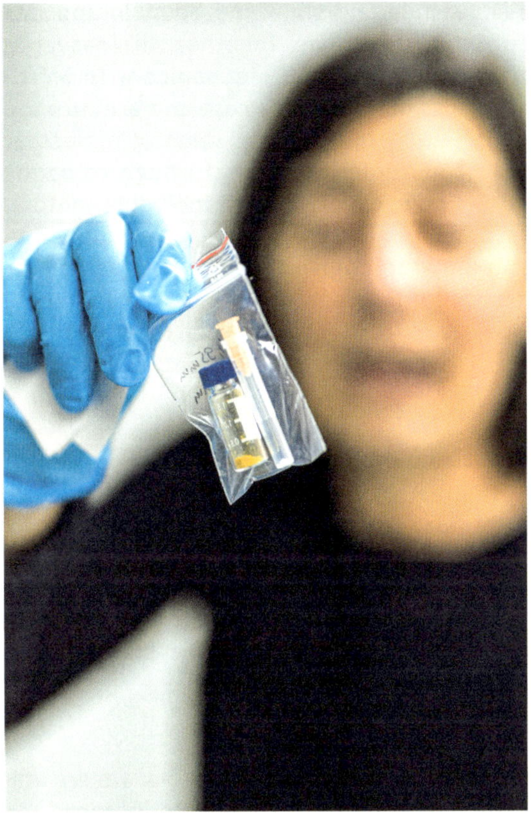

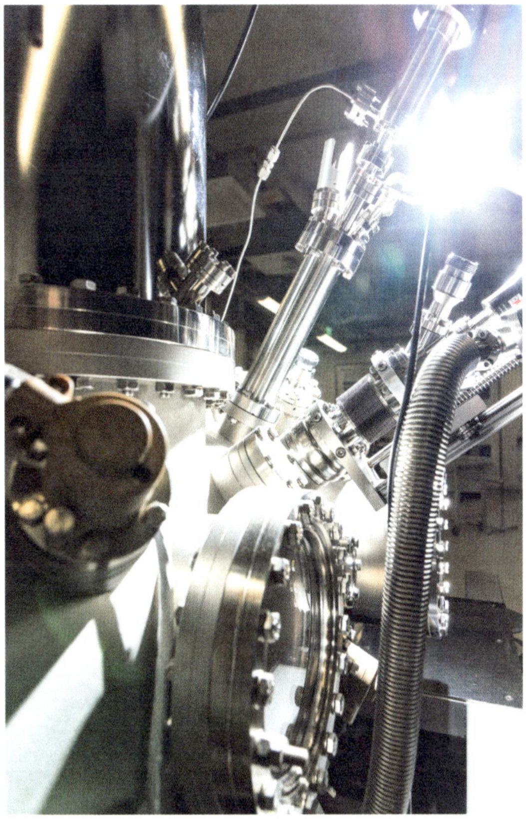

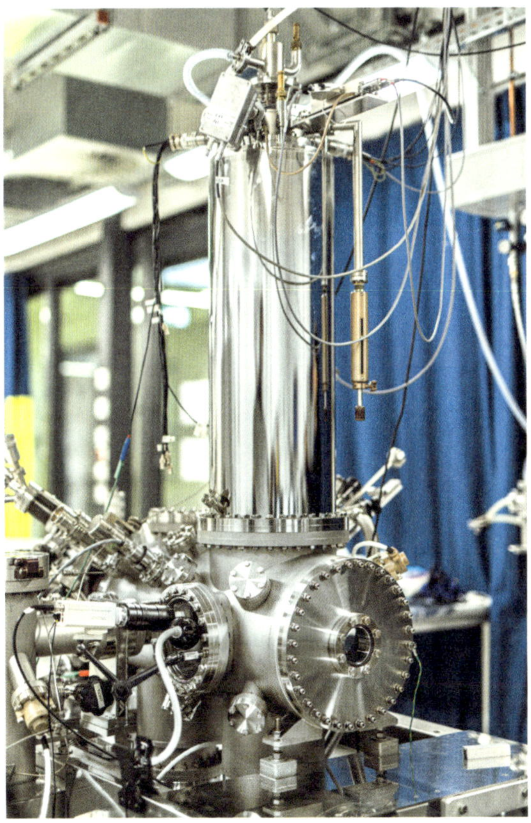

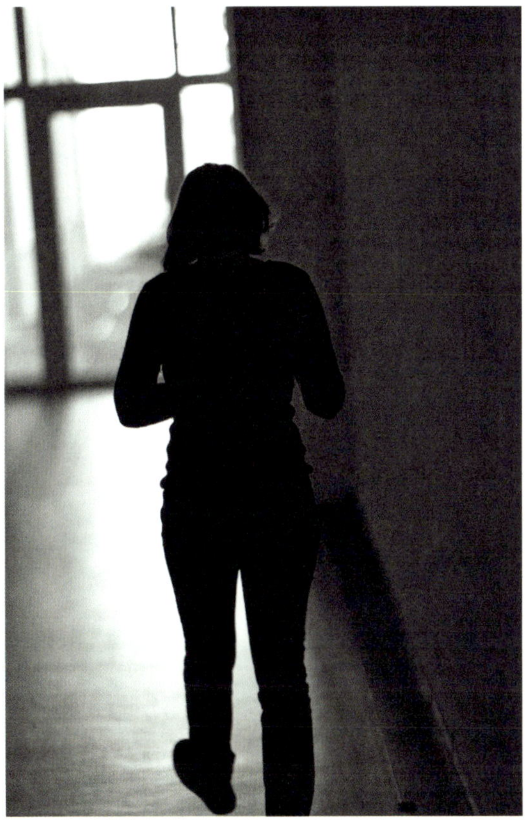

Yes, but we want to make our research more accessible and show the possible applications. Molecular machines are very exciting, but hard to imagine. Together with research groups from six countries we coordinate the European project "Mechanics with Molecules" (MEMO), with the aim of constructing molecular machines that use and execute the collective and synchronous movement of individual molecules. All the steps of molecular mechanical transmission are being investigated, from molecule gears through to nanofabricated solid-state nanogears that use the drive power of individual molecule motors. In order to make our research more accessible to the public we organize, for example, the Nanocar Race in Toulouse. Eight international teams race against each other with their self-developed nanocars. The nanocars must be driven across a one-hundred-nanometer gold plate, indirectly powered by the electrons released from the tip, which means you are not allowed to simply push them along directly.

If speed is the issue, how fast is a nanocar? One hundred nanometers per hour?

Not quite. It's around one hundred nanometers in six hours.

Six hours? Even snails are faster.

Yes, sure, a lot faster, I know that. But you have to remember that every voltage pulse normally only moves the nanocar to the next atom on the surface. On top of that we have to continually map the precise position of the nanocar on the course, so that you can follow the race live around the clock on YouTube. And every image takes five to ten minutes.

The nanocar race is a really good idea. It is a nice form of popularization in such a specialized area that is very abstract and difficult to grasp.

Although we have been researching molecular machines for a number of years, we are still at the beginning. There is still a lot to discover. And we will win the next nanocar race!

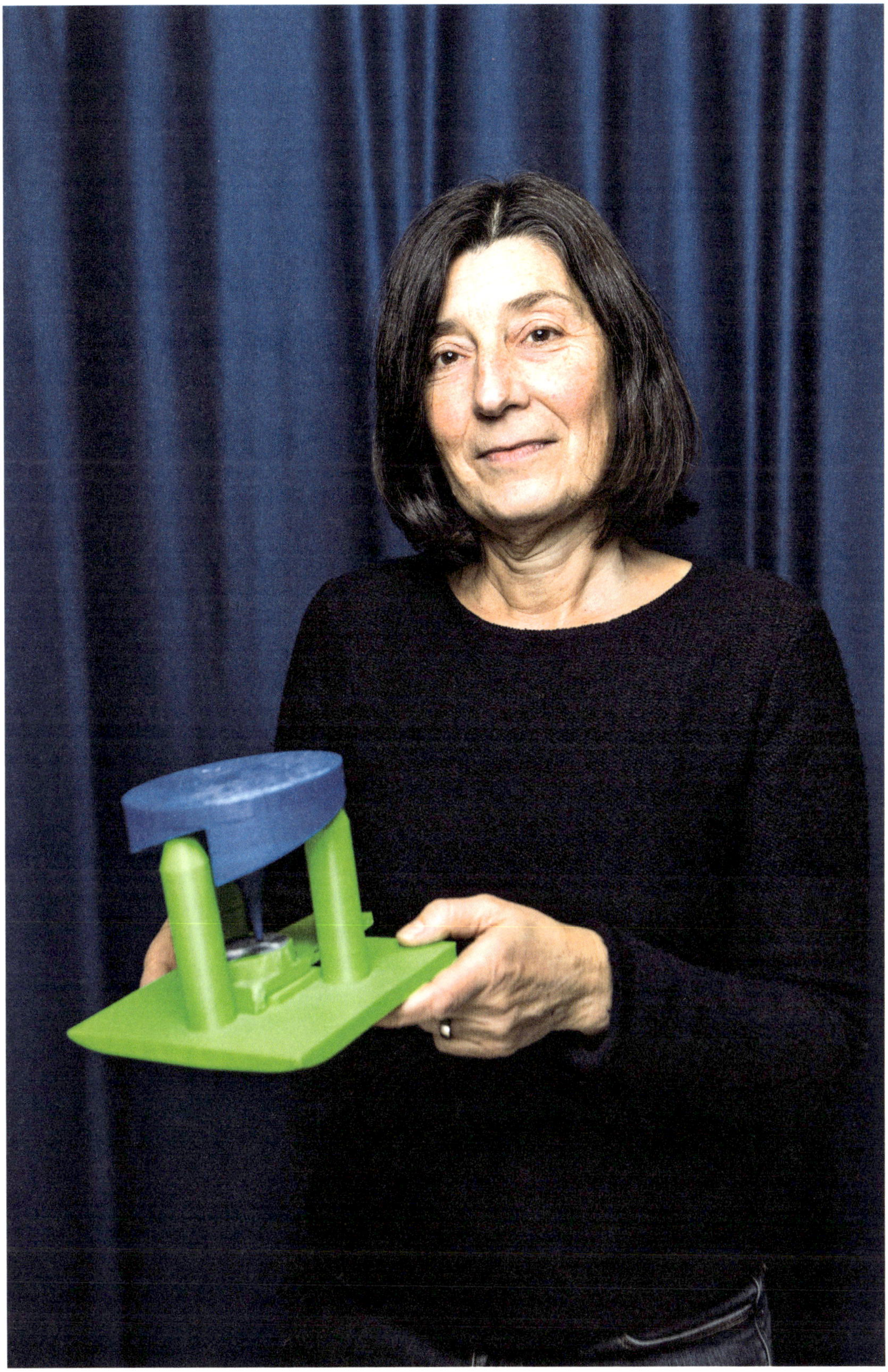

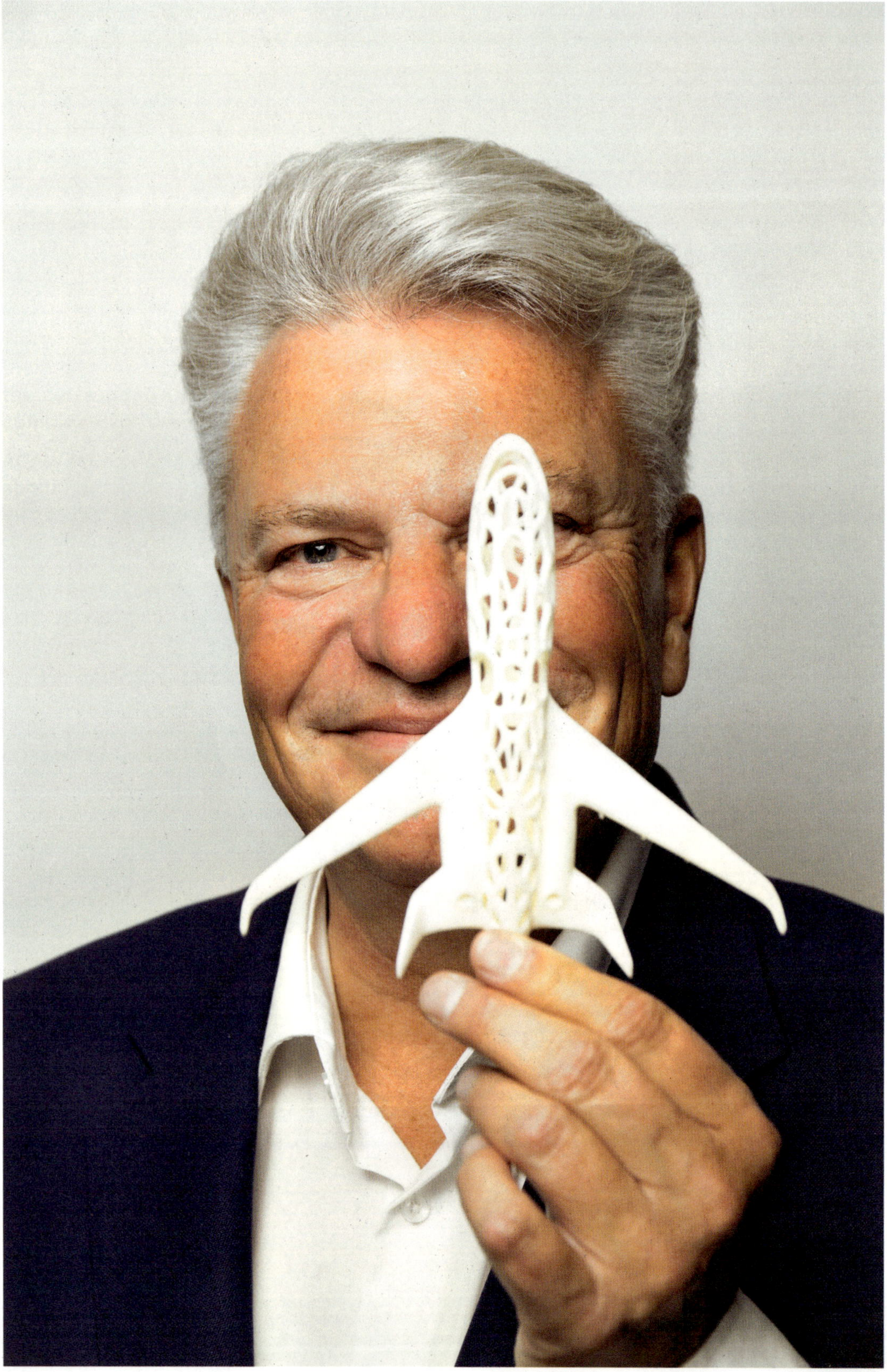

CLAUS EMMELMANN

"In the Future We Will Imagine Products Bionically and Digitally and Then Simply Print Them Out"

Nature shows us the way: what has proved itself over millions of years in honeycombs or bone structure is increasingly being used for the 3D printing of components. Using the additive manufacture process it is possible to produce very complex and filigree structures that function perfectly and are lighter and more robust than conventional components. The director of the Institute of Laser and System Technologies at the Hamburg University of Technology has been working on the bionization of 3D printing with metals for the past two decades. The evolution of nature could soon revolutionize the entire industry.

Professor Emmelmann, in 2014 you were awarded German industry's Innovation Prize, in 2015 you were runner-up in the German Future Prize, and in 2016 you were chosen as Hamburg's Citizen of the Year. What for?

My team and I were the first to focus on the bionic 3D printing of metals, twenty years ago, and we developed a bionic 3D-printed cabin bracket made from titanium for the Airbus A350. That was the first time worldwide that bionic 3D printing of metals was used on an industrial scale in airplane manufacture. Inspired by the structure of a bird bone and manufactured using bionic 3D printing, it was possible to reduce the weight of the three-dimensional, high-performance component by almost fifty percent. In airplane manufacture every half kilogram is crucial. Together with Airbus at the time, we calculated that with similar bionic transformations of other components it would be possible to save a total of 1.6 tons per aircraft. That means savings of roughly one billion euros in kerosene costs over the lifecycle, for example, of a Lufthansa fleet. However, the manufacturing costs for airplane components using the 3D printing process are currently well over one thousand euros per kilogram, while using the conventional stock removal milling process it is under one hundred euros. They are worlds apart, that is, they differ by a factor of ten. The challenge is to reduce the costs so significantly through scale effects that it will be economically viable in the future.

What can the bionic construction method contribute to cost reduction?

On the one side we can improve functionality and achieve weight reduction and greater efficiency. For example, we produce bionic heat exchangers that are more efficient by a factor of four and can therefore save around seventy-five percent of the functional energy conventionally required, while simultaneously reducing the volume and weight by more than fifty percent. We not only save material and costs, but also produce extremely sustainable products with low CO_2 emissions. We not only make the components better, but also cheaper.

What does bionic *mean in conjunction with 3D printing?*

It means that we design components according to the model of nature.

Like Leonardo da Vinci applying bird flight to flying machines?

Yes, exactly. Our scientific engineering competence is that we have a function that we can compute and optimize: for example, heat exchangers through thermodynamic simulations or mechanical systems through topology optimizations. When we use our software to compute our simulations in order to find the ideal structure, we often land at bionics: thus a tree, a tree structure, a hollow construction, or a bone structure becomes the model for our 3D component geometry.

The CNC milling machine can't do this?

No, it can't because it generates the component geometry subtractively and not additively, layer by layer, like 3D printing, and so it is unable to generate complex hollow structures. We always say that we first have to bionize the parts. First, to reduce the costs, but also to increase customer benefits, for example, using our system to produce more efficient graduated and curved boreholes, as opposed to a conventional straight drilled component.

Why are you so successful in international comparison? What do you do differently?

We take a look at the entire value creation chain, from construction and design through to manufacture and quality assurance. The individual manufacturing steps, including material logistics and quality control, can be tailored to each other and digitized. We bionize, design, manufacture, and attempt to digitally automate all the steps in the value creation chain using artificial intelligence. We translate the component construction and manufacture into bionic 3D printing and attempt to make the 3D printing as cheap as possible. Here the laser beam is center stage. It is the essential tool for melting the metal in powder form. Up until now we were very restricted in terms of the laser performance we could employ. It wasn't possible to use more than five hundred watts, otherwise the powder began to vaporize. But we wanted to weld. That is why, in a research project begun five years ago, we started printing with other laser beam profiles that enabled greater welding performance without vaporization.

How should I picture welding using 3D printing?

You should picture it as follows. You have a base plate. A layer of powder fifty micrometers thick is applied to this base plate. And now you melt the powder where the component is to be created and weld it to the underlying base plate.

The laser beam melts and welds at the same time?

That is the most important principle. The computer-controlled laser exposes the component contour into the powder. After every layer the plate is lowered by the thickness of the next layer of powder and the laser welds the next layer of powder. Using this layer-by-layer laser welding application process you can achieve extremely high stability, equivalent to a forged component, which is also extremely resilient both mechanically and thermally.

And what have you achieved in terms of improvements?

With the new laser beam profiles we vaporize a lot less and instead of just five hundred watts can now use up to five kilowatts. That means that the speed of the welding process has been increased considerably.

This also makes larger structures possible, doesn't it?

If we want to be coarse, naturally we can do that. But we want to remain filigree and generate hollow structures with high stability, low weight, and the sustainable use of fewer

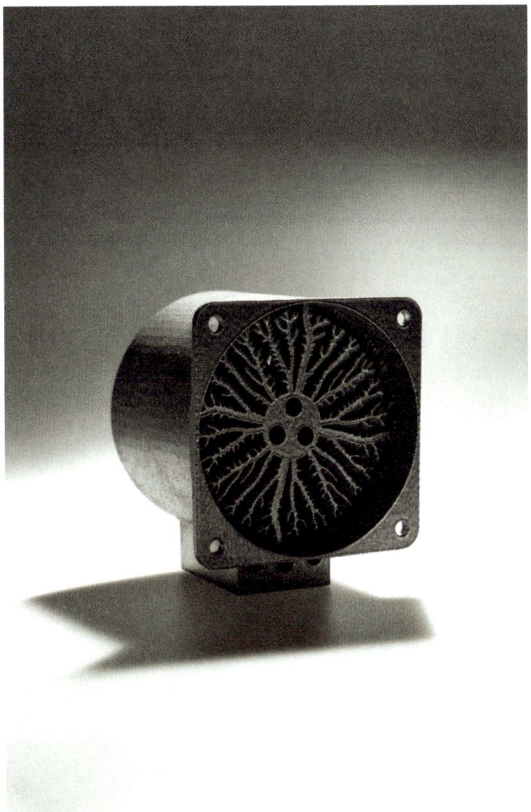

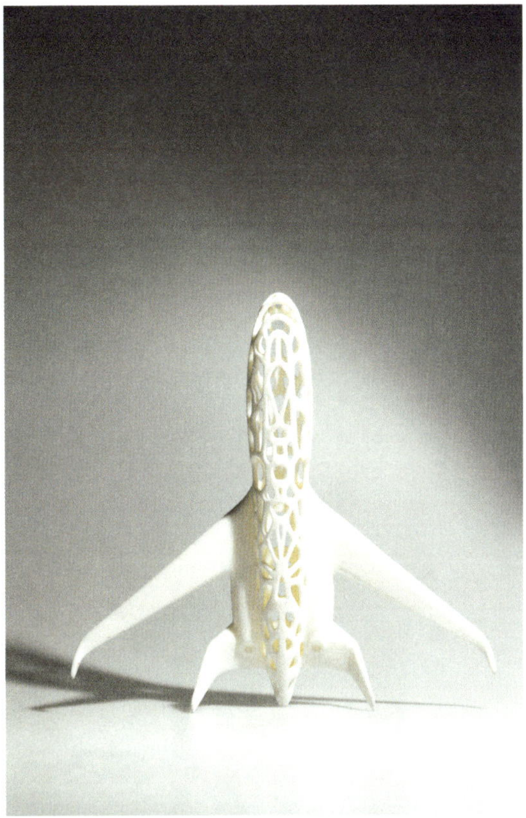

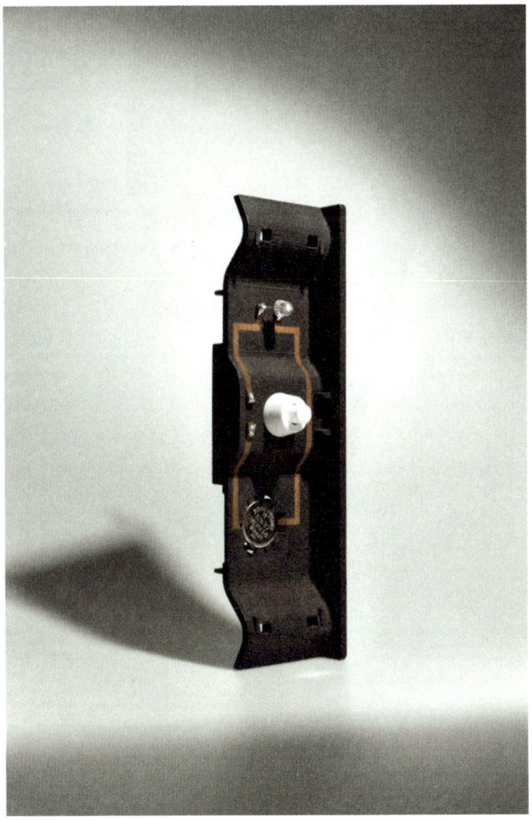

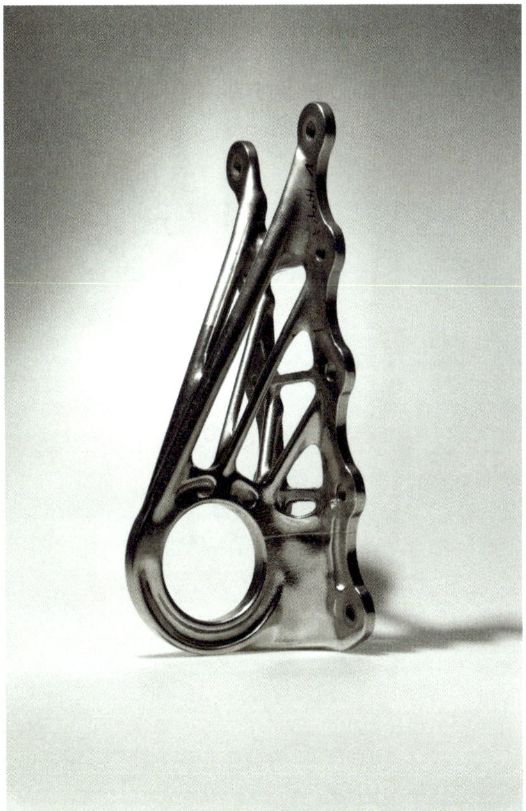

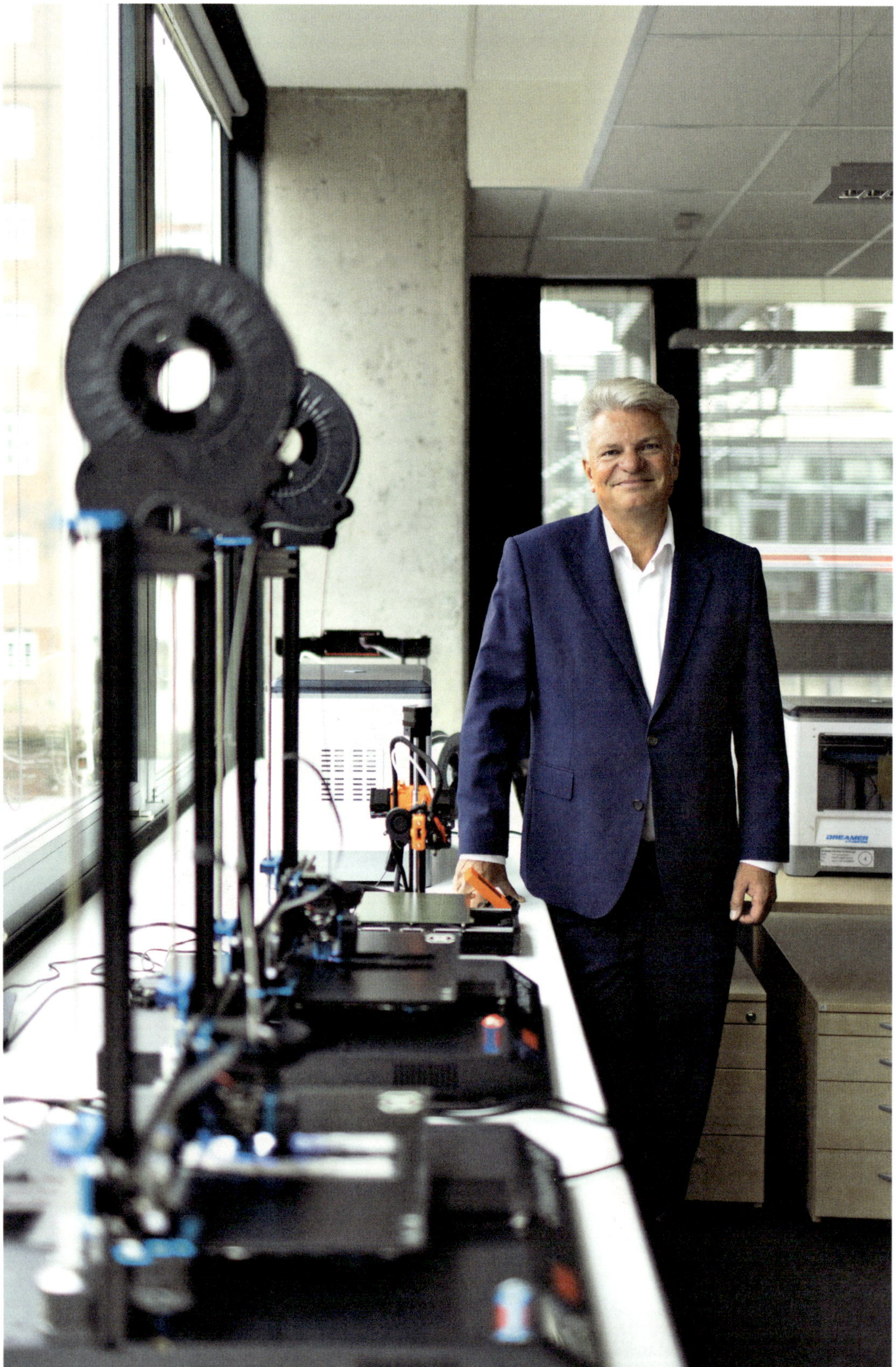
DREAMER

resources for the manufacture and operation of the components. We can now achieve this a lot more quickly and are no longer restricted by a laser exposure process that continually fluctuates between vaporizing and melting.

Aren't new materials needed to strike a better balance between melting and vaporizing?

Different materials naturally have different melting and welding characteristics. Aluminum is ideal and can also be built up very quickly. But it is not possible to achieve high rigidity with aluminum and cover all applications. The materials are generally defined by the individual application. If I require high rigidity, I employ titanium, or high temperature materials for thermally demanding functions, even though it will be expensive. But we are not just getting faster; the lasers are also getting cheaper. This will enable manufacturing costs for 3D printing to be considerably reduced in future. We have calculated that with our new bionic 3D printing method employing so-called ring mode lasers we can even undercut the costs for conventional milling processing. At the moment we are looking for users who, together with other partners, would be interested in developing this new manufacturing method with us.

When will you reach series production with this? The industrial concept of mass customization is based on small series with a high degree of individualization employing the economies of scale of mass production. Every product can be individually tailored to the customer, but the manufacturing processes are not slower as a result. Here 3D printing would be the killer application for the whole industry. With regionalized 3D printers the factory of the future could attain a high degree of individualization and lower transport costs at much lower prices.

I also come from industry and always had the goal of making 3D printing suitable for series production. The industrial experience was important for me, to grasp how it actually functions, select the right technology, and use it to make the customer more successful. That is why metal has such great potential, because conventional metal machining and processing is also expensive. That means if 3D printing succeeds in drawing level with conventional manufacturing methods, then we also have chance of moving into series production. This is difficult with plastic because plastic injection molding is very cheap, while the polymer material for 3D printing is relatively expensive. It is even difficult with three-hundred-euro printers.

In principle, when can a new technology replace an existing one?

Four criteria are decisive for this. Firstly, the function of a component must be met, and where appropriate demonstrate greater efficiency than in the case of conventional manufacture, in order to achieve lower CO_2 emissions. Secondly, the demands in terms of the geometric and material-specific qualities have to be met. Thirdly, the new technology, where possible, must facilitate shorter development, manufacturing, and logistics times. The throughput time must be short, and the individualization has to proceed quickly. The fourth criterion, which is often a key buying factor for the customer, is the cost of the component, that is, the "cost of ownership" throughout the entire lifecycle, including all operating, maintenance, and recycling costs. In the future 3D printing will be the most suitable manufacturing method with respect to all four criteria, if the manufacturing process costs can be reduced by a factor of ten.

Can you give us an example of something that could be successful in the near future?

The aviation industry, for example, is intensively engaged with hydrogen as a future aircraft fuel, to reduce CO_2 emissions. That places enormous demands on the infrastructure. Hydrogen needs to be cooled to below 250 degrees, and then brought back to room temperature as quickly and simply as possible. 3D printing is ideal for the manufacture

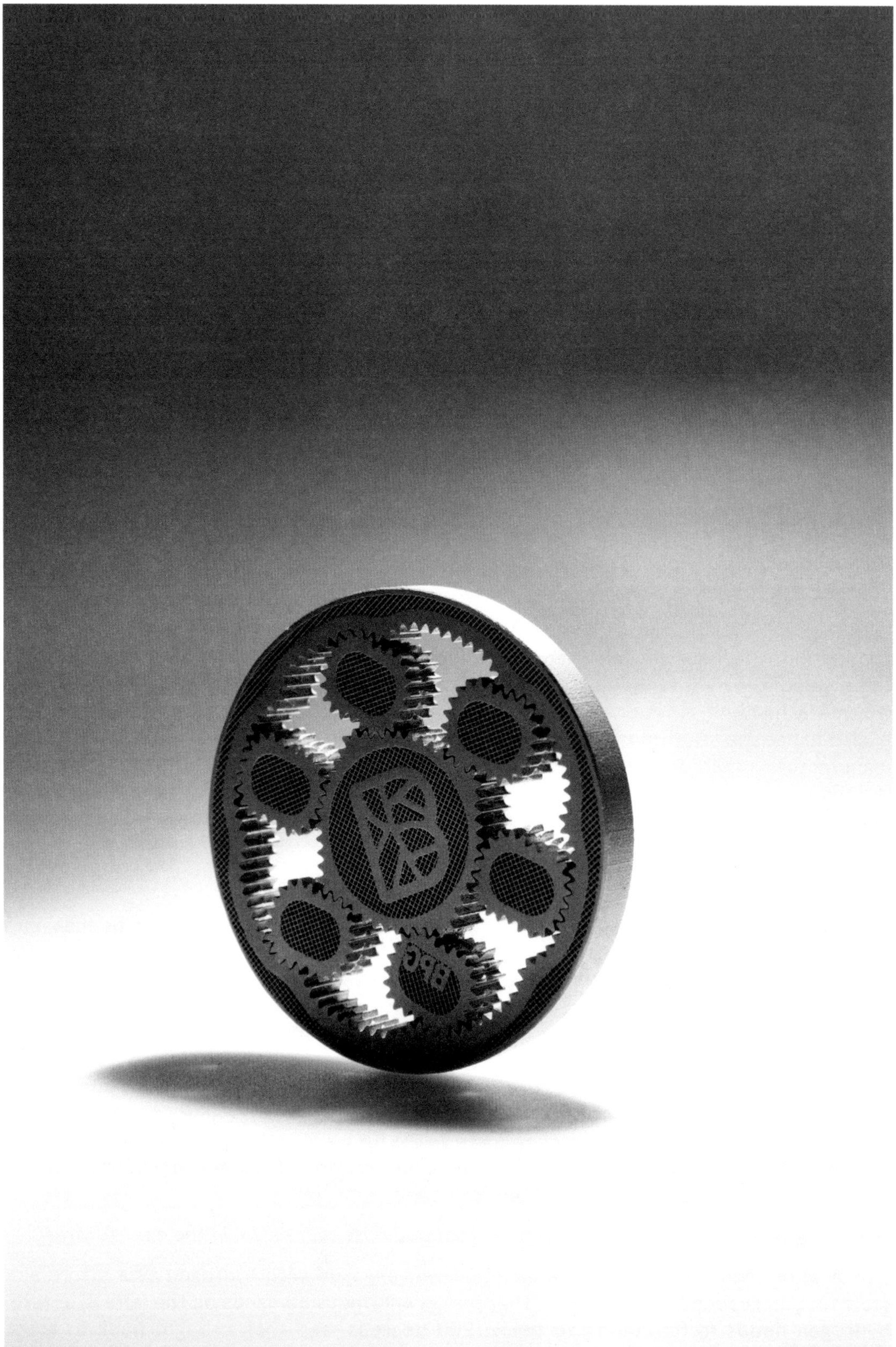
BPG

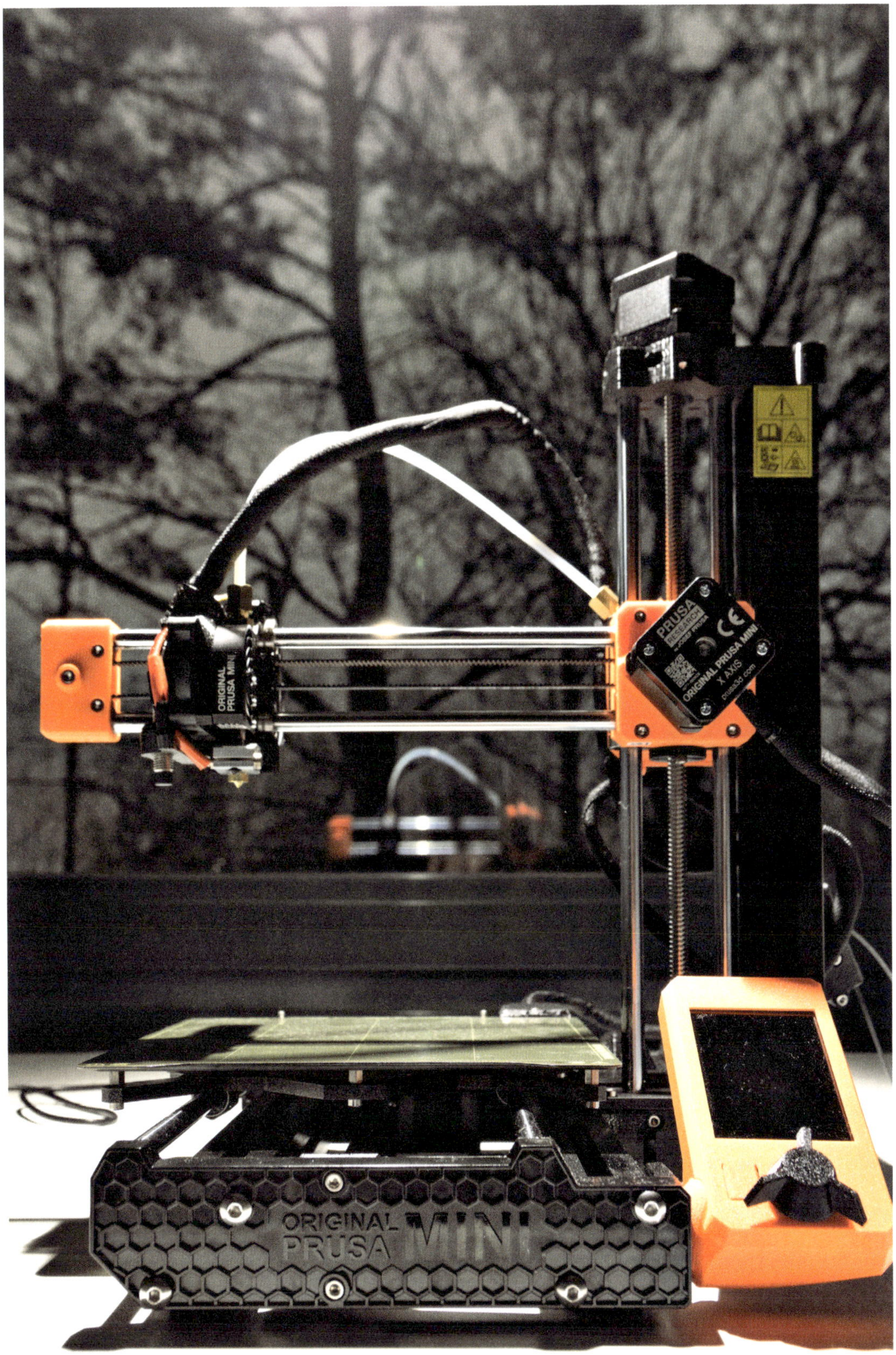
ORIGINAL PRUSA MINI

of this infrastructure, as it can reproduce complex functions and can meet these high requirements with the materials. Hydrogen is a very gratifying subject for us. Just like heat exchangers, which play an important role in the energy supply. We can completely rethink heat exchangers and work with manufacturing procedures that provide cheaper manufacturing costs and higher efficiency, accompanied by a reduced use of resources.

You approach companies with suggestions, resulting in a research contract or manufacture of a prototype. Then the issue is to decide whether it is worthwhile for series production.

This actually needs to be clear beforehand. A research project typically lasts three years and addresses the technical themes such as bionic design, selection of material, material qualification for the welding process, quality assurance, the use of monitoring systems, automation of the upstream construction process, platform development, et cetera. For the heat exchangers we developed a platform where the customer could enter the heat capacity and the manufactured size, and in the space of a few minutes the software had calculated the component that could then be immediately printed out. The vision is to use the digital manufacturing technology so that we can very quickly produce these individual and sustainable components. It is irrelevant to the laser what it prints. It doesn't need a series. Naturally, our vision is to build small modular factories that are close to the customer and are decentrally connected by data cables. To my knowledge, Ottobock is also working on the digitization of the value creation chain, from the patients to the provision of individual medical care. If you can digitize everything from design and construction to manufacture, and then generate it in the medical supplies stores via 3D printer, this has a lot of potential.

Yes, Ottobock is working on this very intensely. Which hurdles do you see? Where can science help in overcoming the current physical limits of 3D printing? With the materials?

There is a lot of potential in material research.

What do experts guess will be the material of the future?

The material must have a high performance profile with respect to function, and at the lowest possible cost. Titanium is a wonderful material, but it is still too expensive here. In contrast, in Australia titanium is a waste product generated in the mining of highly specific materials. On top of this, titanium is very difficult to process because of its hardness. If we could produce titanium cheaply and process it simply, this would be a high-performance material that would fit very well with 3D printing. This is also highly interesting for medical technology.

Titanium can be used for implants.

We have had great success with titanium hips and cups. The titanium hip joint was the first component from the 3D printer produced in large lots worldwide. The bionization of the component also played a central role here, because we oriented ourselves on the human bone structure and the additive manufacturing process of 3D printing allowed the buildup of light, stable, hollow bodies and porous surfaces. These endoprostheses were manufactured according to the model of the spongiosa—a sponge-like system inside human bone composed of fine beams of bone. As a result these titanium hips have a better custom fit, grow in quicker, and accelerate the healing process. Johnson & Johnson and Stryker now have over a hundred metal printers for printing hips or knees in their respective factories.

You say that the excessively high unit costs of 3D printing is a hurdle that needs to be overcome. They can be reduced by improving the speed and performance of the lasers. But this is not trivial.

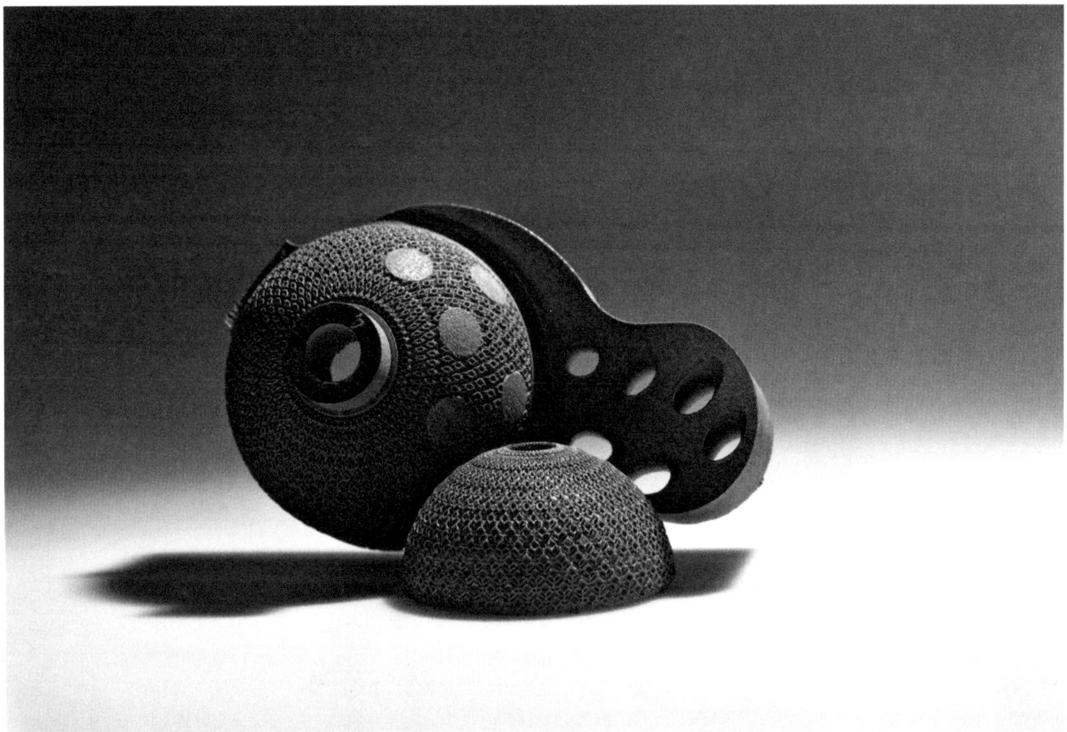

3D printing doesn't just consist of one technology; it offers a whole palette of procedures and materials. The powder bed process is best suited for metals. In the case of plastics one generally works with a filament. Here the thermoplastic plastic is wound onto a spool and heated and liquefied in a nozzle. The liquid filament is then applied to a plate and given its geometry using a CNC-XY table. In principle one can use this to produce a variety of components at a high resolution. There is also laser technology in the plastics field, but this is far more expensive and, similar to the laser melting process, builds up a bed of polymer powder layer by layer using laser exposure. Alongside the faster laser machines, customers now use the smaller 3D printers in series production and increase productivity using numerous 3D filament printers.

But that takes a long time.

Sure, this takes a long time. But if one has a number of components, one can use numerous printers in order to increase productivity. So productivity needs to be improved; we need to find suitable materials and combine the processes effectively. And finally, one has to find the business case in the company: that can be driven by individualization, digitization, CO_2 reduction, or the increase in efficiency. This varies greatly. But actually Germany is predestined to be the pioneer in this field. We are a country of engineers and have mastered especially complex and demanding processes.

The reason that we are export world champions in the first place is because we have mastered the complex industrial innovations.

We also have the opportunity to completely rethink things as a result of the technological transformations brought about through 3D printing. But this is not so simple. When you tell an experienced engineer that he shouldn't mill anymore, then he has to completely rethink. A whole technology is scrapped.

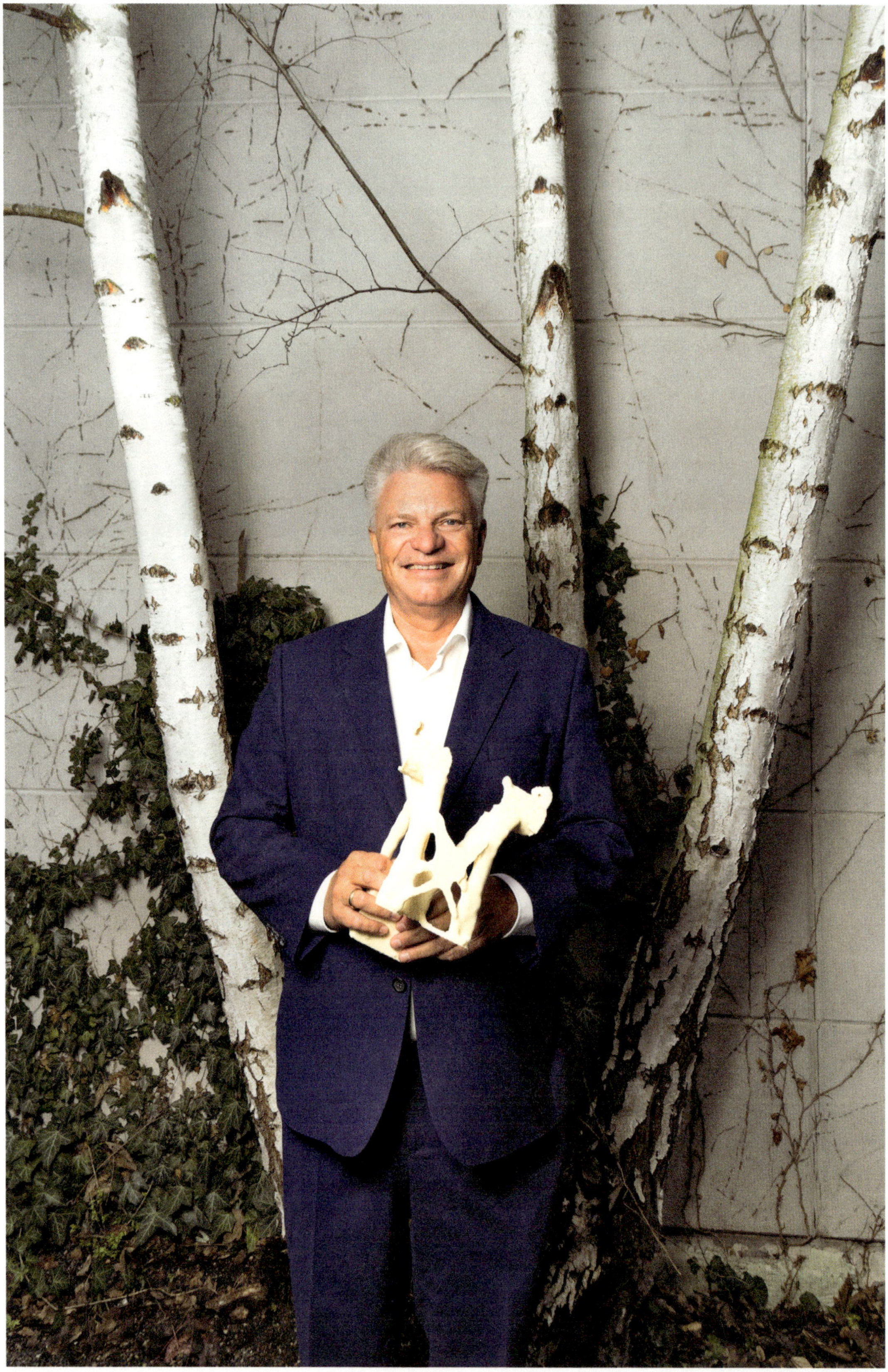

The heavy products can still be cast and milled, but 3D printing is ideal for the digital, the light, the fluid technical systems.

I observe among my employees, who at some point go into industry, that they think completely differently. With an old product it probably doesn't make much sense to change the manufacturing process. But as the innovation cycles are getting shorter and shorter, we can consider whether things couldn't be done differently. Do we always have to screw the parts together and assemble them in such a complicated fashion, or can we develop a single part and then print it out whole? So, in the future we imagine things bionically and digitally and then simply print them out. That is a fantastic scenario for the future.

But how can we reach this point? That we as the German or European research landscape don't just simply supply the blueprints, but also earn something with them?

We need to realize that we will only remain competitive with the application-oriented implementation of these blueprints. And here we actually have a good division of labor in Germany with research, which the state spends a lot of money on, and industry, which needs to cooperate closer with research. We have fantastic talents in research and highly qualified specialists in industry. We need to network these intelligent potentials and get them out of their silos, so that together we can develop genuinely competitive applications. There is still a need for improvement here. As I have also worked in industry it is not enough for me to just gain new insights as a researcher. I also want to take them on the road and see them in action. And naturally, preferably in Germany. This motivates me.

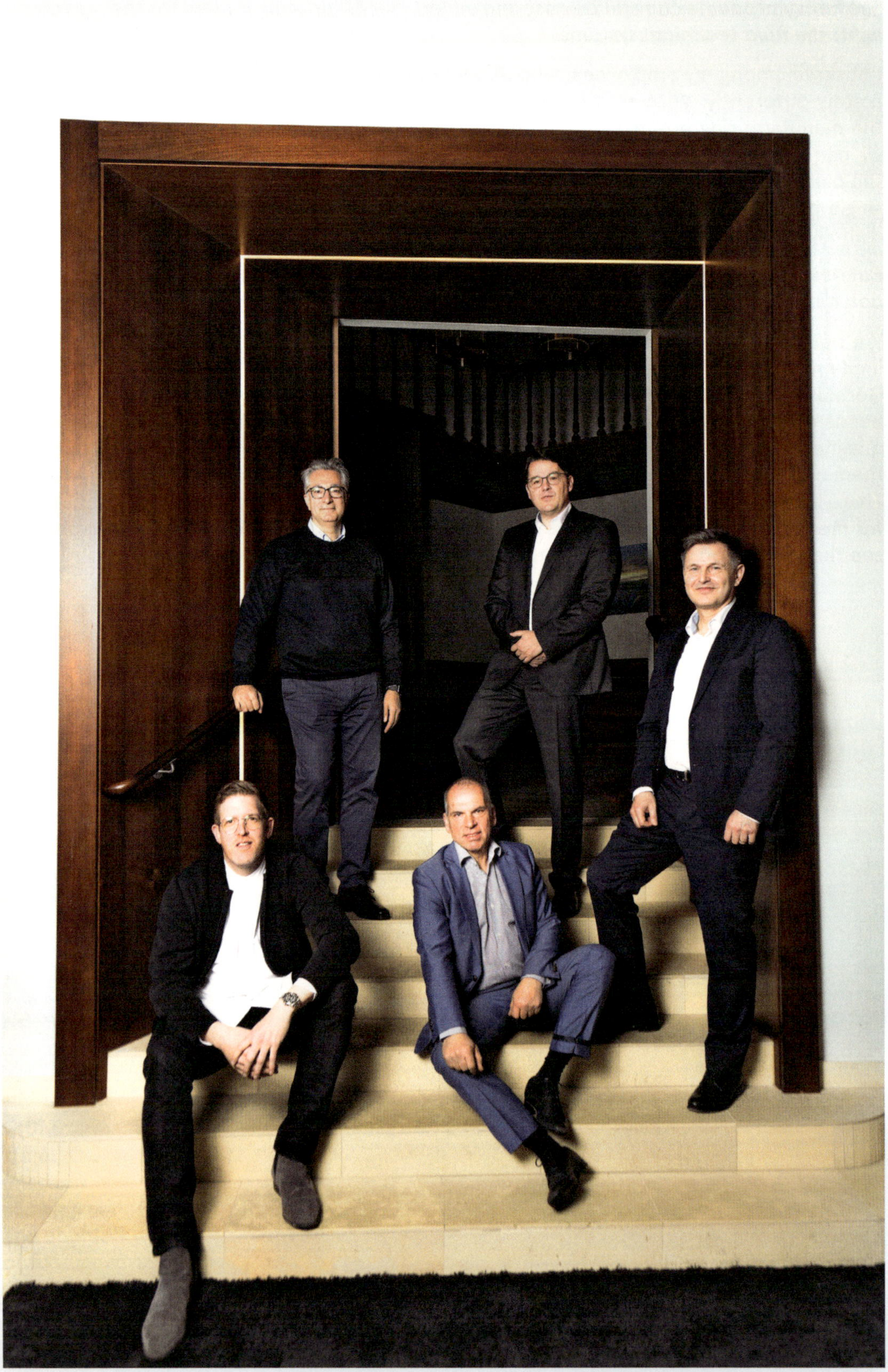

Martin Böhm, Güngör Kara, Hans Dietl, Bernhard Graimann, and Andreas Goppelt (from left to right)

MARTIN BÖHM, HANS DIETL, ANDREAS GOPPELT,
BERNHARD GRAIMANN, AND GÜNGÖR KARA

Reflections on Advanced Materials and 3D Printing

The five experts from Ottobock discuss how the current physical limits of 3D printing can be overcome and how important the research into nanotechnologies and innovative materials is for the development of new application fields and growth markets. With an innovative, patented procedure for the deployment of soft polymers for Rapid Liquid Printing (RLP), Ottobock has achieved a breakthrough that could be significant for a multitude of other industrial applications.

Andreas Goppelt: Francesca Moresco is trying to develop synthetic molecule machines in the form of a single molecule, using nature as her model. In nature, nanomachines already exist, for example, in muscles that contract. Seen at the molecular level, it really does look quite funny: the molecules walk along the substrate like little elves. In plants they capture the sunlight, use it to operate ion pumps, and generate an electric current. This makes nanomaterials particularly interesting because they possess properties that in comparison to their macroscopic counterparts are extremely exciting and fundamentally different. However, they also raise new issues. Nanoparticles can influence cellular structures, and that is why material research also has to examine the associated toxicity. Using nanomaterials, materials can be made extremely light or display electrical conductivity even at high temperatures. The topic of nanomaterials is fundamental—and at Ottobock we need to observe these developments in order to use them for new application fields and growth markets.

Hans Dietl: We already use nanomaterials in coatings, for example.

AG: Nanomaterials can be used in a diverse range of applications such as coatings, textiles, and surface refinement in which a few layers of atoms are vapor-deposited. Furthermore, at a fundamental level the use of nanomaterials can result in batteries and rechargeable batteries that are lighter. This will have a profound effect on our industry.

Bernhard Graimann: Innovative materials have always played a decisive role in Ottobock's innovations such as CFRP springs or silicon liners. Advanced materials will remain a major topic for us in the future—not just at the nanoscale. For example, the University of California, Berkeley, is developing the world's strongest permanent magnets to date. The results are quite promising, and very soon smaller and more powerful electric motors could be available for wearable robotics. This development is very exciting for prostheses and exoskeletons. We are already using conventional electromechanical drives, but if the motors become smaller and more powerful, we can dispense with the gears, resulting in drives that are lighter and above all quieter. Another example is nanomaterial of the sort that is being developed at MIT: polymers that combine the material characteristics of steel and plastic, giving it an extraordinary mechanical stability while remaining very light. High performance and light weight are especially important characteristics for us, since most of our products are worn on the body and less weight contributes to user acceptance. As innovation leaders we have to carefully follow these technological developments and take them into account in our strategies.

I would put molecular machines or nanomachines of the type that Francesca Moresco is studying in the category of nanorobotics or nanobots. In the future there could be medical applications that are actually used as a replacement for pharmacological intervention. After all, it is a popular fantasy that small nanorobots will clean up our blood vessels, close internal injuries, or transport medicines where they are really needed.

AG: Then, of course, there is the question of how these nanobots will actually be able to find the target cell. In biochemistry there are other adequate possibilities for cell-specific drug delivery as one can make use of natural principles such as antigens or antibody interdependencies. Today it is already possible to distribute medicines on a tissue- and cell-specific basis throughout the body using biochemical mechanisms. You don't need small robots for this.

BG: In my opinion nanorobotics is of little relevance to Ottobock. However, advanced materials and nanomaterials are very exciting for us, including for 3D printing.

Güngör Kara: Professor Emmelmann and I are on the board of directors of Mobility Goes Additive, an international network for industrial additive manufacturing. Emmelmann is an expert in 3D printing and has advanced numerous innovations, including in the aerospace field. His strength is that he comes from design and has transferred bionic design into an independent production unit. Bionic production is oriented on the bionic structures and principles of nature, and it enables a complete solution consisting of design optimization and light weight for the entire production line. For Emmelmann there are two essential pillars in 3D printing: firstly, the printer and the laser performance in metal 3D printing needs to have a higher performance, and secondly, it is about bionic design. At Ottobock we have employed bionic design in our lower-leg prosthesis, in order to achieve lightweight construction and high rigidity.

HD: Our basic aim is oriented on nature and the idea of dimensioning the components in proportion to the load and in a configuration that is exactly like our skeletal system. In nature this is continually optimized, because our bone structures adapt depending on the workload.

BG: That is the difference to what we can achieve with 3D printing. A bone is continually rebuilding itself because it is a cellular system. We can't reproduce this with 3D printing. However, with such bionic procedures we can, for example, achieve a reduction in weight. As was already mentioned, lightweight construction is an important topic for us, and topographical optimization by means of modeling and simulation helps us here. Bionics can provide approaches to solutions or inspiration, but it is ultimately an optimization problem. Using 3D printing enables us to produce complex structures in an additive fashion. It would be impossible to produce such structures with conventional, subtractive processes.

GK: Most of all it is a matter of materials. While Professor Emmelmann is developing metal printing for use in producing titanium hip cups, polymer is the relevant material for us because titanium is simply too expensive for use outside the body. The cost factors are a huge challenge with 3D printing. Something has to radically change so that it becomes cheaper. That is why most 3D printing companies that I know are now pursuing a filament strategy because the filaments—that is, thermoplastics in wire form that can be wound to virtually any length—have achieved a far higher cost-performance ratio. The innovation potential is not necessarily due to the better materials, but above all the cheaper ones. We are also experimenting with employing considerably cheaper 3D-printing materials for intermediate products such as functional forms.

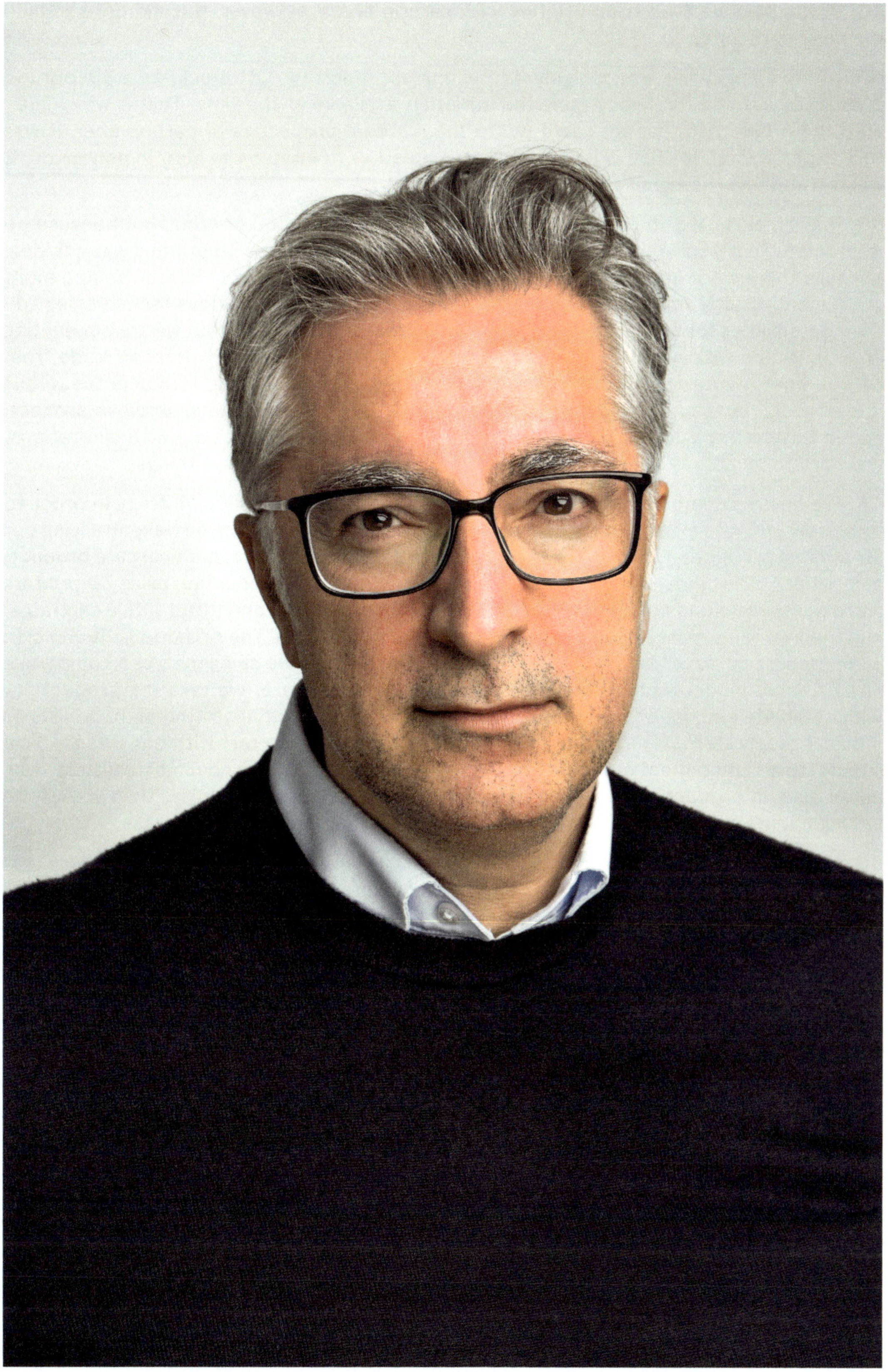

Güngör Kara, Chief Digital Officer, Ottobock

AG: In the past we even tried printing with carbon fibers. However, this did not produce the desired results.

HD: In my opinion this was a highly interesting approach for Ottobock, as a 3D-printed prosthesis was still too heavy using the materials available at the time. That is why I initiated the project: so that we could utilize the material properties of carbon fiber. It was very experimental. We had print heads that enabled us to embed the fiber in polyamide. It worked and had the rigidity required for a socket.

BG: Nevertheless, we don't employ 3D printing for the sake of 3D printing, but because we want to efficiently solve the problems of our users. In this case, the 3D printing was not ideal because there were other solutions that yielded more efficient, cost-effective, and more convenient results. You have to be familiar with and evaluate the various technologies and their possibilities for finding solutions. One of our research projects that we are conducting in cooperation with partners is called "Patch to Patient." It is about patches made from carbon fiber reinforced plastic that are applied in a similar way to a normal process, but using robots. Instead of 3D printing in the classical sense, it is another additive process, and in this case it is a promising alternative. Even when you are dealing with innovations, you have to be able to take a pragmatic and solution-oriented look at things.

GK: We have to combine the innovation strategy with our digitization strategy in order to bring new business models onto the market. We are currently working on decentralizing our 3D printing production to enable our clients, that is, the orthopedic, medical care product, and patient care centers to print out optimized components on location using 3D printers and a complete cloud connection. The printer will be optimized and fitted with a cartridge, and then we will almost be at the level of a Nespresso machine. The orthopedic technician can purchase or lease the printer at an affordable price, and we can generate a completely new business model via the cartridges, scanners, and software. We take the production, with complete end-to-end solutions, from scanning to 3D printing, to the orthopedists or medical supply stores. That is our vision, which we also want to test with our own medical supply stores and patient care centers. Once this infrastructure has been established, new materials and production processes can actually be scaled very quickly. Then we will be talking about completely different markets and growth opportunities.

BG: However, it is not just about an efficient infrastructure. In 3D printing the software that prepares the product's data model is decisive. For this you need modeling and simulation, and high-performance computers for optimizing these structures. In other words, you need good algorithms. There are so many interesting developments in the scientific and industrial world right now. This is the future, and it is enormously powerful, since 3D printing doesn't just make bionic or lightweight structures possible. Instead, through these structures we can integrate functions into these materials. Such functional materials or meta-materials are very exciting for the future.

AG: We are currently experimenting with Rapid Liquid Printing, which, apart from us, has only been mastered by a research group at MIT. It uses soft polymer structures that are constructed in a gel. With this gel it is possible to manufacture the silicon liner used in 3D printing much more quickly and individually. This is a revolutionary technology for all applications in which the soft polymers are anatomically formed or individually tailored. Although there are still a lot of challenges, we have achieved a great breakthrough for Ottobock with our technology, which enables the use of soft polymers in RLP. Our technology, which we have already patented worldwide, enables a diversity of applications extending well beyond our industry. That is a real milestone.

BG: That is a success of our innovation management. In principle this is always composed of three steps: we observe interesting new technologies, identify the relevant ones, and then adapt them for our products. With RLP, we quickly recognized its enormous potential and initially established an internal research project to address it. Then we developed our manufacturing strategies, secured the patents, and now we are bringing this innovation to the market as quickly as possible.

Bionic Regeneration

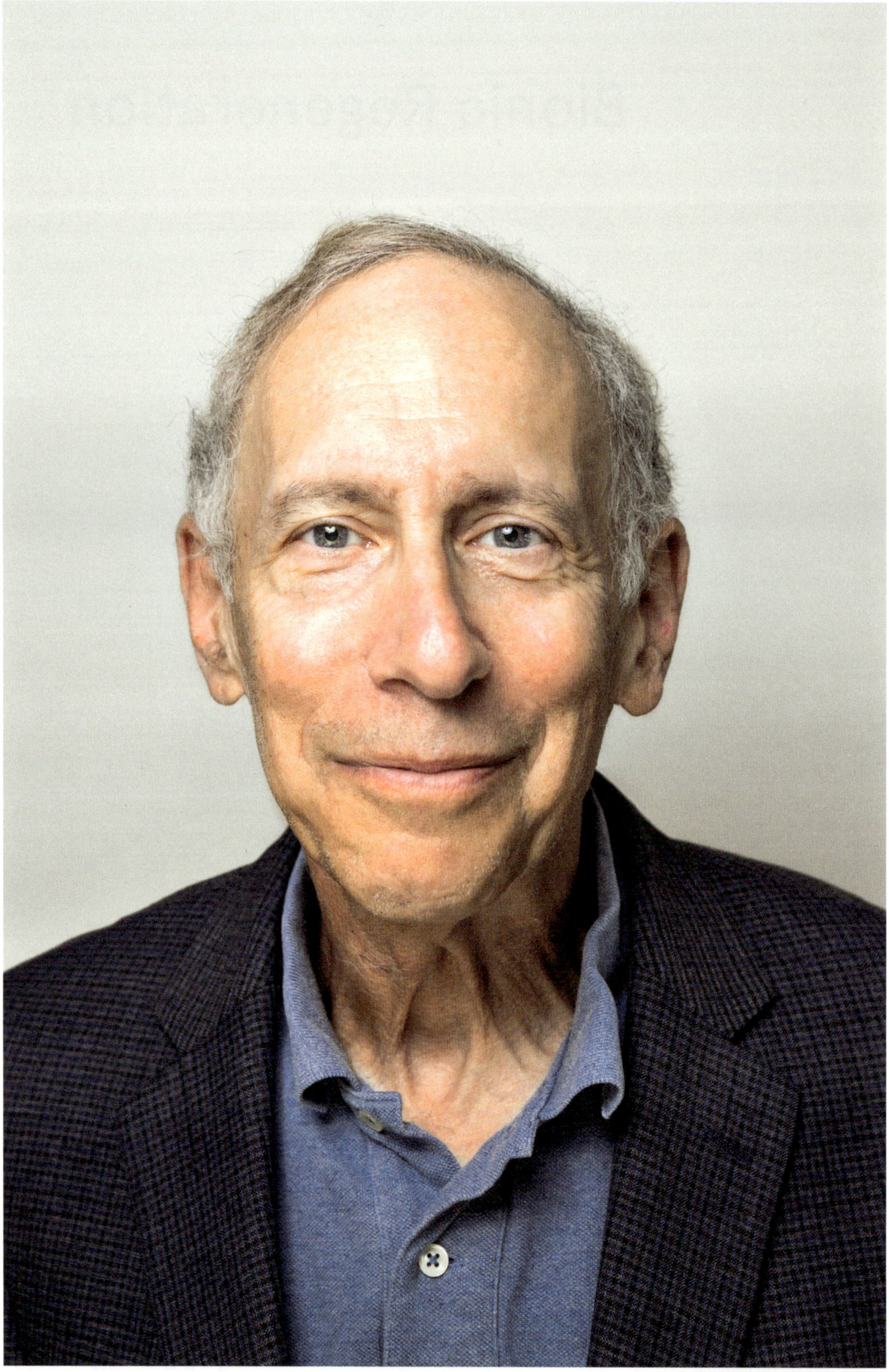

ROBERT LANGER

"The Beauty of Molecular Biology Is That the Body Does All the Work"

The director of the largest academic biomedical engineering research laboratory in the world at MIT is a widely recognized and cited researcher in biotechnology, especially in the fields of drug delivery systems and tissue engineering. He is the most cited engineer in human history with over 1,500 research articles and 1,400 patents that are licensed to more than 400 companies. The research of Robert Langer has saved millions of lives. He has also started numerous biotech companies, including Moderna Therapeutics, a company that has been making headlines with their COVID-19 vaccine.

Professor Langer, you are a science celebrity and the godfather of tissue engineering. You are one of the cofounders of Moderna. Before we talk about tissue engineering, can you tell me a bit about the current status of vaccine development? Your core competency is drug delivery, and one of the most important things about vaccines is that they have to be protected as they travel through the body to reach the cell they have targeted to deposit their mRNA in. You have developed many principles and patents for protective materials in the past, and the mRNA vaccine bundles many of your own competencies. You discovered a law in biomedical engineering that the protein is produced by the human cells, and all you have to do is switch it on—then they can produce everything. Let's start with this miracle.

The central dogma of molecular biology says that the DNA makes RNA, and RNA makes protein. Of course, protein therapeutics have been very successful, but it takes a long time to make them, and there are many diseases they can't treat. So the beauty of messenger RNA is if you pick messenger RNA and protect it, say, by putting it in a nanoparticle, and you make the RNA that will make the specific protein, the body does all the work. You just inject the messenger RNA and the nanoparticle protects it, but then degrades. And then it gives the instruction to the body to make this particular protein, whatever it is, whether you want to make a vaccine, something to cure cancer, or something for heart disease or for a rare disease. That's really the central dogma of molecular biology. What Moderna and some other companies have done is capitalize on that and let the body do the work. You can treat diseases that you couldn't normally treat with proteins, like diseases that are inside a cell or based on the membrane of the cell.

There was a lot of euphoria about the fact that companies such as Moderna were able to develop a vaccine in only twelve months. Of course, Moderna had been working on the technology for ten years, but it was really a miracle to have the first real trial with the vaccines and a whole FDA process and everything in that short amount of time. People were so relieved. But now we are a bit disappointed because the vaccine is helping us, we don't have to go to the hospital, we don't have severe illness, but it's not protecting against the virus itself. We can get infected even though we were vaccinated three times. How can you handle that?

Well, I think a couple things are happening. First, a lot fewer people get inflected—that is if you have the vaccine. That's the first important thing. When you look at the data, it's a huge difference. So when they talk about ninety-five-percent effectiveness, you have to compare it to other vaccines, such as the flu vaccine, which is less than fifty-percent effective. And just like with flu, I think the expectation shouldn't be that it is going to cure you for life. You may very well need boosters. But I think the two important things are: number one, when you look at the data, many fewer people get COVID to begin with. And number two, if they do get it, the chance of them going to the hospital is very low, especially compared to people who don't get it, as is the chance of them dying. So if you look at either getting COVID or hospitalizations or death, they're all much lower. I don't find the fact that you have to give a booster surprising. I mean, you get variants. I think a lot of it may have to do with a specific class of viruses rather than the type of vaccine. So it still seems to me that if somebody gets the vaccine, they're so much better off than if they don't.

What are the perspectives for Moderna? Are they working on new classes of virus now?

Well, Moderna is working on a lot of things. With respect to COVID, one of the things is they have specific vaccines against the various mutations. So there are vaccines that Moderna has made against Omicron and Delta. Moderna is also working on making many more vaccines, because the world is still not as vaccinated as we want. In addition, Moderna is working on other types of vaccines, like for flu, Ebola, and even for cancer. So there are many things that Moderna is working on that I hope will lead to many other treatments.

You are working very closely with the Bill & Melinda Gates Foundation. Gates approached you in 2012, and he gave his famous TED Talk in 2015, where he warned that although we are investing so much money in the military, we will die from a nanovirus. And this is a much bigger bomb than any bomb that we have in the military arsenals. And you developed a lot of other vaccines for other diseases with the Foundation?

I'll break things down into a couple of areas. Moderna makes specific vaccines against nine or ten different diseases. And then there'll be combination vaccines. At our lab at MIT, when we work with the Gates Foundation, the work is much more basic. One example of the kinds of things that we're working on is the problem in the developing world that people often don't return for a second injection.

You developed a medication that releases over time: immediately, then after six months, and then after twelve months. It's amazing!

Yes. That's still in development, but we have created a new process that enables it to happen. So you can give what we call a *self-boosting vaccine*. That's an example of the kind of thing we've done with the Gates Foundation. We are looking at polio as an example too. But like I said, the work that we do at MIT is sort of a platform. Another thing we're working on—some of which is in collaboration with one of my former students, Mark Prausnitz at Georgia Tech—is a microneedle patch, like a Band-Aid, that you could just ship all over the world and make it very easy for people to get vaccinated.

Isn't your lab at MIT the biggest biomechanical lab in the world?

I'd say is the biggest academic biomedical engineering lab in the world.

You are working on a lot of issues. How did you develop your own interest since the field is so broad? Tissue engineering on the one hand, then drug delivery, which is the most important. It's a big development for the whole pharmaceutical industry. You save millions of lives with that alone. How did you develop this very broad interest?

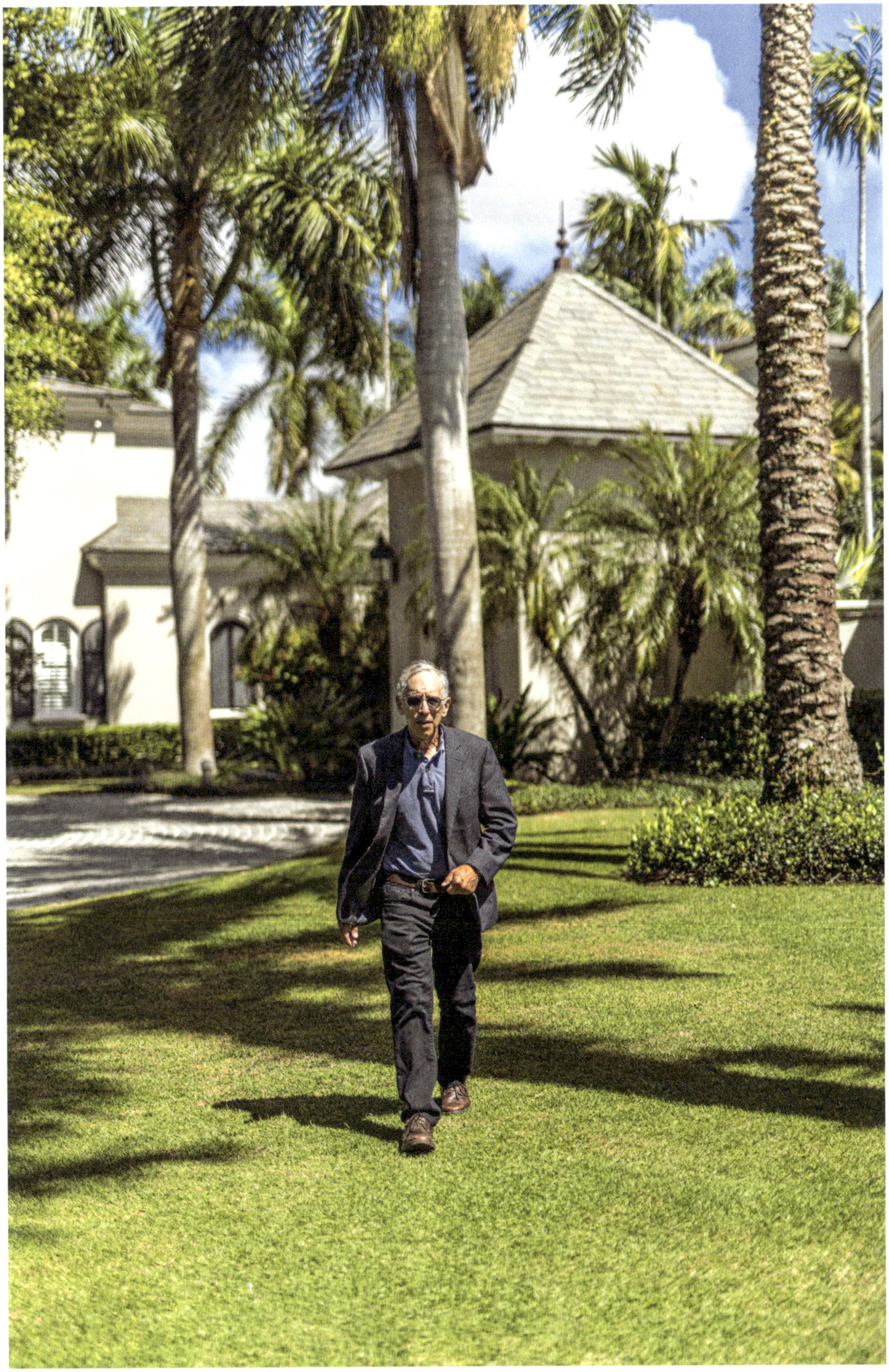

Well, engineers like to solve problems. I started getting interested in chemistry when I was a little boy. I got a Gilbert chemistry set.

Really? I got it too, but I didn't choose chemistry as my career.

Well, you seem to be doing fine. But it got me interested. I liked magic: I would mix chemicals, and they changed color. But the way I got involved in a lot of this was when I got done with my PhD at MIT. I ended up working in a surgery lab at Harvard Medical School and Boston Children's Hospital with a man named Judah Folkman. We were working on a new way to think about treating cancer and discovered what are called *angiogenesis inhibitors*, molecules that could stop blood vessel growth. We had to invent a way to study it. We needed a way to deliver large molecules through tiny particles, and so I got involved in doing that as well. In 1976 we published two papers. One was the first encapsulation of delivery of macromolecules through small particles, published in *Nature*.

Were these small particles nanoparticles?

Well, they were probably microparticles, but if you sliced them a little bit more, they'd be nano. I mean, it was the same.

It's the same system.

Yes, it can be. There are different ways to make nanoparticles. But it was the first time that anybody had delivered molecules like DNA or RNA, nucleic acids. So we did that. We also used that approach to isolate the first blood vessel inhibitors. That's how I got started. A lot of people were very skeptical of that work. They didn't believe blood vessel inhibitors existed, and almost no one felt that you could deliver large molecules. This was in the mid-1970s. So I began trying to understand the science of it better, and then I also got other ideas of how you could use materials in medicine. Being at Boston Children's Hospital, as a chemical engineer, was a little bit like being a kid in a candy shop. I saw all these things that medical doctors did. And I thought, well, you could use engineering to solve this or solve that. That's also partly led to tissue engineering.

So you're basically an engineer?

A chemical engineer.

How did you come to tissue engineering?

Well, there was Joseph (Jay) Vacanti. Jay was a surgeon, and he and I worked together on the angiogenesis problem when he was a resident. When he became a physician at Children's Hospital, he was in charge of the liver transplant unit. He wondered if we could make livers from scratch or any tissue from scratch. We started talking about it and came up with different ideas about how to do it. It was because of him that I got involved in it. Of course, the issue for him was that with liver transplants there's a huge donor shortage. So he wanted to have another way to make livers.

I think you started with blood vessels, engineered blood vessels, and then the liver. What exactly did you do?

Jay and I wrote an article for *Science* a number of years ago. The way I would explain it is you take a material, like a scaffold, maybe that material is a polymer, and you take a cell—it could be a stem cell, a liver cell, or a blood vessel cell. Sometimes you need more than one cell. But these cells are smart, so you give them the right chemical signals, and then the right template with just the polymer. And they kind of organize, when they have those correct cues, and then they will make a tissue from scratch. I mean, sometimes you could call it an organoid, but they'd make the tissue from scratch. And depending on the cells you use and the conditions, you can theoretically make almost anything. Some, of course, have gone further into the clinic than others.

How do you control the cell? Let's take a liver cell. How do you implant the command that it should now produce this or that?

Well, there's two different things. If you take the liver cell, hopefully it will do what it's normally done. There are five types of liver cells that you might use, such as hepatocytes, Kupffer cells, and so forth. Sometimes you may only pick one, but it'll make hopefully what it normally makes. If you wanted to change that, and people are looking at that, one could use some type of genetic engineering to instruct the cell to make another type of molecule.

You didn't have genetic scissors in the old days when you started, did you?

Well, there may have been ways to do it, but they certainly would have been in very early stages in the 1980s. Basically we would just take individual cells.

And where would you get the RNA into the cells?

You could do one of a number of things. You could take the cells from the patient themselves, and it would depend on the situation, you could take cells possibly from a close relative, you might take a cell line like neonatal skin fibroblasts, but you would have a cell type that would do what you would wanted.

The liver is one of the organs that can regenerate on its own?

Yes.

What about other organs like the heart? Aren't kidneys simple as well?

Well, I don't know that any of them are simple, but people have worked on all of them. We haven't done so much work on the kidney. With Tony Atala, we did some work on ureters. With Gordana Vunjak-Novakovic, who was one of my fellows and now a professor at Columbia, and Milica Radisic, who's a professor at the University of Toronto, we made heart tissue. You can theoretically make anything. I mean, again, you might use it for different things. We've not been putting the heart tissue in patients but we are using it for drug testing—so you could make a heart on a chip. Milica and Gordana started a company, and I'm involved in it too. They're working with about thirty companies testing drugs, because you can do it much faster. You don't need to use animals and things like that.

So in the very distant future we won't need operations anymore, because we will be able to stimulate the heart to regenerate and perhaps even to repair itself?

Well, I don't ever want to say that you won't be done with those things completely. I think it will take a lot of time. But my expectation is that someday we may be able to do heart repair better because of things like this. I hope so.

We will live longer because of these technologies you're working on.

I hope so. Healthier and longer, I hope.

Can you tell me about the regeneration of limbs? Because if you and Dr. Atala and other people are successful, Ottobock won't need to produce prostheses anymore. So what do you think about the regeneration of limbs?

Well, I think that's obviously a lot more complicated. I think when you try to create one cell type or one structure, it's hard enough, but if you need to do a lot, it'll take a while. But I think at some point in time, sure, all these things will happen, but that could be hundreds

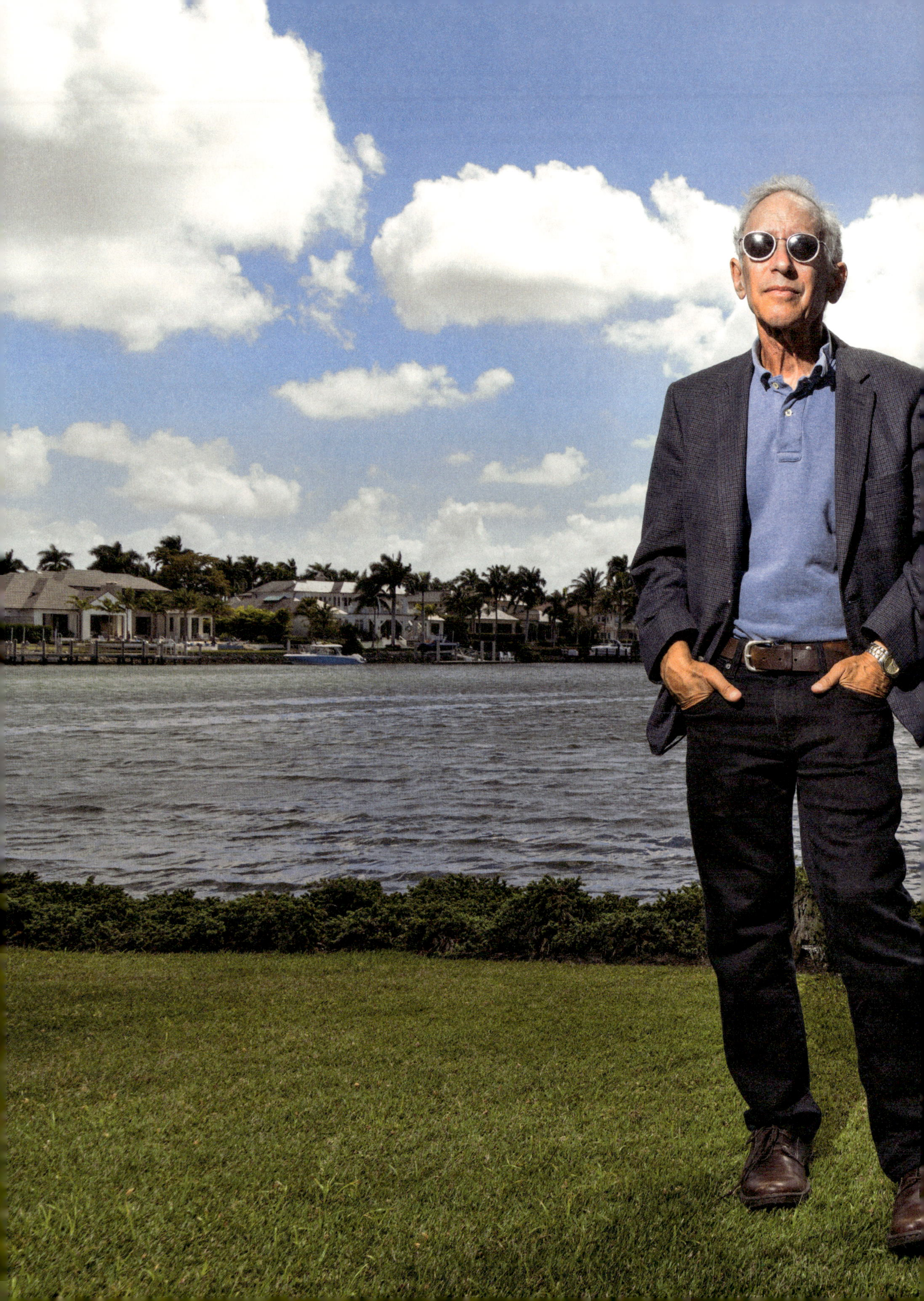

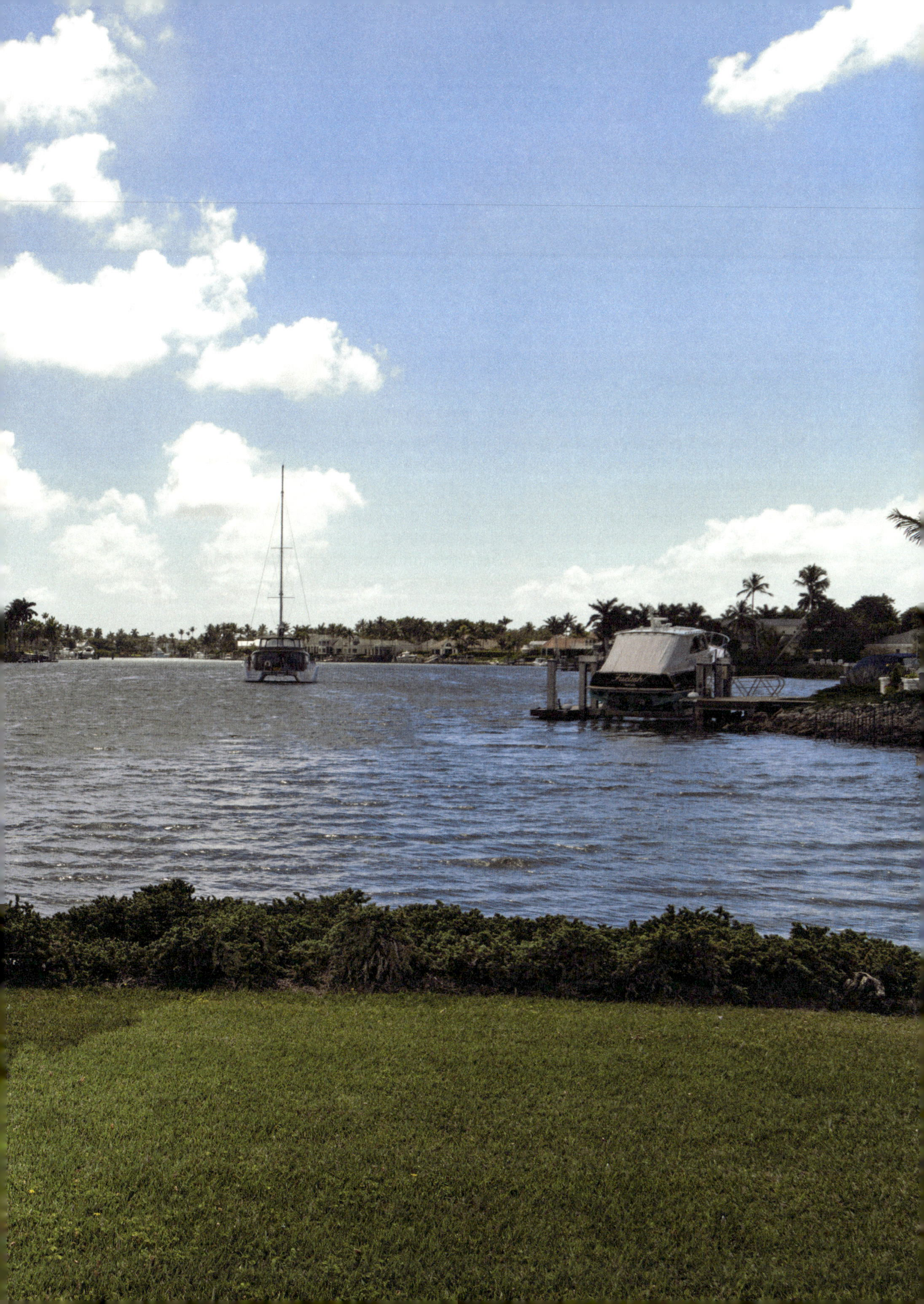

of years off. So Ottobock may be in business for a while. And by the way, these things could probably help.

It's very important, because they are already experimenting with tissue engineering to connect the body better to the prosthesis.

Absolutely.

One of the major pain points of patients is that prostheses are very uncomfortable to wear. The more comfortable the design, the longer the patient will wear it. The reality of prostheses is that a lot of people don't use them because they are so uncomfortable. They only use them when they really need to, because it's still so uncomfortable to have an artificial body part. If they can be better connected and integrated into the body so that patients have the feeling that they are really a part of their body, maybe they will accept them better.

Well, obviously, when you lose something, that's a hard thing.

Yes. There is a lot of progress. Dr. Atala showed us in his laboratory how they produce tissue with 3D printers. It was really amazing. What would you say is the perspective for these kinds of technologies?

Well, I think it's a big part of our future in several different ways. I think that over time we will see more and more tissues that use tissue engineering. I view bioprinting as part of that. I also think we'll see it used in drug testing. I've mentioned the heart on a chip, and at our own lab we're doing work with Li-Huei Tsai on a brain on a chip. With Gio Traverso, another one of my former fellows, we've developed a gastrointestinal tract on a chip. I think it'll have a big effect on speeding up research and hopefully minimizing the use of animals and people. There are some companies that are using this technology to make meat and also to make leather. So there are all kinds of applications that people are exploring.

Yes, of course. You mentioned the brain. Will it be possible one day to cure Alzheimer's and Parkinson's by stimulating the brain? We are now producing tissue outside of the body in the laboratory, but how can we stimulate the human body to do it by itself?

Well, I think the first thing is that you need to understand some of these diseases better. One of the biggest problems with Alzheimer's has been the lack of scientific understanding. I think understanding what really causes it and having better ways to test it are the first steps, because we don't know yet what the right treatments are. That's why there's been such a disaster of clinical trials. I mean, most people have tried the same thing to get rid of amyloid plaque, which may or may not be the right way to do it.

Because the vaccine just stimulates the human cell to produce everything it needs to fight the cancer or the disease or the virus, maybe we can do it with the brain as well?

Yes, but you need to know what it is that you want to produce. That's really the problem. If you look at the COVID vaccine, we know what you're trying to produce. You're concerned about, say, what's called the *spike protein*. I think once you understand what you're trying to achieve, then you can do it. I think the problem with a lot of diseases is that biology isn't well enough understood. And I think that personally is the biggest problem with Alzheimer's. There's may be others too.

Looking back on your rich career, you have 1,400 patents, you are the most cited engineer in the world, and you have incredibly broad fields of competencies and experience. What are you focusing on now? You told me that you're not retired. You are still in the MIT laboratory. What do you want to achieve in the next three to five years?

I'd say three things. We're working on a number of new ideas that I hope we'll bring further along. I mentioned some of them, like the brain on the chip. There are other things that we're trying to do in terms of figuring out good ways to stabilize the vaccines and microneedle patches. With the Gates Foundation, we're working on better ways of giving vital micronutrients to people in the developing world. And then we're also doing a lot of work on new oral methods of delivering insulin and new ways of delivering gene editing and gene therapy agents like CRISPR and messenger RNA, and solving problems or helping solve problems, like targeting them to specific tissues. So there are a lot of things in the lab. That's the first thing. I think the second thing is to take the things that we've already done scientifically in the lab and get them to patients.

Making them scalable in the industrial sense.

Right. That may come through companies or through collaborations that our lab does. But to get them into clinical trials and get them approved by the regulatory authorities. The third thing is training: training the best future chemical and biomedical engineers in the world and having them become leaders in their fields.

What do you think will be the next big leap?

Well, there's two levels of leaps. One is a clinical leap, where you take things that have already been researched, and then they get approved by regulatory authorities and get widely used. In that respect, one of my thoughts is Laura Niklason's work at Humacyte. She started in our lab many years ago to try to make new blood vessels, and they're in very advanced clinical trials now. In terms of basic work, that could be any number of things about engineering cells to make them less likely to get rejected and to make sure that they stay in the correct differentiated state, but there could be many types of things.

When will we be able to finally fight cancer?

Well, I think we're fighting it, and I think some advances have been made, but we haven't won that fight. Scientists have made a dent in it, in the case of immunotherapy and CAR T cells. But I think there's a lot further to go. And similar to Alzheimer's, we don't understand cancer that well.

Is the role of the immune system underestimated in cancer research?

I don't think that we underestimated it, but I don't know that it's that well understood. I think a lot of these things really come down to not understanding the most fundamental biology of some of these diseases. The more we understand, the better basic research gets, I think the better the chances are that we'll come up with good treatments.

Can you tell me more about your future vision of the tissue engineering?

Well, I think what you'll see someday is all kinds of tissues and organs. I don't know if it'll be like a factory, but I think that many things are being worked on, some by us and a lot by others. In our own lab, as I said, we started doing stuff on blood vessels, spinal cord repair, and restoring hearing. Jeff Karp and I have discovered some molecules, and that's in clinical trials now. Again, a lot of that has to do with cellular regeneration of a certain type of stem cell. I think over time we will see more and more tissues produced. I think we'll also see organs and tissues on a chip, which will help drug testing and basic understanding of biology.

What do you think about people like Elon Musk, who founded the company Neuralink to make a patch in the cortex of the human brain with a lot of microneedles to get all the signals in order to help people regain sight and sensations? Is this progress? Or do you think the biomedical engineers will make the race?

When I look at science in general, a lot of times one person gets credit, but it's many, many people that make the race and that deserve credit for it. I don't know how it will end up, but I do feel all these things are good. All these things will hopefully lead to progress. And as that leads to progress, we all build on other people's work. I think Isaac Newton said that. If he's done well, it's because he stood on the shoulders of giants. When you see real products that help people, it's the culmination of the work of many people, hardly just one.

There will be a time where other scientists standing on your giant shoulders. What is the message to them?

When I give commencement speeches at universities, my usual message in terms of science—but it's not limited to science—is to dream big dreams that can change the world and make it a better place. But a lot of times if you then try to do things like that, whatever your field, a lot of people will tell you that it's impossible, it won't work, and you shouldn't try that. And I tell them, do try that and don't give up. I mean, keep trying to make your dreams happen. I don't mean you don't listen to people, but you keep trying and hopefully good things will happen.

We can certainly say that you made this world a better place. Thank you, Professor.

Well, that's very kind. Like I said, I had a lot of help in terms of what I did, and I've had great students and great people to work with.

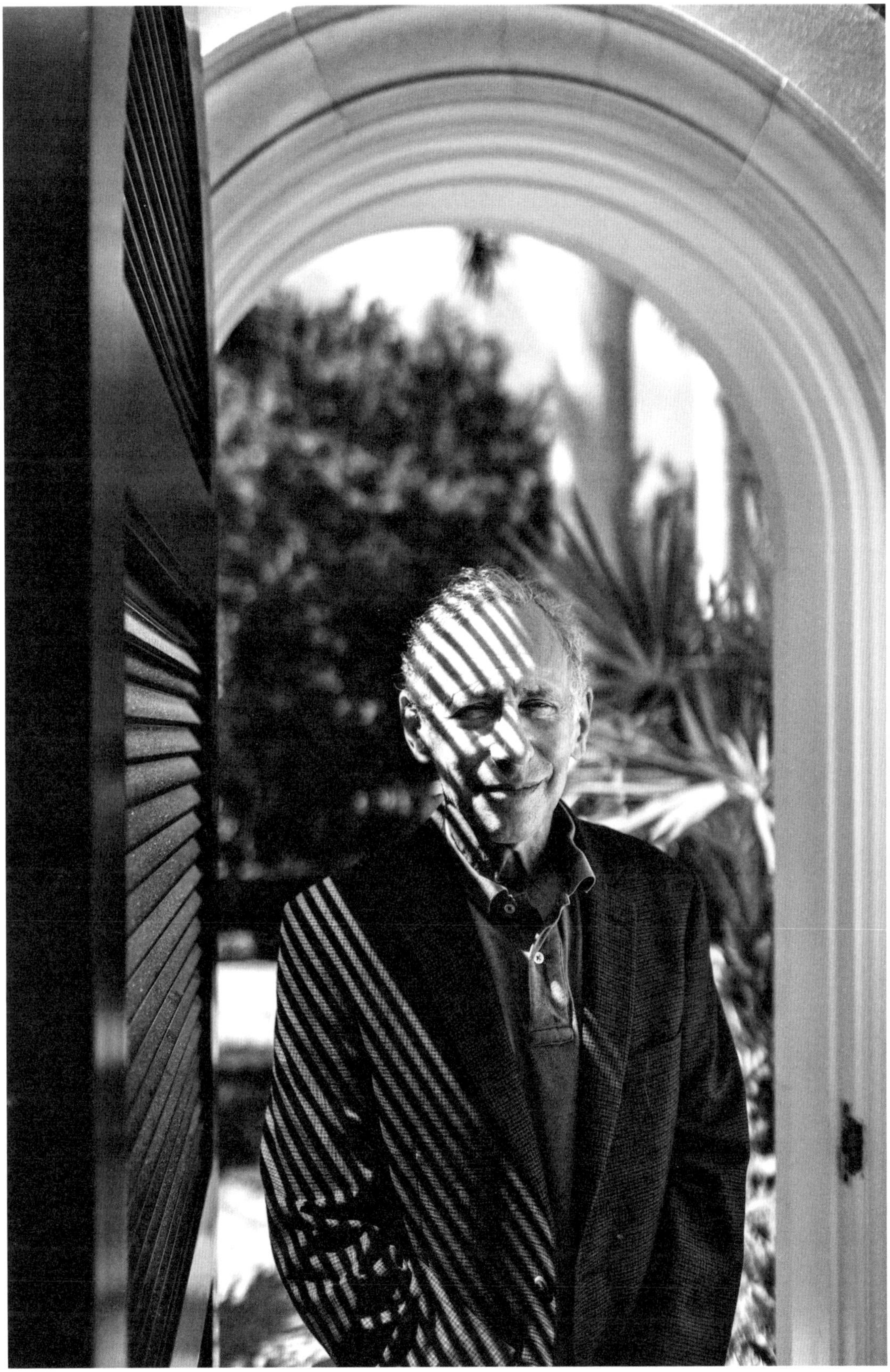

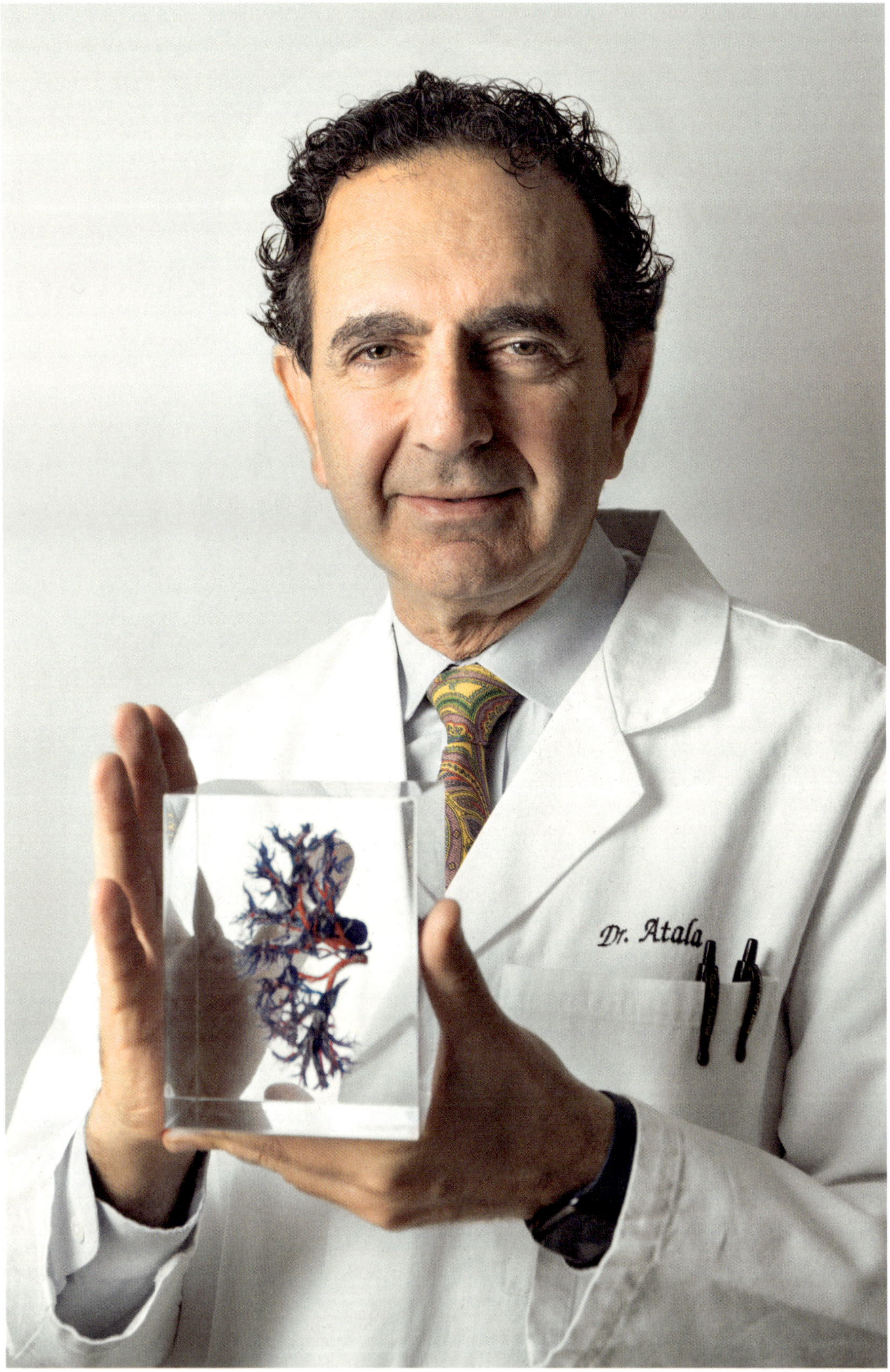
Dr. Atala

ANTHONY ATALA

"It Wasn't a Wonderfully Designed Experiment—It Was Really an Accident"

Boston was always a place for tissue engineers to be. Ioannis Yannas had tried to grow skin at MIT in the late 1970s using expanding cells outside the body and putting them on materials. So the concept was already there, and Robert Langer made a big breakthrough in the late 1980s before Anthony Atala took over the pioneering role in the 1990s. Today the bioengineer and urologist is founding director of the Wake Forest Institute for Regenerative Medicine with a mission to improve patient's lives through regenerative technologies. When he is not cultivating artificial organs in his laboratory, he continues to practice medicine. He is a warm-hearted and empathetic doctor one would gladly confide one's problems to.

You are still a practicing physician, helping people with bad injuries, lost limbs, or serious burns in a very concrete way.

I trained as a surgeon, and what really inspired me was helping people as a physician. When I got into research, many people asked me why I didn't stop practicing. The fact is that practicing medicine is very critical to what we do, because if you're not there on the front line interacting with patients, you don't really know where the needs are. The technologies change so rapidly. It's interesting how people think that medicine advances slowly, but the fact is that we are doing things today in our practice that we never thought possible just twenty years ago. A lot of what we learned in medical school three decades ago no longer even applies, because things keep changing rapidly. When you're working in research, you cannot be working on yesterday's technology. You have to be working on the technology that you need. So clinical practice has been very important for me.

This is the main problem of research: how to adapt it to practical things and products in real life. It's very good that you are on both sides of the table.

Being with the patients is what really keeps inspiring me, but it also helps guide our next direction for research, because it helps us identify the needs in real time.

Managers call that customer centricity. *You listen to your customers and change your innovation process according to the needs of customers so that the engineers who are building the product know what is needed.*

I'll give you an example. We were working on an engineered heart valve, and then we were chatting with the cardiovascular surgeon next door who said, "Well, we will not be doing heart valve implant surgery much anymore. It's going to change with the new technology that's coming out. We're going to be able to put the heart valve through the blood vessels, noninvasively." We had been working on our technology for years, and we had to quickly shift because that's where the field is going. It's about being able to see where things are going instead of where they are.

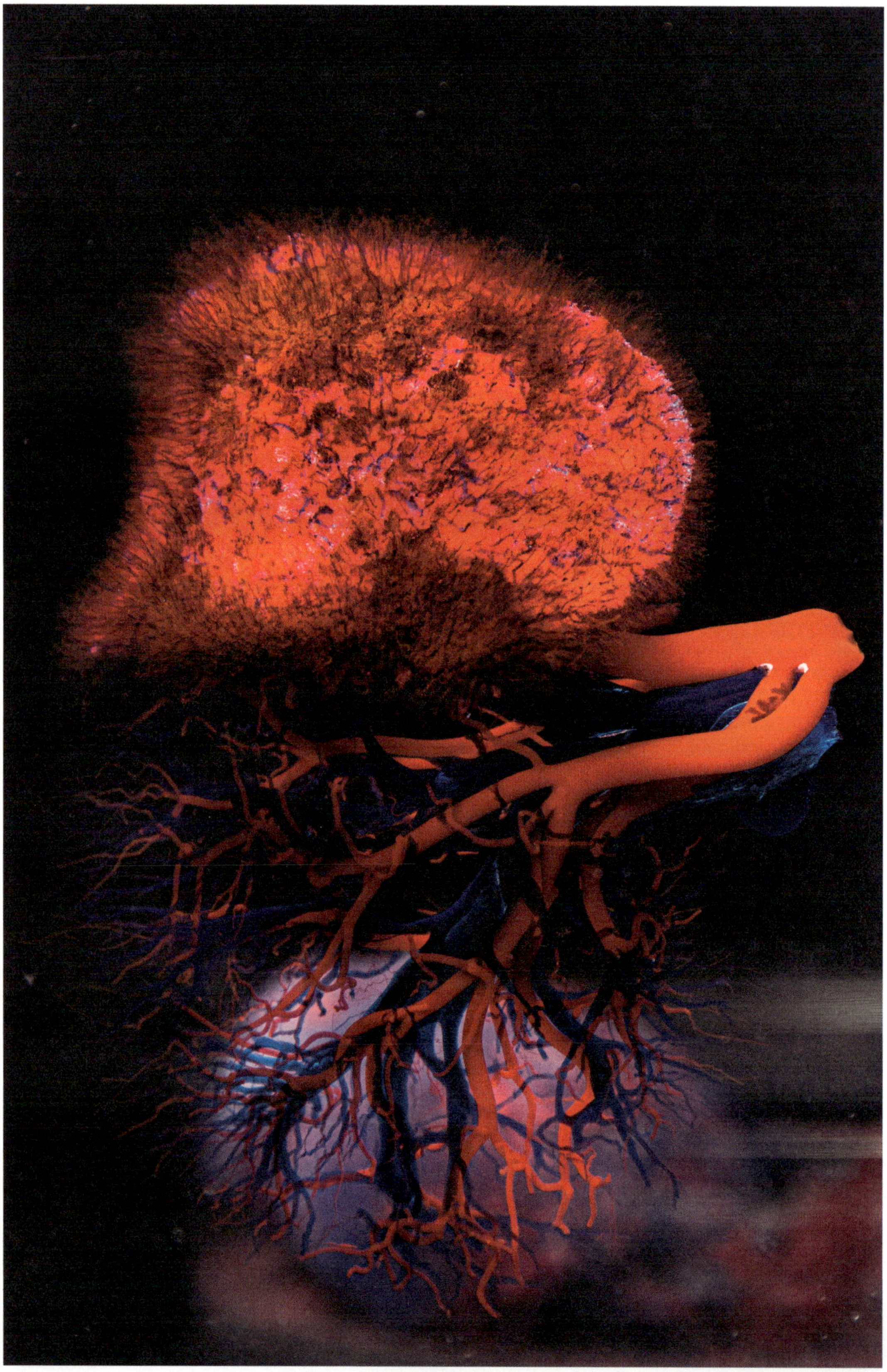

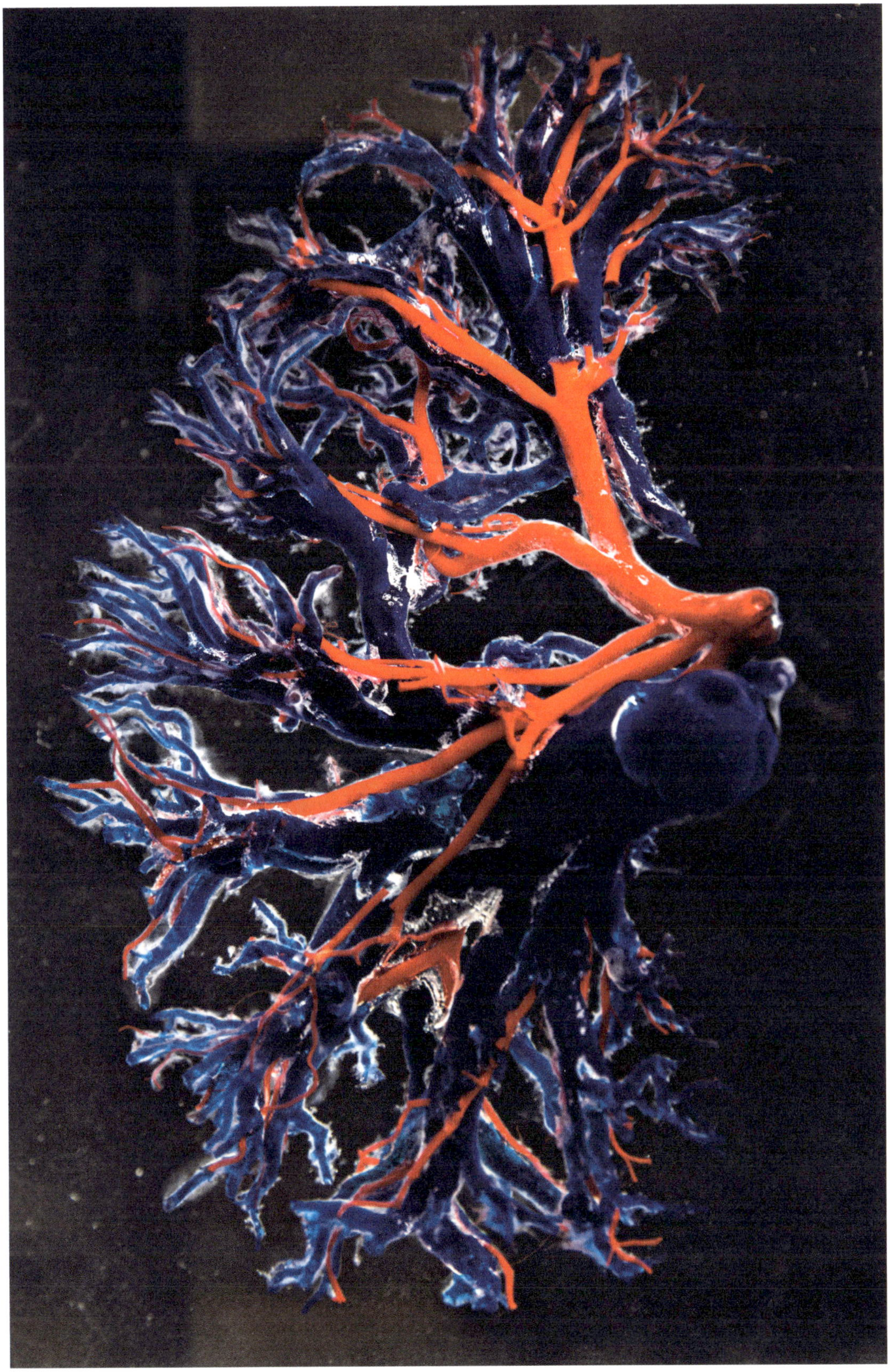

Kidney casts: two different depictions of the internal vascular system of a human kidney (left and right)

It's so much better for the patient when the procedure is noninvasive.

It's just amazing. Before we had to crack the chest open, cut the bone, and do a bypass to put in a heart valve. I'm amazed at how the technology has changed.

My cardiologist told me that I will never have a heart attack because I have a genetic guardian angel. Even though I smoke, drink a lot of alcohol, get very little sleep, and work hard, he said that he cannot see any reason for me to change my life. So I will never end up on your operating table.

[*Laughs.*] Oh, you're a lucky man! You know, I'll tell you something: if you drink, chances are that you won't have heart disease. I probably shouldn't have told you that! Alcohol is very good at preventing heart disease.

Really? How much alcohol a day?

The data is a very interesting. Drinking the equivalent of one to two glasses of wine per day is ideal in terms of preventing heart disease. However, if you drink less than one glass, it doesn't have that positive effect. And if you drink more than two glasses per day, it can be detrimental. So it's okay to drink—just not more than two glasses.

That's not very good news.

It's good. I'll start drinking more now!

You were the first person in the world to grow a bladder, so you are a real pioneer. At what point in your career did you decide to artificially grow organs and tissues?

I was actually a very reluctant researcher. In fact, I didn't want to do research. I just wanted to do my clinical practice. And when I went through the last stage of my training, my boss told me that they were adding a research track to the training. Since they were changing it, he said that I could choose. I told him that I just wanted to do the clinical track. And he said, "Really?" This was Dr. Retik, who was a surgeon-in-chief at Boston Children's Hospital. And he said, "You know, I really think you ought to do it. You should think about it." So he gave me a week to think about it. And a week later he came back to me and asked me if I had thought about it. And I said, "Yes, I have thought about it. And I still only want to do the clinical track. I don't want to do research." He said, "Really?" I said, "Yes." He said, "Well, is your wife home? I'd like to say hi to her." So he got on the phone with my wife. My wife told me, "You need to do that extra research here." When I asked why, she said, "Because he knows what would be good for you." So I started my training with Dr. Retik, but the scientist that had been hired to be there had not arrived yet. He was not coming for another six to eight weeks. There was research going on in terms of trying to get liver cells to grow, so I just thought that maybe I could get urothelial cells to grow—the cells that line the urinary tract.

Are they stem cells or specialized cells?

They're actually already specialized cells. The challenge was to get cells to grow. That was really a challenge back then, we're talking about a time when the term *regenerative medicine* had not yet been coined. Just the concept of growing a tissue was like science fiction, and the challenge with the bladder was that you really couldn't get those cells to expand very rapidly. So that's where I started to focus. The research was basically trying to get those cells to grow so we could try to implant an engineered tissue in a patient. If I had known then what I know now, I probably would've never gone into such a project.

So we have to thank your wife that you did?

Absolutely!

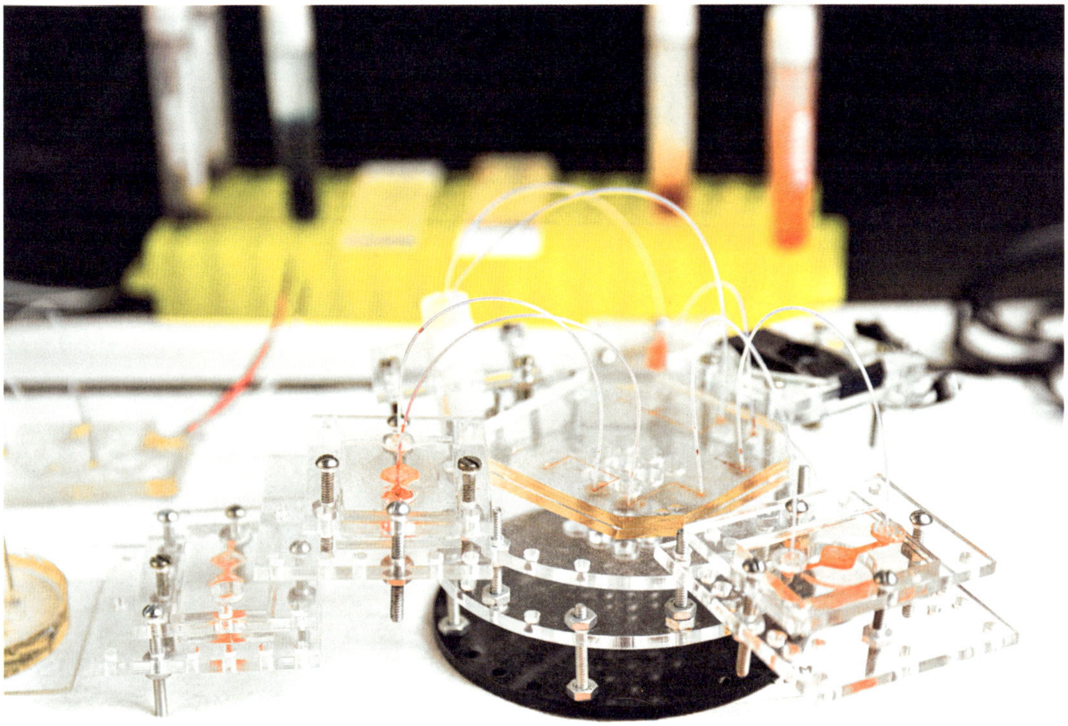

How do you manipulate the cells and get them to grow in this way?

Basically, it was just a growth factor biology. In fact, our very first federal-funded grant was on growth factor biology: how to basically identify the best growth factors that would allow these cells to grow.

Wasn't there a Nobel Prize for that in the 1990s?

There was in terms of growth factors. Absolutely.

So we are talking about the late 1990s now?

It was actually around 1990. Of course, an engineered tissue had not been implanted yet. It was really a pie-in-the-sky type of thing. But it worked, and we were able to get the cells to grow. There were many challenges: getting the cells to grow, getting the right materials in the body. You wanted materials in the body that would not degrade but also not have an inflammatory reaction. The idea was if we could get tissue to reconstitute at the bench, then we would have a solution that one day could help us avoid organ transplantation and its rejection challenges.

Is rejection the main problem? If I need a bladder, you grow one with my own cells, and my body will not reject it?

Exactly. For the concept of tissue engineering to really work, you had to work where it was going to be a benefit. It's not just doing the research. If you're not able to get the cells to grow and expand, then you're not going to be able to put them into a patient. So the whole point was how to really get under stem cell biology. The concept was to go to a patient like you, take a piece of tissue from your body, grow and expand the cells from your body, and

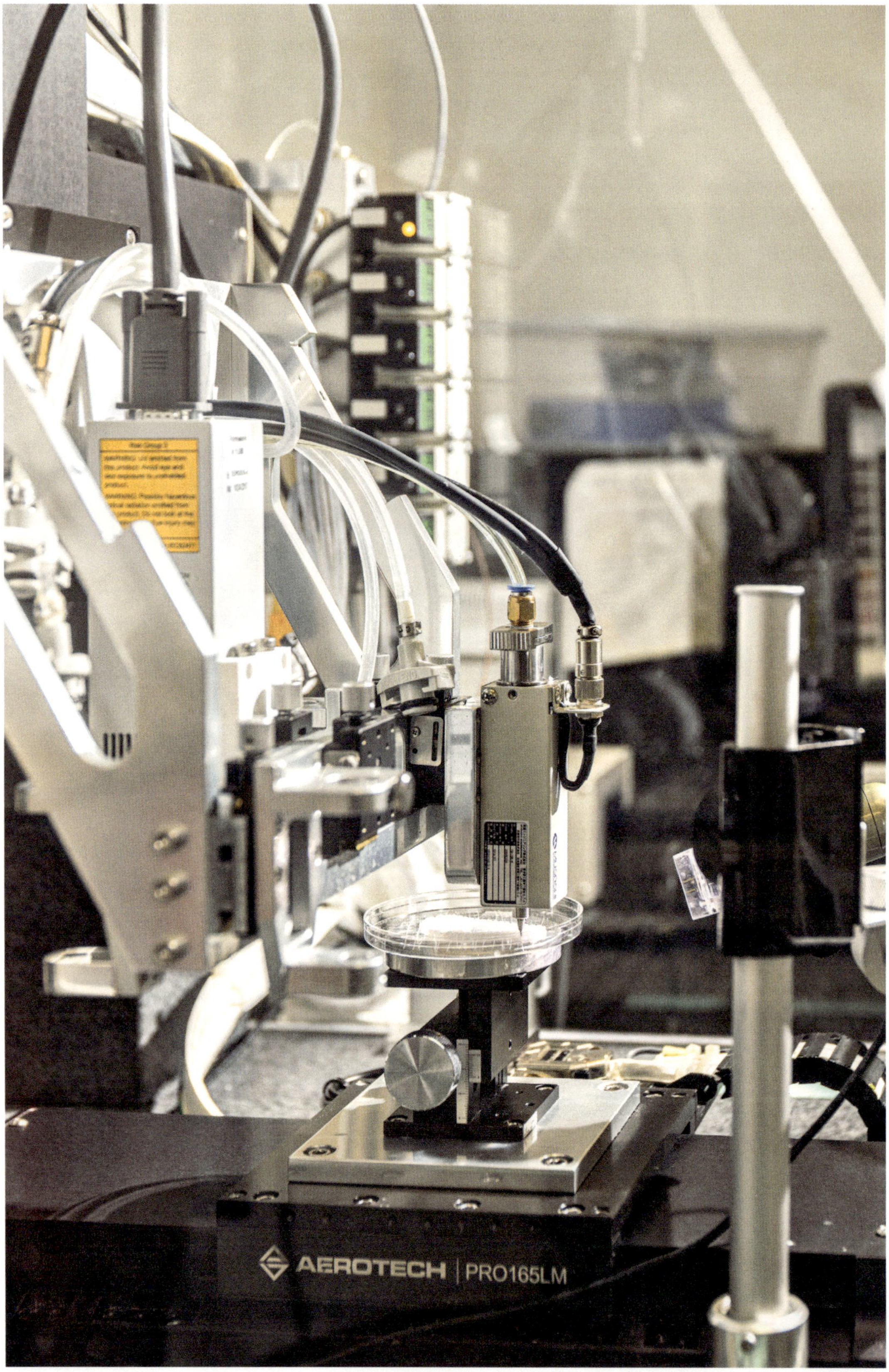
AEROTECH | PRO165LM

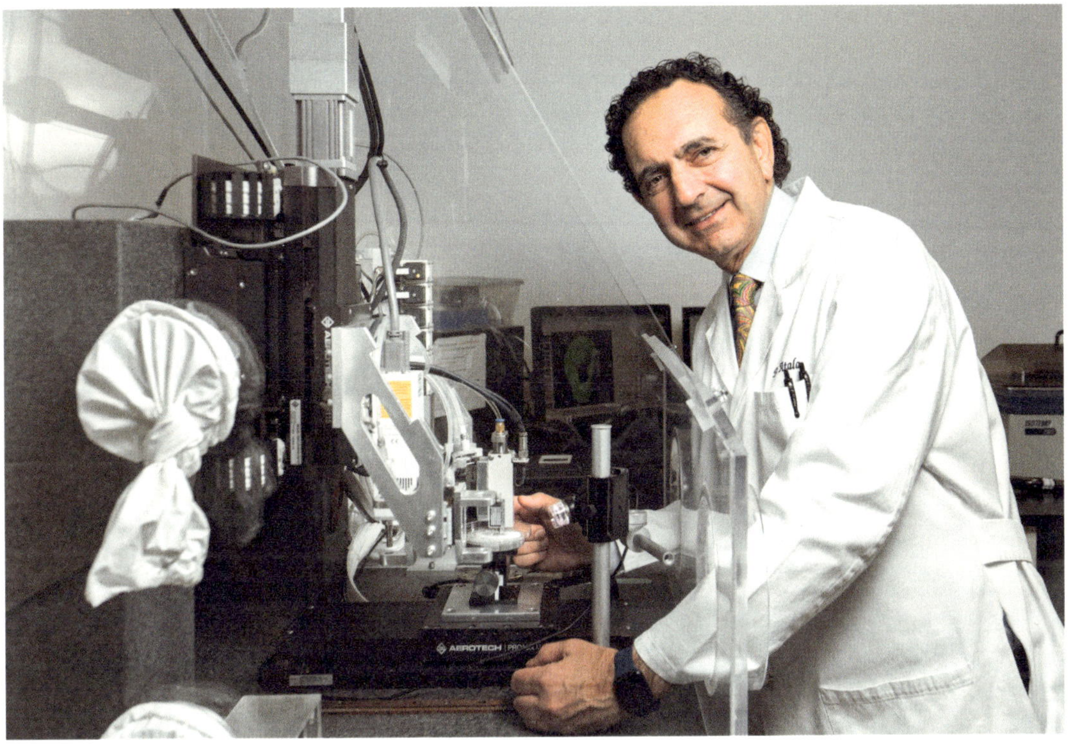

then put those cells back into your own body. It was a good concept. But if we didn't have that understanding of the actual cell expansion, you could never do it in a patient. That was the problem with many cell types in 1990. And the challenge in 1990 was that most human cell types could not be grown and expanded outside the body. You could grow them in terms of actually harvesting them and having them survive in a tissue culture plate. They would survive for a few weeks, and you could experiment with them all you wanted. But you couldn't really expand them because you needed the expansion stage to put it back into the patient.

How did your breakthrough occur?

Honestly, I wish I could tell you that it was a wonderfully designed experiment that had all the facets behind it, but that's not really what happened. We did have a grid of all these different conditions to get these cells to grow and expand. But at the end of the day, we would keep getting these samples from the operating room that patients would let us have because they were just going to be discarded. It would take hours to process these samples. One day the sample came out very late: it came like at six o'clock at night. And you know, we decided to just scrape the cells and put them in the growth media. That's how we created this impromptu simple process that was not in our initial grid. That's what worked. It was really an accident.

Everybody wanted to go home because it was late.

You know, it took hours to process each specimen. We had just processed one the day before, and now we didn't want our team to stay again. It turns out that the technique was a much more simplified process, and it ended up giving us this huge expansion that we needed. The concept of actually putting cells on a material had been there since 1970—twenty years before we'd even started to work, fifteen years before the term *tissue engineering* ever got created, and before papers had been published on this. But it was basically getting the right material and right cells. There was another scientist in Boston—it's interesting how a lot of this happened in Boston—Ioannis Yannas from MIT, who also in the late 1970s had actually tried to grow skin using expanding cells outside the body and putting them on materials. So the concept was there, and then of course, Bob Langer in late 1980s, and then us in the 1990s. Boston was really an interesting place.

You were working close together because it's a very small community.

Ioannis Yannas was a wonderful scientist working many years before us. I did work with Bob, and I worked with Ioannis as well. I mean, we communicated and exchanged ideas a lot, but the person who published a paper in the 1970s was Dr. Eric Tanasha at UCSF in California. The interesting thing to realize about this field is that it is a process of people coming up with these thoughts and then it takes another ten years to advance it further and it takes another ten years to put things together.

You put ideas together from papers that inspire you to do research of your own. So you put it together and it worked?

It worked. We had to really play with the materials a lot. There were two schools of thought back then: Ioannis was talking about natural materials such as collagen being the best, and Bob was talking about using artificial materials that you engineer and manufacture, such as polyglycolic acid, which is the artificial material used for stitches in surgery. So to create our first bio materials, we actually combined the collagen with the artificial materials to make our own tissues because we found that both gave it good properties. One alone had good properties, and the other one alone had good properties, but both together had better properties, so we ended up doing that. We are standing on the shoulders of giants who came before us, and we keep advancing the science. And hopefully others will use our work to advance the field. We're not giants, but hopefully they'll stand on the shoulders of the work that we do.

How was the story perceived?

We were finally able to test the materials, and eight years later we put our first engineered organ into a patient.

This was the bladder.

Right. It was a very anxiety-provoking experience. You never know, right?

I read that it was a bit of a fight due to the ethics of putting it into a human body.

There were ethical issues around other technologies, but there were no ethical issues around this one because we were using the patient's own cells. And we put them right back into the same patient.

You didn't use embryos.

Exactly, there were no embryos. The whole stem cell debate started at the same time when we implanted our first organ, but thankfully we didn't have that ethical debate with this technology. However, it was still extremely anxiety-provoking because you can do all the testing you want, but when you finally put it into a patient, you never know. You can do the testing over and over again to show that these technologies work preclinically, but you never know until you put it into a patient.

Do you use genetic techniques like CRISPR?

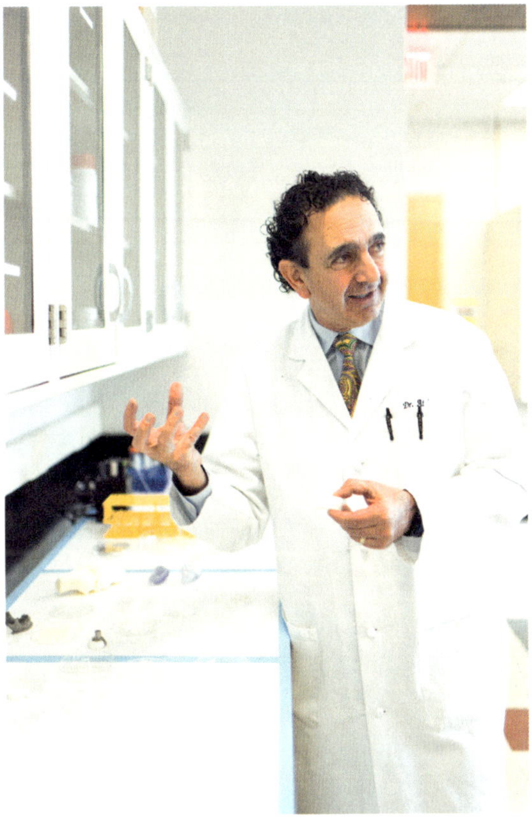

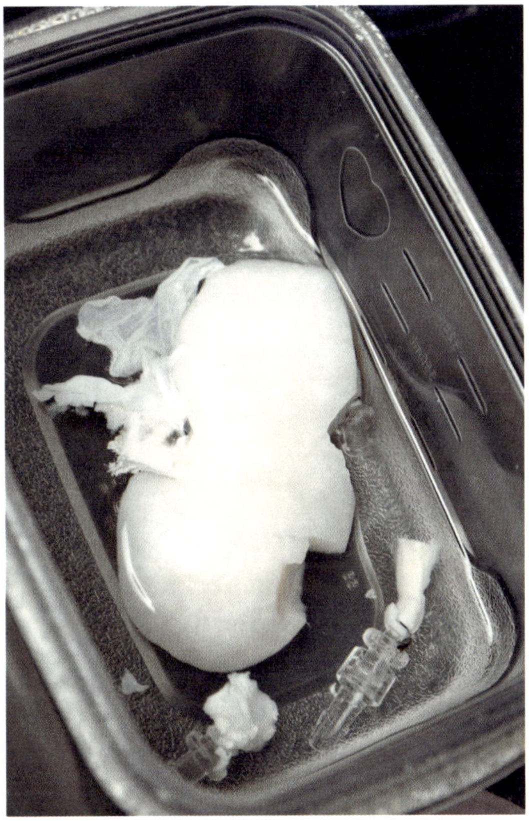

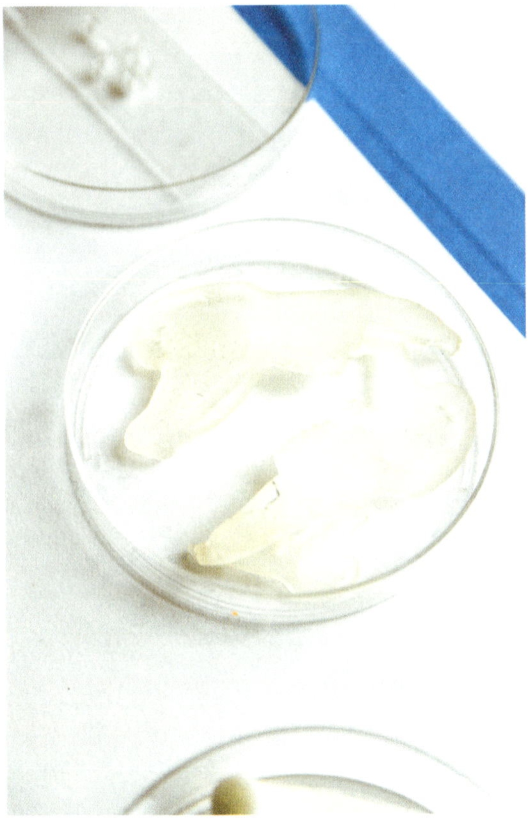

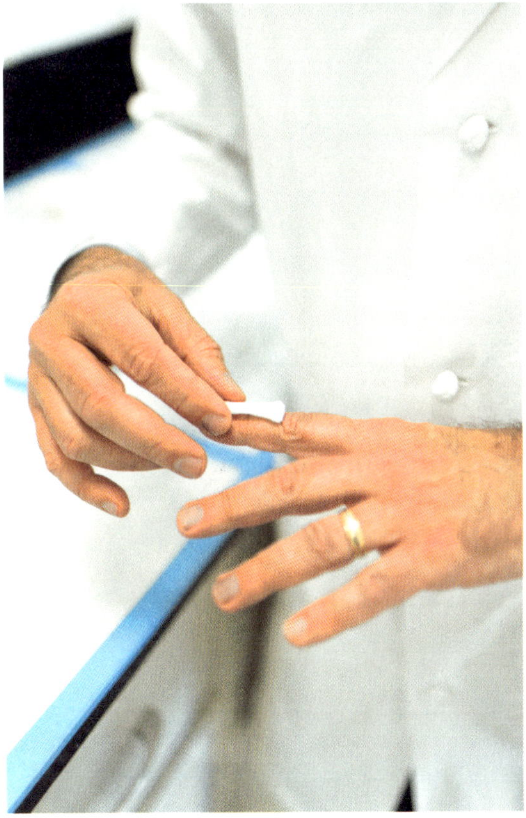

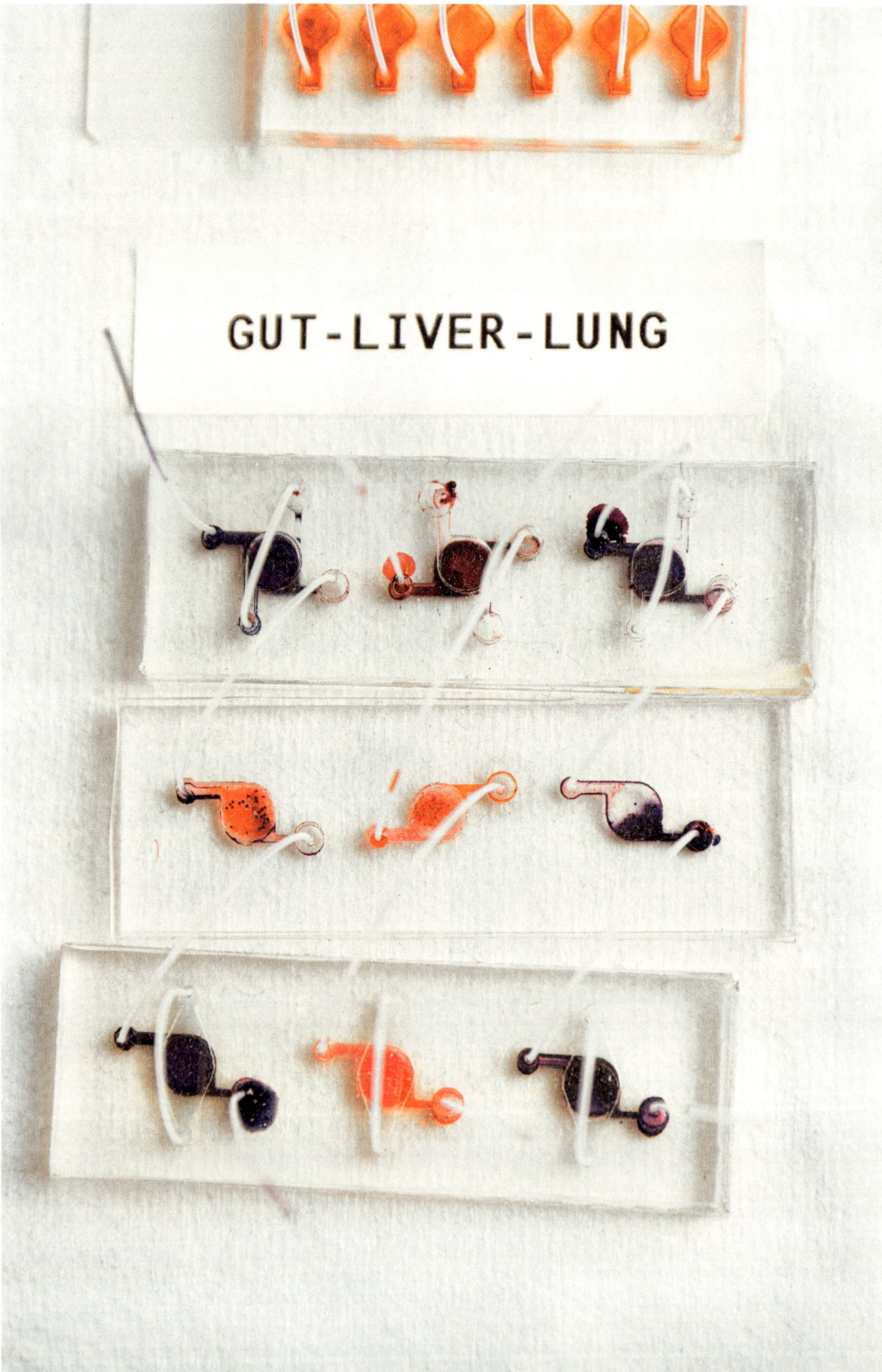

Body-on-a-chip: micro-sized 3D organ structures on chips for testing how organ systems react to aid drug development process

Now we do, but back then we did not. Now, of course, there are many different tissues that we put into patients. We have over fifteen applications of our technologies that have gone into patients at this point. And we use all kinds of things. Our team is working on using cells alone, using materials alone, using materials and cells together, using small molecules to induce regeneration, 3D printing—a lot of different technologies and strategies, but, of course, that's how we got started.

Can you describe the most important technologies you are trying in patients?

We have a number of technologies that we have put into patients. For example, we've put skin into patients, we've put in a urethra, which is the tube connecting the bladder to the outside of the body. We've put in cartilage into patients, we've put in vaginal organs for patients who were born with congenital absence or defect of the organ.

Does this have applications for promoting gender diversity?

At this point, we have to use the cells from the patient's own sex-specific organ. That's what we have done, but it could be applied in the future to gender diversity.

So one day you will be able to print a penis?

One of the technologies that we have applied for at the FDA to start clinical trials on is to engineer penile tissues. It's one of the technologies for which we are recruiting patients right now. We are also working on kidney cell therapy in patients with kidney failure. So we have a number of technologies out there that are already in patients and some that we're recruiting currently to start clinical trials.

You started in the urology field because it was your field from the beginning. Is that why you started with the bladder?

I started with a bladder because of the urological field. But we expanded to other tissues such as skin, cartilage, muscle—so both urological organs and nonurological organs. Another clinical trial we currently have is the use of skeletal muscle for patients with rotator cuff injuries, for example, with shoulder injuries. At the institute we're working on about forty different types of tissues in organs. And fifteen applications are currently have been applied to patients, but our goal is to really keep increasing the number of tissues and technologies that we get into patients in the future.

What are some of the techniques that are typically being used to develop lab-grown organs?

Much of what we're seeing right now in terms of organ development is focused on strategies that enable greater precision and scaling up, and 3D printing is a technique that can be used to add those features to the field. At our institute, we're using 3D printing to tackle the development of a number of organs, such as skin, tubular structures like blood vessels, urethras, hollow nontubular organs, and miniature solid organs. With 3D printing, we can print both the cells and the scaffolds together. And while we print the cells and the scaffolds, we're also printing the channels to bring the vascularity or the blood flow into the construct.

Can you explain how the intended patient's cells are able to multiply inside the printed scaffold?

The first step in printing an organ or tissue structure for a patient would be to expand the patient's cells in the laboratory to obtain the millions of cells required to engineer the replacement tissue. The printing process itself is designed to keep these cells alive and provide the conditions for them to continue to multiply. The printer has small nozzles to precisely deposit both biodegradable, plastic-like materials to form the tissue "shape" and water-based gels that contain the cells. This water-based "ink" is optimized to promote cell health and growth. In addition, a lattice of microchannels is printed throughout the structure to allow nutrients and oxygen from the body to diffuse into the structures and thus keep them alive while the tissues develop their own system of blood vessels.

Then you need to get approval for a printed organ from the FDA?

Yes. We have a team here that works with the FDA on a regular basis. We work very closely with the FDA to make sure these technologies get to patients. And we have our own FDA-compliant manufacturing facility here. So at the institute we basically have the concept of the idea. We do the proof of principle on the benchtop. We do the preclinical testing. We also do the conversion of the technology to manufacturer product through our process development unit. We have a regulatory team that takes care of quality assurance, quality control, filings with the FDA. We have a manufacturing team that manufactures the product, and we have a clinical trials team that actually coordinates the clinical trials. And it's all done within this building. The only thing we don't do in this building is to actually treat patients. We do everything up to the point where we coordinate the production of the product. And then we deliver the product to the patient.

What will be the next organ you work on?

We're working on a number of them. The next one to go into patients and the clinical trial is penile tissue.

This is a breakthrough that is extremely important for many people.

Yes, you know, it's interesting, because if you are a patient that has a particular condition, any organ becomes the most important organ. That's why our goal is really to try to expand the indications. So more conditions can be treated, and more patients can be helped.

What is your vision?

The mission of our Institute is to improve patient's lives through regenerative medicine. Our vision is to create a global transformation where patients worldwide can be helped with these technologies.

How do you make this technology scalable? Will you send blueprints to Africa so that people there can build a factory or 3D printer to produce locally? Or will you give the patents to a company that will manufacture them for the whole world?

It's both actually. We're big on collaboration at the institute. In fact, we promote collaboration, because we really do believe that the competition is the disease. It's not our colleagues across the world but actually the competition that is the disease. That's what we want to solve. We all have to work together to bring these changes.

You make other people rich with your ideas.

That's okay, as long as people benefit. Of course, there is intellectual property.

Do you have your own company as well?

We basically license, so the technologies themselves belong to us.

And the institute benefits. Do you personally benefit?

If you have patents then you can. But basically the way that it works is we have a lot of collaborators. Every year we do an inventory of how many collaborations we have. As of this January, we had approximately four hundred active collaborations of which about over three hundred were national and about a hundred were international. And in a lot of these collaborations, we're actually making sure that we exchange, that we are providing

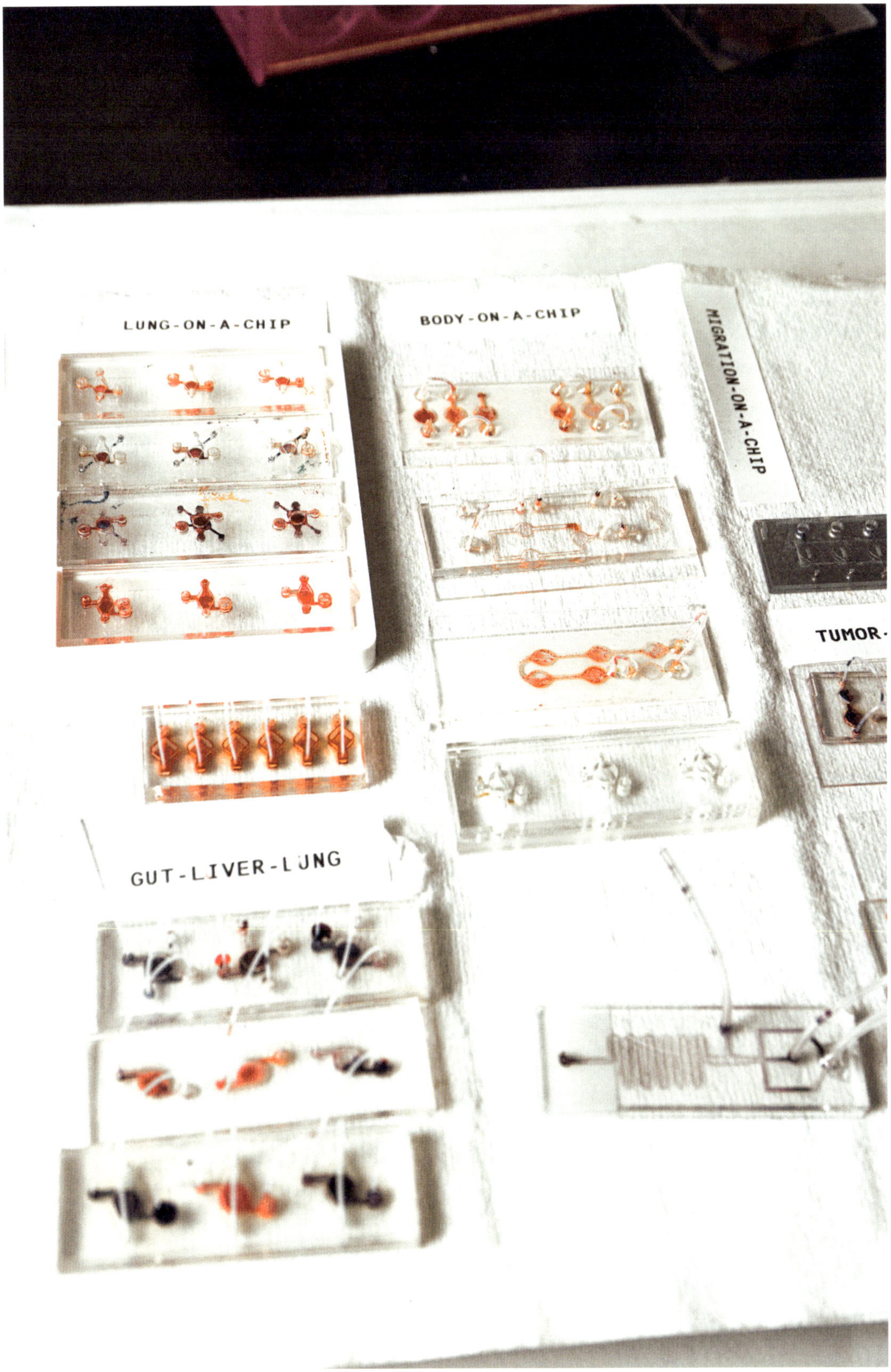

Body-on-a-chip: micro-sized 3D organ structures on chips for testing how organ systems react to aid drug development process

information, reagent cells, and know-how. Because we honestly do believe that these technologies should be available to everybody. Of course, there are other methods to make that happen: through industry, patents, et cetera. But at the end of the day, this is bigger than all of us know, right? We want to make sure that patients get the benefit they need.

So the areas that link you to Ottobock are tissue engineering and muscle engineering.

Yes, we do grow muscle. We have clinical trials with muscle at this point. The nice thing about muscle is that you can use it for different indications. For example, you can use it for the rotator cuff injuries we talked about, because when people get a shoulder injury, they get muscle weakness. We can also use muscle as therapy for patients with urinary incontinence, for example, or fecal incontinence, by buttressing up the urinary or fecal sphincter. The nice thing about it is that for some of these tissues you can really apply it to different areas as well.

Can you tell me about your work with stem cells?

There are human embryonic stem cells and induced pluripotent stem (iPS) cells. Both are very powerful because they can grow very rapidly, and you can turn them into anything. But unfortunately, by definition, human embryonic stem cells and iPS cells form tumors. In fact, you cannot call a cell pluripotent unless it forms tumors. Because they're pluripotent, these cells can turn into anything and grow rapidly. On the other hand, you have adult stem cells, such as bone marrow and fat, which will not form tumors but unfortunately cannot be expanded very well and have a limited ability to turn into other cells. So, approximately twenty years ago, we postulated that perhaps there would be a stem cell population from the amniotic fluid and the placenta. When you take the human embryonic stem cell, you take it at twenty days of gestation. You take an egg and a sperm, mix them up, and put it in an incubator. Twenty days later, when it's the size of a pinhead, you take the stem cells out—you have human embryonic stem cells. That's it, week three.

So you can grow stem cells in the laboratory?

You can use that egg for in vitro fertilization, or you can take the cells out, which is only about a one-hundred-cell cluster. You can start growing those cells, which are human embryonic stem cells. But they are so primitive that they form tumors.

Really? I didn't know that.

Yes, that's why you have not seen human embryonic stem cells widely used. It's because, by definition, they're so primitive that they can turn into anything. Those cells want to become bone, cartilage, everything—so they're unchecked. If you put them in the body, they form tumors. It's the same thing with induced pluripotent stem cells. You take a cell from skin, or from any tissue, and you revert the genetics to take it back to a human embryonic stem cell state. But again, they form tumors. That's why even though the human embryonic stem cells have been around since the 1980s, you're not seeing too many of these cells used clinically.

So it was the wrong way in the science.

No, it's not the wrong way. This study really allowed us to learn about stem cell biology and know what to do. We thought it would be easier to control the cell's primitive nature, but it's hard to control it. Then you have the adult stem cells, like bone marrow and fat, and those cells will not form tumors because they're already adult.

They are already specialized.

They're specialized, so they won't form tumors, but because they're specialized, they don't have the potential to become many other tissue types. And they don't have the potential to grow and expand in large quantities. So we postulated back then that just maybe there would be a stem cell population in the womb. You're taking human embryonic stem cells at week three from an incubator, so if you just wait an extra seven weeks, you can take a chorionic villus sampling, a commonly used diagnostic technique where you put a needle in and take a small sample of placenta.

Just to check if there's any diseases?

Yes. So we postulated that perhaps there would be a stem cell population in this tissue that would be a powerful stem cell as well. In fact, it was a fishing expedition. But we did find a stem cell population in that tissue that goes into all three germ layers, just like in embryonic stem cells. You can create endoderm, ectoderm, mesoderm, you can grow them and expand them in large quantities just like human embryonic and iPS cells. But because you're further down the development, they don't form the tumors.

So you just have to find out how to stimulate them. If you nudge them in the right direction …

That's exactly right. So unlike human embryonic and iPS cells, amnion-derived cells don't form tumors and unlike adult stem cells, you can grow and expand them, and they will go into all three-germ layer. This is a stem cell type that we've been working on.

Regarding the mechanism for growing cells, is there a possibility that in the distant future you will be able to take cells from the patient to the laboratory? Can you skip some genes, or can I grow it myself? Is that part of your vision?

Absolutely. With the current concept, we go to the patient. We take a very small sample of tissue from that patient's own tissue and organ. We grow and expand those cells outside the body. We can then print those cells, or we can make them by hand, or we can put them on a hydrogel and inject them. But the work that we're doing right now also involves inducing your own stem cell regeneration by creating small molecules. We have a small molecule program where we are actually creating small molecules that you can just take as a pill that will activate the cells for regeneration. That's actually something that we're currently doing in the research stage. It has not been applied to patients yet.

Your organ changes when you take a pill?

I'll give you an example that's easy to understand: inducing hair to regenerate. This is a problem for patients who have cancer treatments. For women who get radiation and their hair never comes back. That's a problem not just for cosmetic purposes. It has a real medical indication, which is very serious for some patients. We use a small molecule formulation that induces your hair stem cells to grow. That's exactly what that is: it's a small molecule that has a few applications. It induces stem cells to regrow, and once you induce those stem cells, you may not need to keep taking the pill. Once we induce those stem cells, those stem cells will generate new tissue, and the tissue is there.

How can the artificial molecule use the stem cell to produce more hair?

It has to do with knowing where the stem cells are. We know where the stem cell populations are for every organ. We know the molecular pathways that lead to the formation of these stem cells and to the differentiation of these stem cells to create what they're supposed to do. So if we can create small molecules that effect the change on that pathway and reignite the pathway, you switch the stem cell on again.

How do you do that?

Because you know what it does to start with. It needs a boost. You know, it's basically like starting the engine and driving your car. If you run out of gas, you need two things. You need more gas, and you need someone to start the engine. So those are the paths that you need to take. You need to give it more gas.

Like hormones or messengers?

They're basically small molecules that activate those pathways again. A small molecule that will generate more fuel. A small molecule that will turn that key ignition again.

Does it do this in a biochemical way?

Yes, biochemically you're able to do it because that's what your body knows how to do. It's just that at that point, it no longer does it. Because genetically your system shut off that particular area.

Now I understand. So you need my stem cells now?

No, we induce a patient's own stem cells to grow.

Do you restart the area of stem cells responsible for the growth of hair?

Yes. You have a lot of hair, so you don't have that problem. Your system keeps going, and you still have that fuel going. That's what you're doing genetically. But somebody else doesn't, either because they were born with that predisposition or due to an injury or disease.

So if you found out the mechanism how to restart the stem cells of hair growing, you can provide it to everybody who has the problem? Or is it still an individual problem that requires building the molecules for that specific individual?

That's the nice thing about it. These small molecules are ubiquitous. They're the same.

So you can produce them on a large scale, and everybody can benefit.

Yes. It's basically a genetic inducement that you're doing with a small molecule that can be provided as a pill.

That's great! I would like to try it, but I am happy that I don't need it.

I know, you really have a lot of hair. It is not because of the drinking and smoking. You're a very healthy person, and you have been blessed genetically.

Thank you, doc.

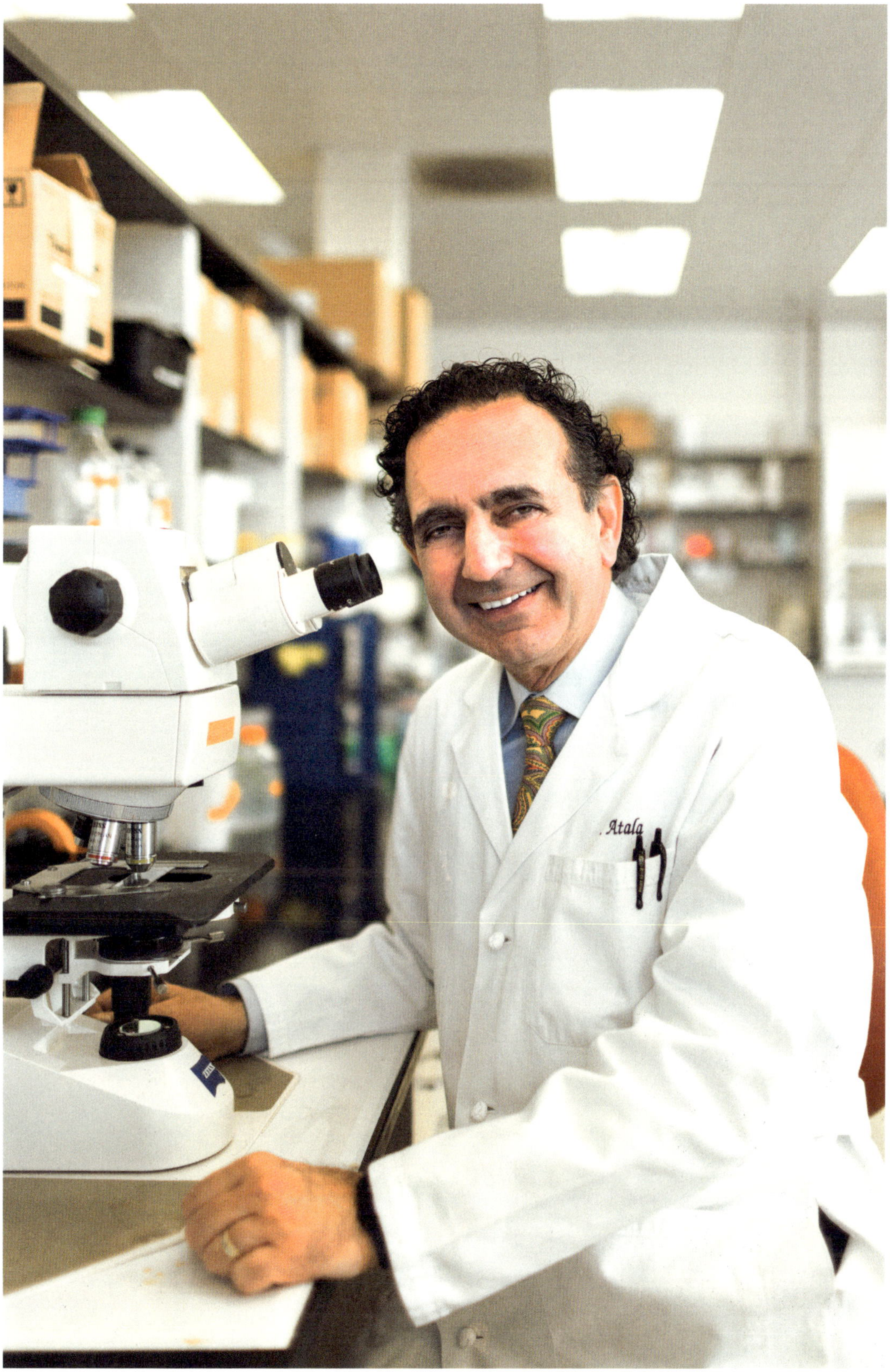
Atala

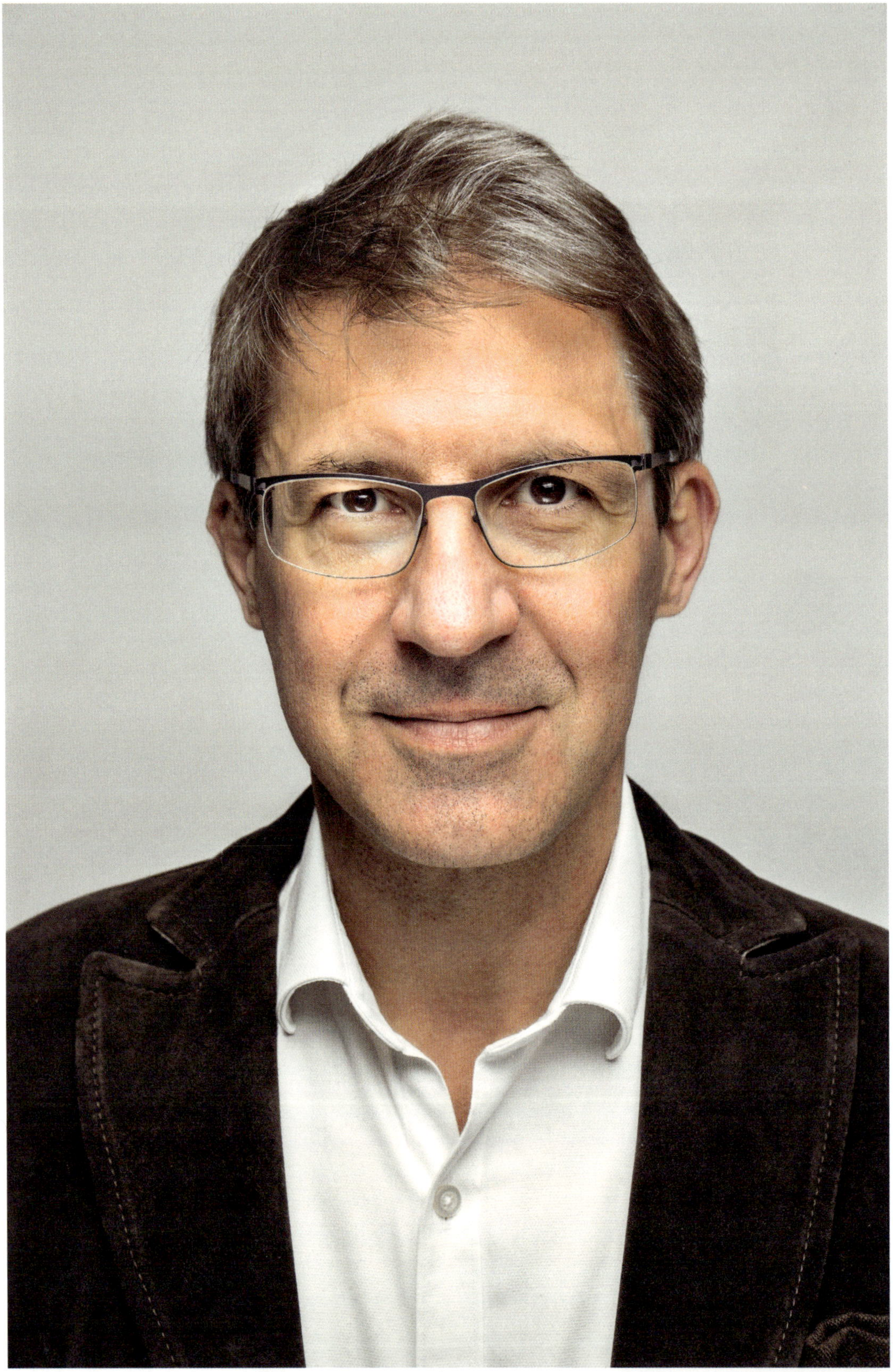

MARK KOTTER

"If All Cells Could Become Medicines, Everyone Would Potentially Have Their Own Repair Kit"

The neurosurgeon heads a regenerative medicine laboratory at the University of Cambridge, where he researches stem cells and cellular reprogramming. As founder of the human synthetic biology enterprise bit.bio, he is successfully challenging fundamental dogmas in biology to translate biology into technology that combines synthetic and stem cell biology, and to manufacture human cells at scale and thus make possible a new generation of cell therapies. Human cells are also valuable for pharmaceutical research and drug development, and they can be used to control of advanced biological circuits for biomanufacturing. So far Mark Kotter and his partners have been able to raise 250 million dollars. In the distant future, bionics and biology may come together to form completely new sorts of interfaces.

The kind of stem cell research that you do is forbidden in Germany.

That's actually not true. We use induced pluripotent stem cells that can be generated from a simple blood draw. These are based on a synthetic biology paradigm that now underpins my two start-ups, bit.bio and Meatable.

You are a trained neurosurgeon—what brought you to start two companies?

I initially started out studying math and physics. Before entering medicine, I worked as a nurse assistant in a ward with patients suffering from spinal cord injuries. Such injuries are life-changing events, and they also pose a huge challenge from a research point of view. So I got really interested in spinal cord repair. Now, a few decades later, I'm a neurosurgeon, working and treating patients with spinal cord injuries and running a lab. The reason I ended up working on stem cells was because it was very clear to me that a pill won't be able to heal a spinal cord injury—but cell therapies may make a difference. Cell therapies are based on the concept that cells themselves form a new class of intelligent drugs. The first generation of such therapies are currently in the clinic. My research eventually led me to spin out a number of companies.

We are familiar with cell therapies as a way to treat leukemia.

That's correct. These treatments are really about to transform medicine, so it's quite exciting. When you look at the history of medicine, we started off with somewhat airy concepts, with quacks inspecting urine and treating people with leaches. Over time science was brought into medicine, and treatments were tested more systematically. The scientific approach led us to the first generation of medicines, small molecules—aspirin, for example, and antibiotics. One of the disadvantages of small molecules is that they are not very precise, and they distribute across just about any part of the body, which leads to side effects. Nevertheless, a lot of very helpful drugs that continue to be in use are based on small molecules. In the 1980s our understanding of biology evolved to a point where we

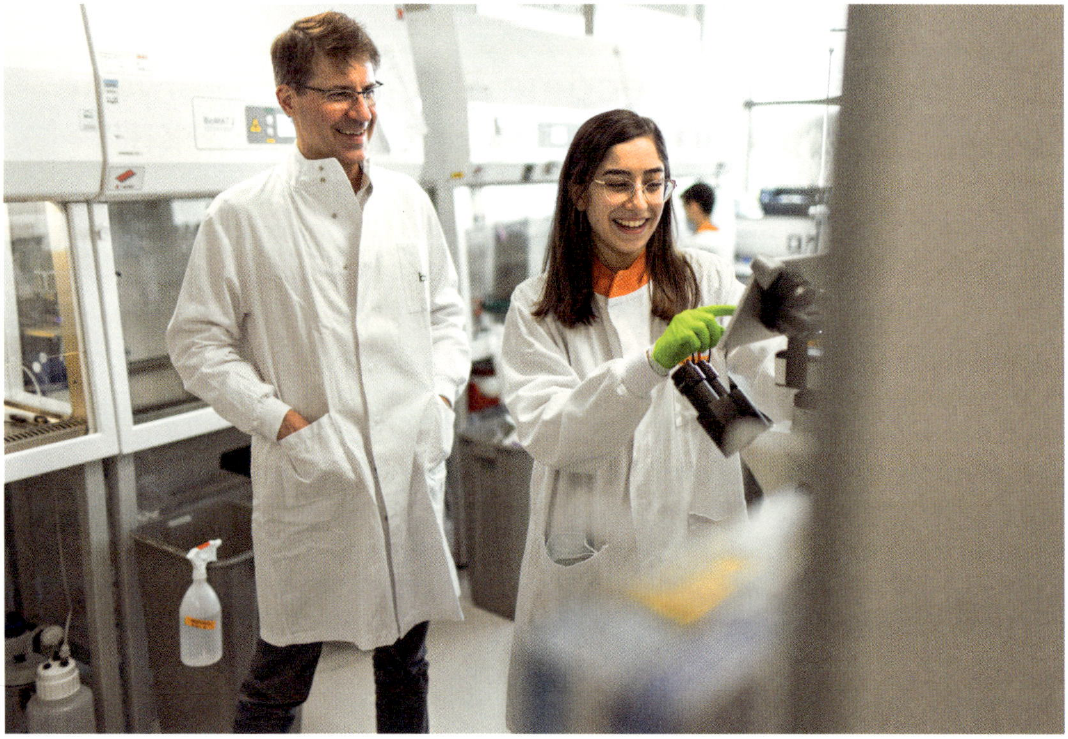

Sarah Hussain, research assistant for cellular phenotyping, and Mark Kotter

understood DNA, and how genes are transcribed into RNA and then translated into protein. In this new paradigm of molecular biology we learned to clone genes, and this opened up a second wave of medicines based on proteins, which is called *biologics*, including drugs such as insulin and antibody drugs. These are very precise drugs based on large molecules that have enabled us to tackle a number of difficult diseases. Sir Greg Winter, who is on the board of bit.bio, received the Nobel Prize in 2018 for discovering a mechanism by which you can develop such antibody drugs. Today we are developing the third generation of medicines, a revolution that uses cells themselves as drugs. Apart from being able to replace lost cells, these living medicines are able to react to their environment to effect change. If the right cell is transplanted in the right context, it will know what it needs to do, and this can be highly effective. As you mentioned, the first approved cell therapies, which are called *CAR T cells*, provide transformational cures for deadly blood malignancies. For this purpose, T cells are engineered with CAR receptors so that they recognize cancer. But this is only the beginning. The concept of using a cell as a therapy is much larger. It is sometimes referred to as *regenerative medicine*. Scientists have been working on, for example, neural stem cells for spinal cord injuries. This was my motivation for entering the field.

There have been a number of studies using stem cell derived cells in the clinic.

Yes, so far studies have, for example, tested the use of retinal pigment cells in the eye to treat loss of visual acuity, and recently we heard of a breakthrough when scientists replaced cells secreting insulin that are lost in the context of type 1 diabetes. The first patient treated with a pancreatic islet cell product showed an extremely promising response. So cells are becoming medicines—that's the wider context of why this is important. If you had the right cell, you could cure the disease, which essentially means that all cells could become med-

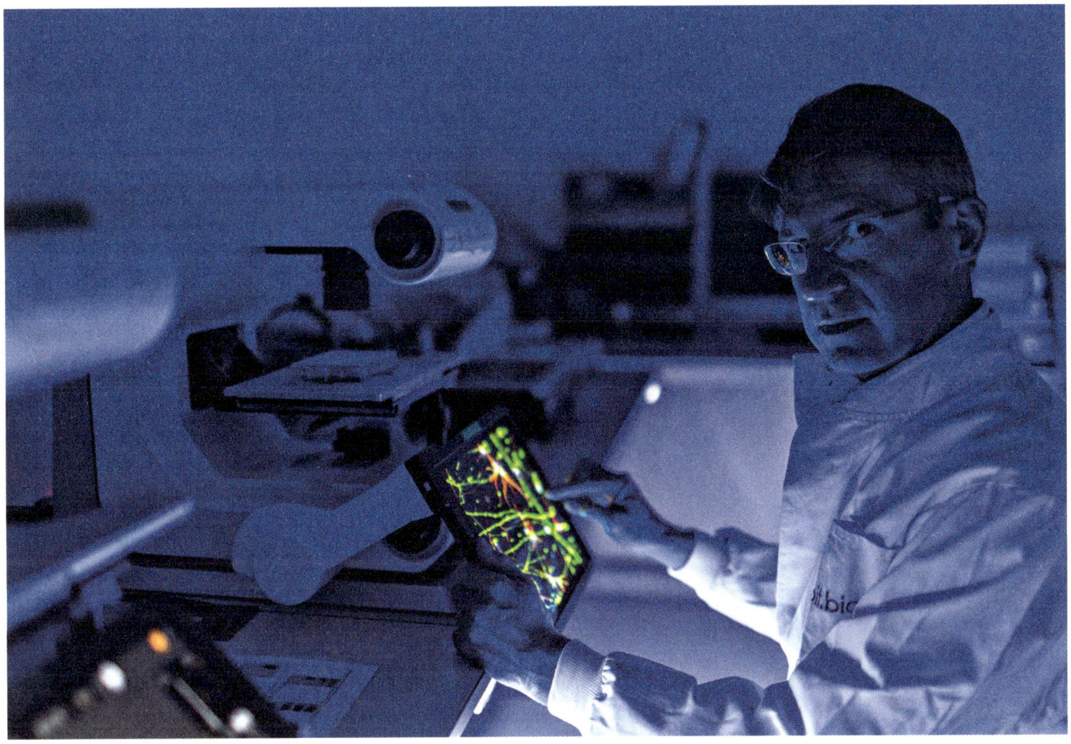

icines. We're just at the beginning of this incredible revolution. And once we produce the cellular components, we can assemble them into organs, as in printing a liver. That's the trajectory we're on.

We've known about pluripotent stem cells for twenty or thirty years. The initial wave of human stem cells was extracted from embryos. This, of course, led to ethical concerns, which is why this approach was heavily criticized, resulting in the US shutdown and the high regulation in Germany.

But the important thing that happened was in 2006 or 2007, when Shinya Yamanaka discovered a way to produce stem cells from skin cells using a cell-programming approach. With this technology, no more embryos need to be sacrificed. In addition, this technology makes it possible to create stem cells from each of us—meaning that everyone potentially has their own repair kit. It's a phase shift in science, and I think we're just grasping the consequences of that. Yamanaka received the Nobel Prize for his discovery in 2012. By the way, he also showed that you can rejuvenate cells—another fascinating facet of biology.

We already have a self-healing mechanism that kicks in if we cut ourselves.

We're not as good as axolotls, though, and the brain and spinal cord in particular do not regenerate very well. But stem cells are just the starting point. If someone transplanted a stem cell into your body, you would just grow a tumor. It wouldn't have positive effects because a stem cell is so early in development and its function is to create all cells in the human body. When you transplant pluripotent stem cells, they form teratomas—essentially creating the tissue components of an embryo. So while pluripotent stem cells are not helpful for treating a disease, they are an excellent starting point for producing the cell

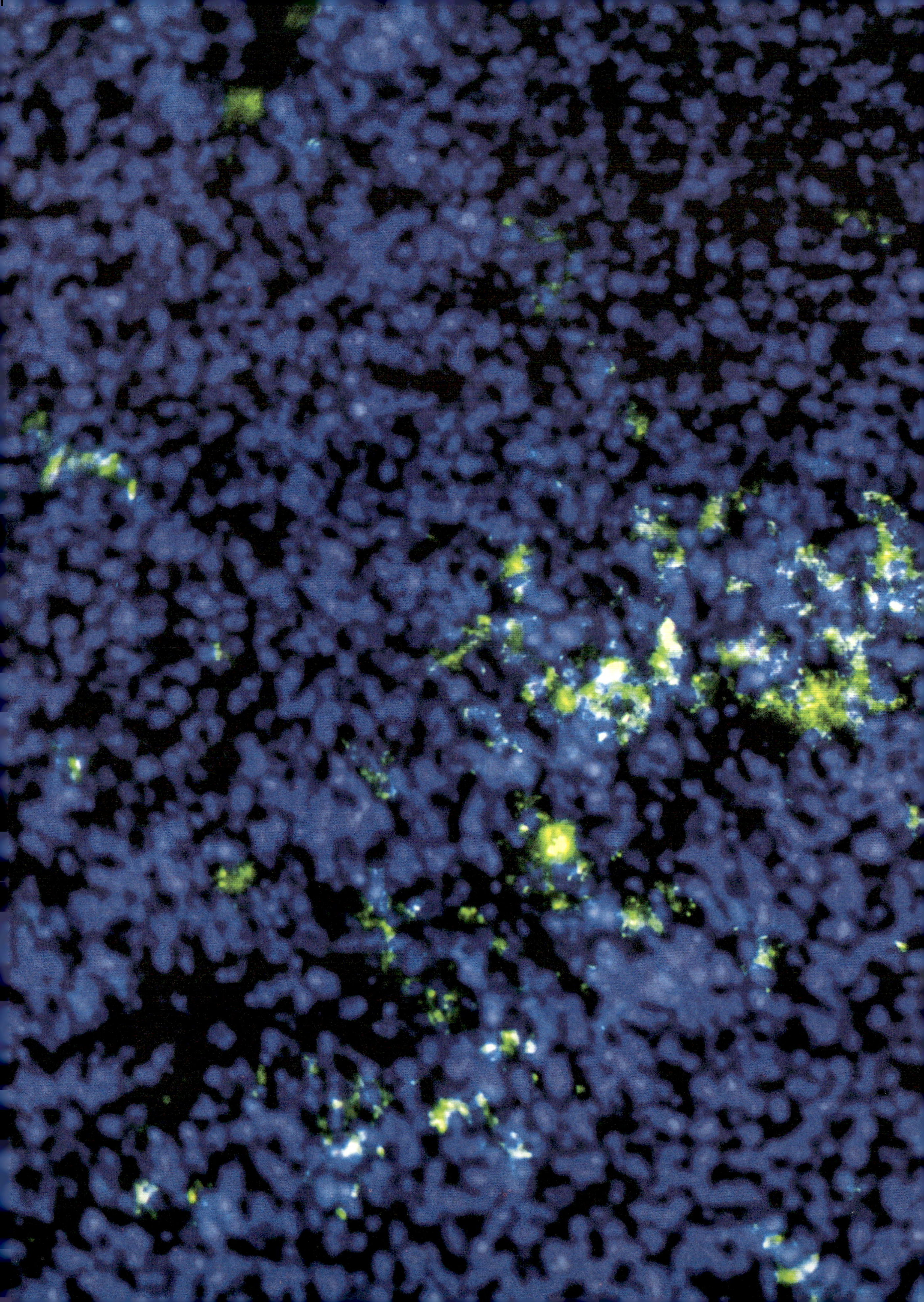

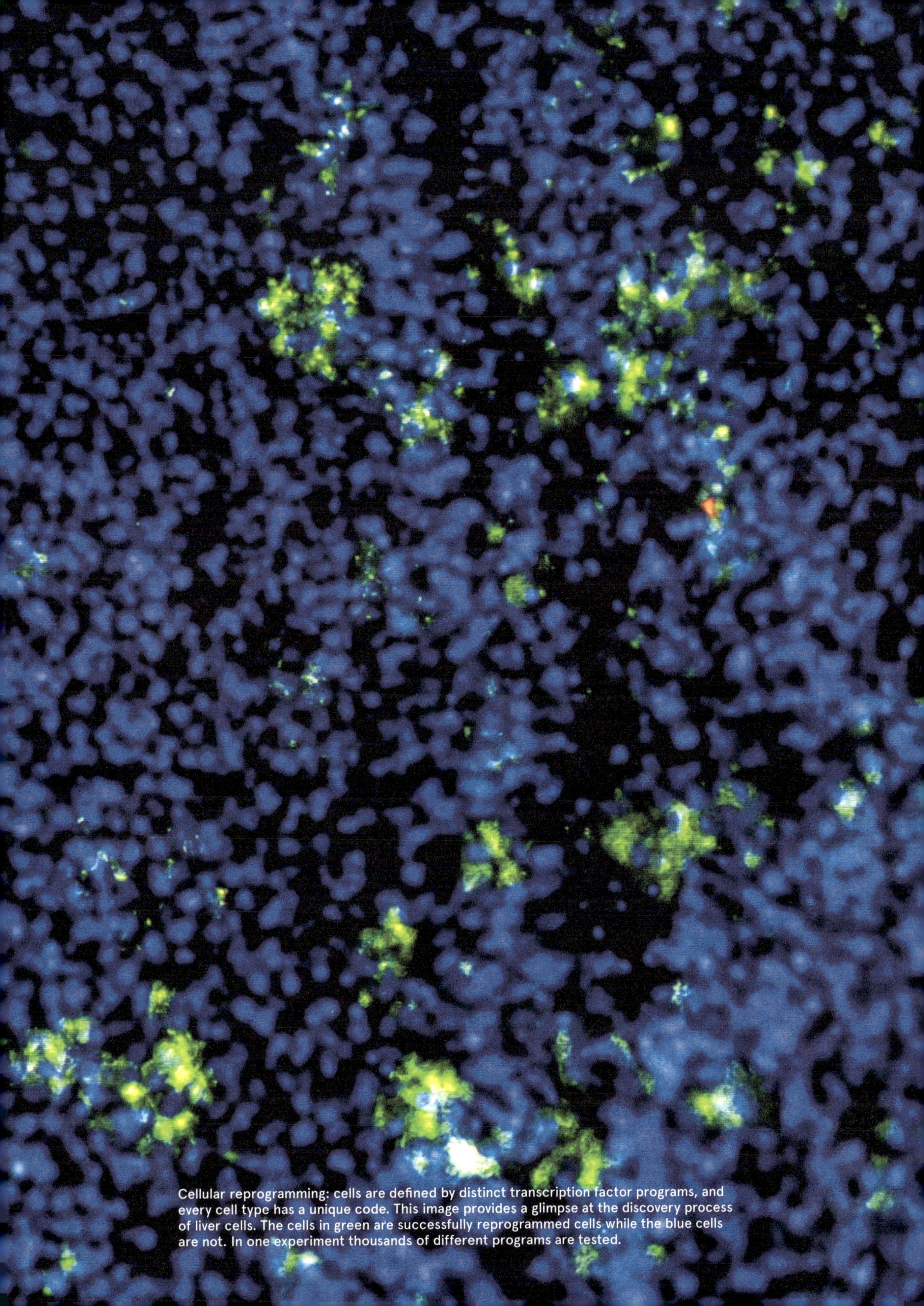

Cellular reprogramming: cells are defined by distinct transcription factor programs, and every cell type has a unique code. This image provides a glimpse at the discovery process of liver cells. The cells in green are successfully reprogrammed cells while the blue cells are not. In one experiment thousands of different programs are tested.

types that are useful as a treatment because stem cells can self-replicate, grow rapidly, and have an open epigenome. To treat a cancer, for example, you need specialized cells, such as immune cells.

T cells can kill cancer.

Yes, T cells constantly survey our bodies and, when they detect cancer cells, they kill them with chemokines. Similarly, you need pancreatic islet cells to treat diabetes, liver cells to replace the cells in cases of liver damage, or a muscle cell to build up muscles.

Can you produce stem cells on a large scale?

Pluripotent stem cells grow like weeds and are relatively easy to scale. The problem is turning stem cells into the cells that can actually serve as a treatment. To achieve this, most scientists use a process called *directed differentiation*. This approach tries to model embryonic development in cultures. Embryonic development starts with a fertilized egg cell that creates pluripotent stem cells, and then these differentiate into increasingly specialized cells to create every cell in the human body. Pluripotent stem cells start by creating three more limited stem cell populations: mesoderm, ectoderm, and endoderm; these go on to create slightly more specialized cells at the next branch point, and so on—until they end up as fully specialized cells—skin cells, immune cells, brain cells, et cetera. The goal of directed differentiation is to use the chemical cues that during development instruct these cell fate decisions to nudge stem cells down particular paths. There has been tremendous progress in the field, and we are now able to create a lot of different cell types, but there are two problems. The first is that this process takes time. For example, generating neurons, brain cells that can signal to each other using electrochemical impulses, can take more than one hundred days. The second problem is that directed differentiation entails multiple intermediate steps, and cell fate choices are based on random events; when cells choose to create their offspring, they often roll the dice! This ensures that really all cell types in the body are created during development.

How do you get a muscle cell?

It's difficult with the traditional stem cell approach. The trajectories that stem cells take are like branches on a tree. Every branch leads to a particular cell type, one of which are muscle cells. Following these trajectories is a long and complex process. Manufacturing protocols based on directed differentiation are long, complex, and riddled with inconsistencies. They are also difficult to scale. This is the reason why there are no stem cell–derived treatments available in the clinic, even though we've known about it for a while.

This is where synthetic biology comes in.

Correct. Yamanaka's approach is synthetic and has nothing to do with embryonic development. He demonstrated that if you activate four particular transcription factors in a cell, it turns into a stem cell. This was completely unexpected, and people wondered if it was even possible. The dogma was that cells are defined by their history. But Yamanaka's work suggested that they are defined by a transcription factor program. Yamanaka was not the first—in 1987, thirty years earlier, Harold Weintraub's group discovered a gene, a transcription factor that they called *MyoD1*, and found that when you switch on MyoD1 in a fibroblast, a kind of connective tissue cell, it turns into a muscle cell. Weintraub must be credited for discovering the first programming protocol. Unfortunately he died before he turned fifty, and hence his work was forgotten—until Yamanaka's discovery. A few years after Yamanaka, Marius Wernig demonstrated that the concept of cell reprogramming can be applied more broadly. He showed that it was possible to program a skin cell into a brain cell, and then he took stem cells and liver cells and programmed those into brain cells. This made me realize that this new cell programming approach can potentially be generalized and that we needed to rethink how a cell type is defined.

This is where the technology of bit.bio fits in.

Yes, bit.bio is based on this synthetic biology paradigm; the technology developed in my lab makes reprogramming very robust and scalable. It allows us to shorten the timelines of cell production by a factor of ten, so instead of forty to fifty days, we can produce brain cells in four days. We can also create cultures that are pure, and the process is highly reproducible. In fact, the scientific community didn't believe that this level of precision was possible. Most people believe that biology is inherently fuzzy—but in fact, our technology enables us to engineer biology, and so cells do exactly what you want them to do.

This seems unique ...

In theory, we can apply this approach to recreate every human cell type. We can scale manufacturing and have no problem reaching industrial scale. This is very different from the traditional approach. Finally, reprogramming creates cells that are much closer to the cells we actually have in the body than the traditional approach. We have recently demonstrated that reprogramming can overcome another bottleneck that has held back the field: the embryonic state of cells created by directed differentiation. It seems that we can literally program the maturity of cells.

I'm curious about Yamanaka's reprogramming approach. A stem cell contains a lot of information, with many possible differentiations—so how can a skin cell be programmed into a stem cell if all the other information is missing?

Fundamentally, every human cell has the same genetic information, regardless of whether it is a skin cell, a brain cell, or a stem cell. The difference is which genes are active.

So it's a question of activating the right genes?

That's right. You can look at a cell as if it were a piece of hardware or a biological computer. The nucleus is the hard drive; it contains the DNA that stores the genes. Genes interact with each other to form programs called *gene regulatory networks*. Of the twenty thousand genes, only about ten thousand are active in a particular cell type. You can change the cell type by changing the active program. The crazy thing about biology is: if you change the program, the physical components of the cell change. If you change the program of a computer, you only get a new function. In biology, the activated genes translate into proteins, which then form lipids, sugars, metabolites, et cetera—the physical components of a cell. So the program dictates the hardware!

So RNA determines which proteins are generated?

Correct. RNA encodes the active proportion of the genes, and subsequently RNA is translated into protein, which form the physical components of a cell.

So when you reprogram, you use transcription factors to activate the segments of the DNA that are relevant for a particular cell type? How do you activate these segments?

Genetic programs are controlled by a set of genes called *transcription factors*. For example, if we activate the transcription factor NGN2 in a stem cell it will turn into a neuron. NGN2 activates all the genes that are important for a cell to become a neuron. Often transcription factors bind hundreds of different other genes. A whole new gene network starts to form in the cell. If you think about the computer analogy, the genes in a cell together form the operating system of life—you could call them "Life OS." There are about twenty thousand genes in the human genome, and about ten percent are transcription factors. Transcription factors are like code words; they are the genes that you need to understand to program cells with a new identity. From experiments we know that combinations of one to six transcription factors determine a cell type. Every cell type has a unique code.

You can impress me with science fiction. Is there any connection between your technology and artificial intelligence?

The operating system of life is complex: twenty thousand genes, two thousand transcription factors. How do they all interact with each other? And how do we even comprehend what's going on? At the moment we can already conduct experiments with thousands of cells and then read out the activity of thousands of genes from each cell using a technology called *single cell sequencing*. A human brain cannot process this amount of data, so you need systems that help you interpret the data. That's where artificial intelligence comes in. You can also start learning about the structure of the data, which is where other mathematical approaches can play a role, for example, to identify structural redundancies.

Artificial intelligence is very good for recognizing patterns and detecting wrong patterns.

Yes, and it is very good at handling large datasets. The more experiments we conduct, the more data we generate, the more intelligent the AI becomes. Consequently, the number of additional experiments goes down over time. If you close the loop between wet lab and intelligent data systems, you can create exponential growth in terms of knowledge.

This could also be helpful for developing drugs because you can configure it endlessly.

That's right. Our focus at the moment is creating new cell type programs and getting faster at it. But when you read out the operating system of life, you can apply this AI to other aspects such as drug discovery. This is complex and requires partnerships.

Who do you collaborate with on this?

We collaborate with the London Institute for Mathematical Sciences. I am a big believer in interdisciplinary approaches between biologists, mathematicians, physicists, and computer scientists. Such collaborations open up entirely different views of the problems we need to solve.

So this is the new operating system of life?

It's the old operating system that we're now discovering.

You discovered this operating system of life, so it's like you found God.

We are discovering the programming language that nature invented to create human cells. We're learning; we're not creating.

You know, this sounds suspicious. It's megalomaniacal.

Well, flying to Mars is a similar sort of bold idea. I strongly feel that to solve earth's problems we need to embrace the possibilities technological development offers. But you are right—we also need to find out what to do with this technology and how to safely use it. That is why we created an ethics board at a very early stage of the company. Professor Marie-Claire Cordonier-Segger is a well-known lawyer who works on ethical issues and sustainability. One of the many things she did was to compile a report on what could go wrong if we develop certain aspects of our technology. This knowledge kept us from going down certain paths. For example, if you developed cells of the nose and the lungs, a terrorist could use them to create an effective virus. Innovation always has two sides. The question is how you apply it. We take this very seriously and want to make sure that our technology is not used in the wrong way.

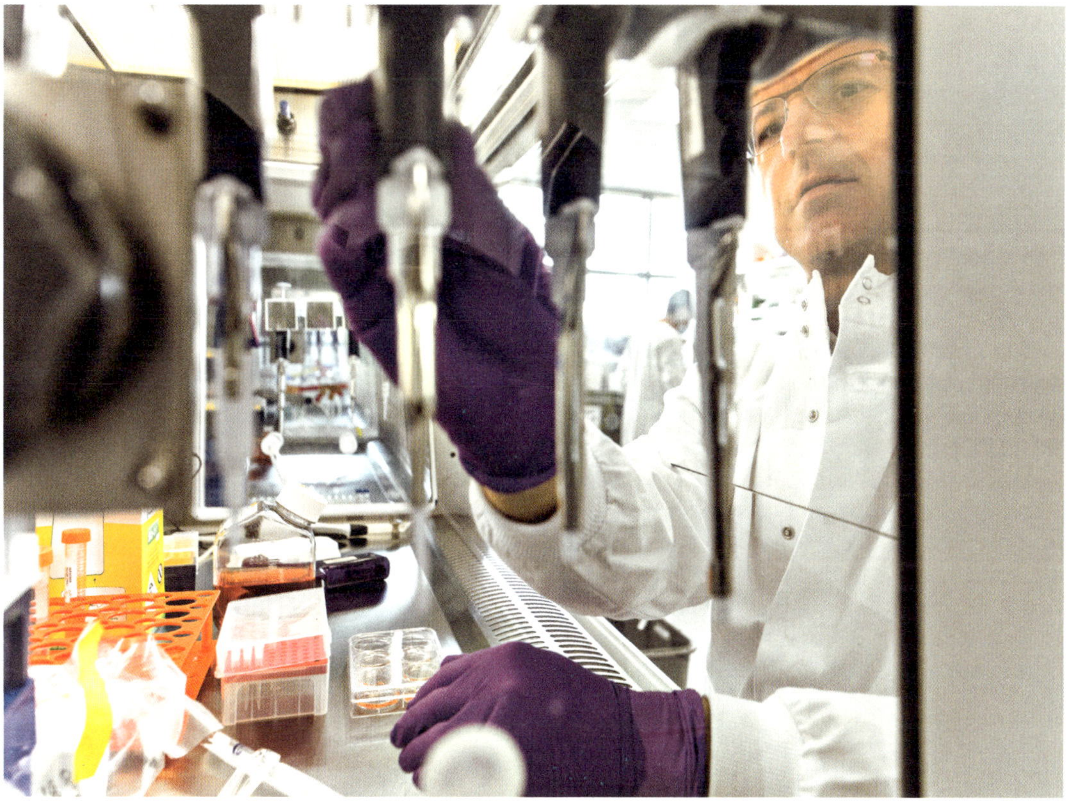

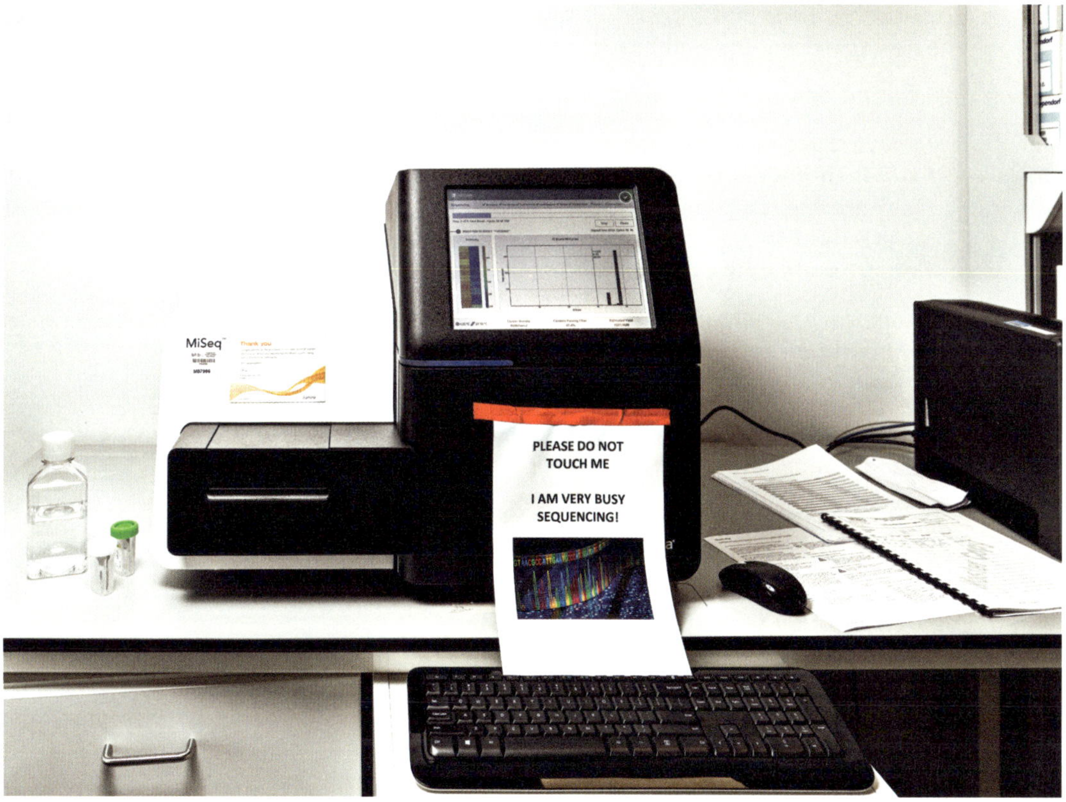
MiSeq
PLEASE DO NOT
TOUCH ME
I AM VERY BUSY
SEQUENCING!

So you could also create embryos and switch off any genes you don't want or engineer a new program into embryos?

Anyone who has a CRISPR tool can do that—in theory. At bit.bio we have set a strict boundary: we don't engineer germ cells—neither sperm cells nor oocytes. This is off-limits. Bit.bio will not engineer any cell type that can be passed onto the next generation.

That would open up Pandora's box.

Yes. I don't want us to play any role in this. And quite frankly, the opportunity to cure diseases such as Alzheimer's and cancer—all of this is sufficiently interesting.

How do you technically activate transcription factors in a cell?

Most scientists, including Yamanaka, use viruses such as Sendai or lentivirus. The issue is that these vectors insert randomly into the DNA. This causes stochastic problems—sometimes the genes activate, sometimes they don't. There are also other kinds of biological vectors such as transposons, but these also randomly integrate into the genome. All these vector-based approaches have one major problem: the cells recognize the newly introduced genes and try to switch them off. It's like the cell's immune system is saying, "I'm a stem cell, not a neuron. Why are you activating this neuronal program? I'd better switch it off." So, to operate this cell programming paradigm, we needed to find a hack to force the cells to switch on new cell type programs. This is the opti-ox technology developed in my lab.

How does it work?

We found that to make sure the new genetic programs are executed, we had to insert them into specific locations of the DNA, called *genomic safe harbor sites*. These sites protect the integrity of the cells as well as the program. Our technology creates a form of control system, which turns cells into a piece of hardware that you can control. The moment we switch on a new program, the cell has to execute it.

So why are you the only one who can do that?

Now that we have discovered the use of genetic safe harbors, other people can do it too. However, the IP is of course patented. Why genetic safe harbor sites exist is really a kind of miracle. They are the ideal landing pads for foreign DNA, and they exist across different species. They are hard to explain from an evolutionary standpoint because it is not clear what selection pressure exists to protect these sites across species.

Can you explain how you activate the specific sequences? Do you use genetic scissors such as CRISPR to implant a sequence in the DNA? Isn't this a very common procedure now?

Correct. Gene engineering is sort of commoditized now. Our hack is the locations in which we engineer the new program into the genome. Programs inserted into genomic safe harbors allow you hijack the program of a cell.

You're using so many words with bad associations like hacking *and* hijacking.

The reason I do is because we operate a software paradigm of biology. This is not the traditional stem cell approach; differentiation is a bit like growing plants. Reprogramming is based on a transcription factor code, hence the language.

It's also not classical gene therapy.

Actually, it's a bit more like gene therapy. The difference is that we engineer stem cells in a culture dish; gene therapy engineers cells in the body.

How do you activate the program once you have inserted it into the safe harbor sites?

Our opti-ox technology consists of two elements: a genetic switch and the new program. The switch is based on a peptide, a small protein, that is sensitive to a small molecule. It is inserted into the first safe harbor site; the transcription factor program that it controls is inserted into a second safe harbor. When the program is too large, you can extend this to other genetic safe harbor sites. To activate the program, all one needs to do is to add a particular small molecule, and within a very short time the cells respond. If you activate the right transcription factor combination that encodes a new cell type, this leads to the activation of a new set of gene regulatory networks that ensure the cell turns into the new cell type.

Are these landing pads the holy grail?

We think they are extremely important, and our investors do too. So far we have raised more than 250 million dollars for bit.bio and Meatable.

When did you figure out that you could also use them for human cells?

About 2015. It all started with our plan to create a particular cell type, a brain and spinal cord stem cell that plays an important role in regeneration called *OPC*, or oligodendrocyte precursor cell. We first set out to discover the transcription factor combination that defines an oligodendrocyte. Within a relatively short time—approximately six months—we found two transcription factors, Sox 10 and Olig2, that are able to program stem cells into OPCs, but the problem was that only about one in a million cells converted. The conversion was rapid—it took seven days, as opposed to 140 to 170 days using traditional differentiation methods. So we knew that we had discovered something important, and we wanted to know why it was that only a few cells turned into OPCs. The dogma at that time was that if the efficiency is low then you will need to improve the transcription factor program. But our data indicated something else. It seemed that the cells switched off the program during the process and that this limits the conversion.

They switch off the sequence or the cell?

The transcription factor program that we wanted to activate. The cells recognize that this program shouldn't be active, and they simply switch it off. The technical term for this phenomenon is *gene silencing*. Like everyone else, we started with introducing these transcription factors using viruses. From there we tried everything out that we could think of—all sorts of viruses, different forcing functions, and selection criteria such as using antibiotic resistance genes and transposons. We brought together the entire expertise of the Stem Cell Institute in Cambridge, and we just iterated. Over three or four years we only made incremental progress. It was actually quite scary—I risked our reputation, and we nearly ran out of money. It was really a moment of crisis.

Then the creator of all of that sent you an angel from the sky.

Yes, in the form of Roger Pedersen, who founded the human part of the Stem Cell Institute in Cambridge. He proposed that we investigate genetic safe harbors. Matthias Pawlowski, who did his PhD with me and now runs his own lab in Germany, worked on this project for years. One morning he came into my lab and showed me a film in which stem cells were turning into brain cells within a few days. The crazy thing was that every cell in the culture turned into a neuron cell at this lightning speed. I could not believe my eyes. This was entirely unexpected. Traditional differentiation takes forty to fifty days and only about ten to twenty percent of the cells convert. So I thought this video must be photoshopped.

You couldn't believe what you saw?

I couldn't. Then Matthias took me down to the lab and we stained the cells, demonstrating that the video was actually true. I thought, "Maybe it's a glitch; maybe the stem cell line that we have used isn't quite right. So let's take another cell line and test it." You know, what this video showed stands against every dogma in biology. Three months later Matthias had a new stem cell engineered, and when he started the program, the same thing happened. At this point I got goose bumps. We decided to try it out with another cell type program that defines muscle cells. This took another few months, until we had two new stem cell lines engineered with a muscle program. When Matthias first turned it on, all cells died. We were ready to accept that our technology didn't work. But then another colleague named Daniel Ortmann, who now is a leading scientist at bit.bio, said, "Let me tinker with it." He improved the culture conditions, and a few weeks later Daniel and Matthias showed me another video in which all the stem cells in the culture turned into muscle cells. It was really mind-boggling. We summarized our findings in a paper, and when we tried to publish it, we didn't get into the journal we wanted—the reviewers said there was nothing new. I spoke to the editor and said, "Are you kidding me? Nobody has ever seen anything like this before." She basically responded, "Look, I don't think people believe that it's really true." [*Laughs.*] Now we know that it's absolutely true. We have been using our technology to manufacture many different cell types.

What was the key moment in the discovery process?

It was an experiment that we did that allowed us to monitor the activity of the OPC program during the reprogramming process. This suggested that there is a problem with gene silencing. The way we did it was to add a red color protein that stains the nucleus of the cell together with the transcription factors. As long as the nucleus was red, we knew that the program is turned on. And we discovered that at the beginning of the process nearly all cells are red, but after a few days, the staining is lost. Then, after ten days, we found that only about one cell in a million reprogrammed successfully. The successfully reprogrammed OPCs still had a red nucleus, while the rest of the cells didn't. That's when we understood that the problem is likely to be gene silencing. It took the rest of field five years to get to that point. By then we already had our technology. That's the quantum leap that we made.

There's a lot of things you can do with this technology. If a hospital sends you some cells from a patient with cancer in the liver, would you be able to build cells for this patient in the future?

In theory we can produce any cell type. In your example, potentially an advanced immunotherapy could help. These could be either derived from the patient's own cells or off the shelf. When it comes to cell therapies, there are two paradigms in a ten-year horizon. The first paradigm is based on highly individualized therapies, using autologous, patient derived cells, and the second paradigm are off-the-shelf therapies. Both paradigms can address the immunological rejection of the cells following transplantation. If I take your cells and engineer them and then transplant them back to you, your immune system will be fine with them. If I take someone else's cells, these need to be matched. Matching is sort of a halfway house. In addition, there is new technology out there called universal cells, where you can take cells from a donor, engineer them to remove the immunological surface molecules, and make them invisible to the immune system. These cells can be applied to any patient. When cell therapies become mainstream in ten years or so, some of them will be based on a very efficient way of creating the patient's own cells, or a universal cell technology that works for everyone.

Your approach can be applied to all cell types.

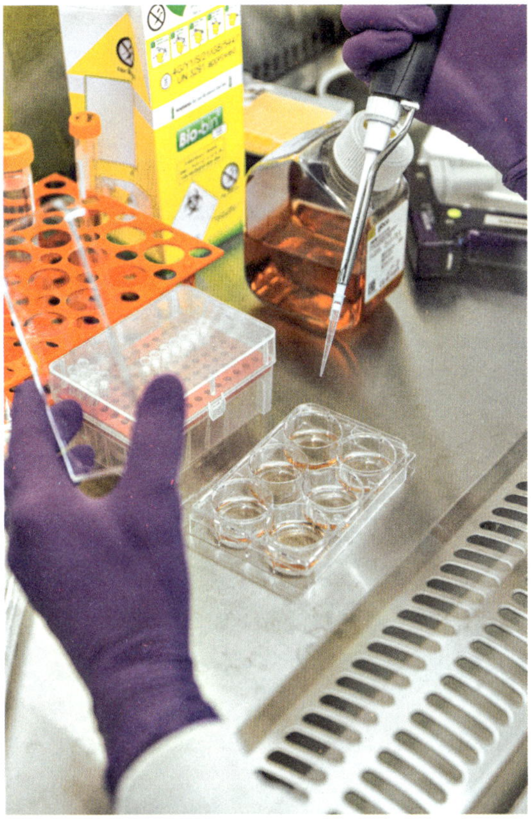
Bio-bin

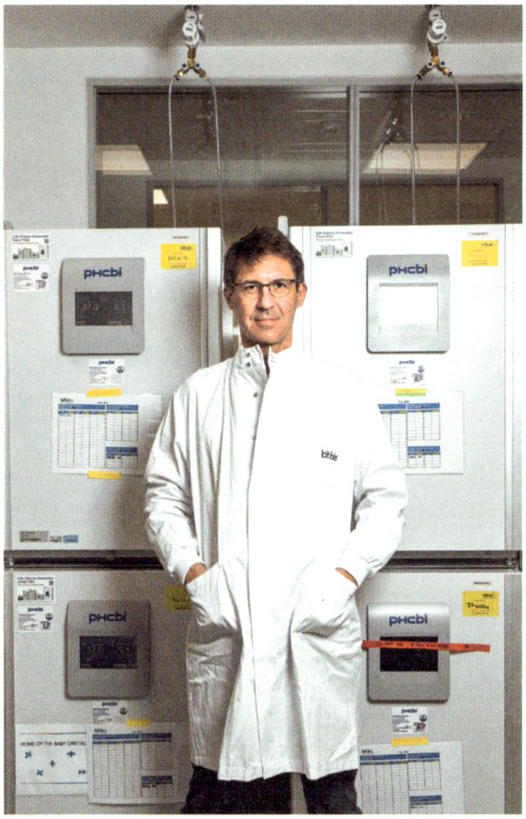
PHCbi
PHCbi
PHCbi
PHCbi

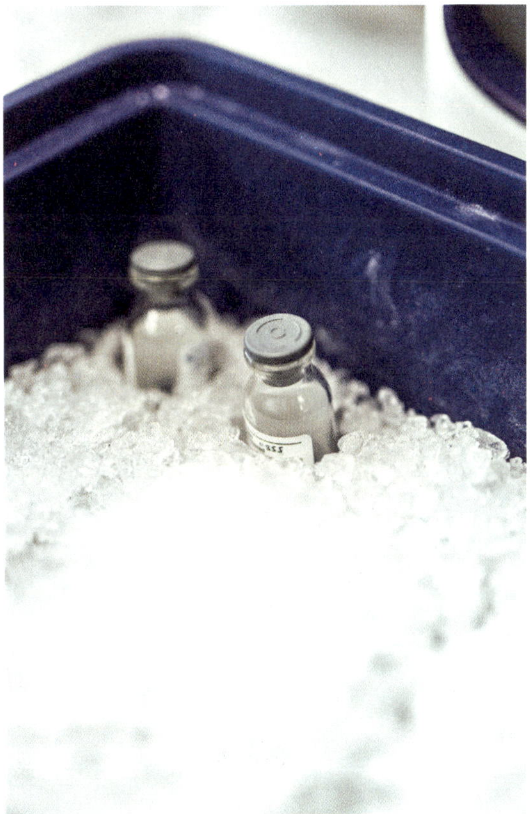

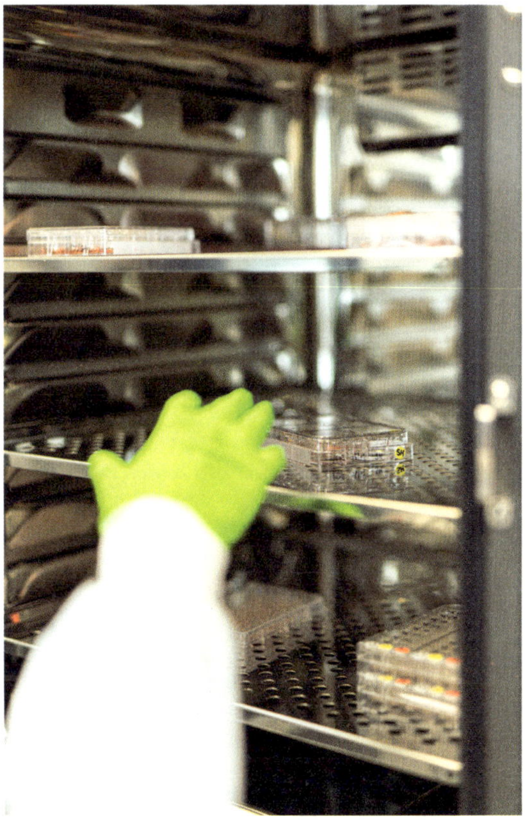

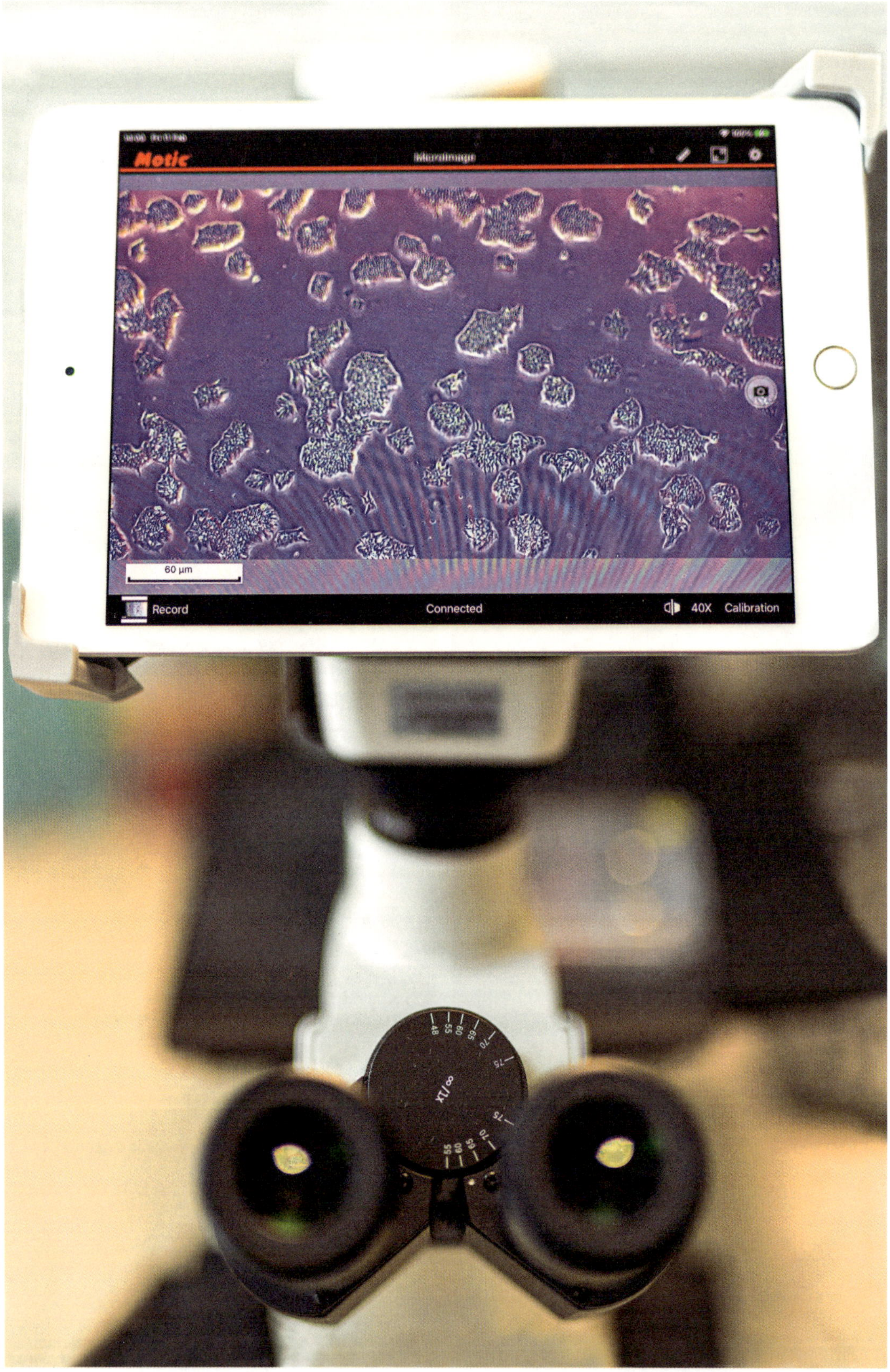

Colonies of induced pluripotent stem cells (iPSCs) in a tissue culture dish

Hence our bold ambition is to be able to recreate every human cell type. And once we have the relevant cell types, we can use them to create new medicines in the form of cell therapies, or assemble them to print organs. But cells have even wider applicability. An example of something that might be relevant for Ottobock is that we can interface cells with implants. Our collaborators Damiano Barone and George Malliaras at the University of Cambridge are working on seeding our cells onto neuroelectrical implants to better connect the implants with the brain. Right now, brain implants such as those used for Parkinson's basically consist of a wire. This is a crude way of interfacing with the brain. Moreover, these implants cause foreign body reactions that further limit the interface. Imagine if you cotransplant neurons that interface with the electrodes and then connect to the neurons in the human brain. This kind of interface could be much finer and more granular, and it will reduce the scarring. This is just one example; we can think about similar concepts in other parts of the body. So in the distant future, I think bionics and biology will come together to form a completely new sorts of interface.

But the brain consists of billions of cells ...

That's where bionics could play a role. For example, in the context of spinal cord injury, cell transplants alone only have a limited effect—they are likely to restore hand function in particular cases, but this not unlikely going to make patients walk again. But imagine if you were to connect bionic implants with cells above and below the injury, you could bridge the injury. This may be able to overcome some of the limitations. The convergence of bionics and biology is very exciting. It's going to take a while to get there. My mentor, Rick Klausner, says that "We always overestimate what science can do in five years, and we always underestimate what can happen within a ten-year period because progress is exponential."

Could you build an artificial muscle from muscle cells that you can connect to a steel prosthesis?

Theoretically yes—but there is a lot more work to be done. We already can produce muscle tissue, but we also need to connect it to the blood system and to the relevant nerves. Still, it seems possible—it would be very interesting to hear what Ottobock would do with a muscle in this context. As a surgeon I can think about reconstructive surgeries.

I can see so many possibilities. If there's cancer in the liver, there are two possibilities: you can mark the wrong cells so that the immune system eats them, or you destroy them with chemicals or radiation.

Correct. Both approaches are theoretically possible, but you would probably do this with another approach. However, you could also grow a new liver. Or both.

How would you do that? Would you transplant new cells?

We're very interested in the regenerative part.

So the liver is the wrong example, because it's already regenerative.

It's a highly regenerative organ, but not in all contexts. If you have fibrosis, suffer from genetic defects, or acute liver failure, the regenerative capacity is reduced, and cell therapies could provide an important solution.

So you bring in new liver cells to restore the function of the liver?

That's something the field is working on. We are taking a close look at liver diseases. When you think about developing your first product, you want to minimize complexities. Perhaps some of the most complex applications would be cells that connect to certain circuits in the brain and therefore need to be in a precise position. Simpler applications would be, for example, immune cells. They can be infused into the bloodstream and then they go where they need to go. Another excited application are pancreatic islets cells as these can be transplanted under the skin.

This is wonderful! You will be able to cure diabetes.

Yes. The first proof of concept study happened last year with cells generated using the old-fashioned paradigm, directed differentiation.

Could you print an entirely new pancreas?

A new pancreas, or a new liver. I think that's going to happen in the next ten years.

Ten years? Not five. Let me ask you why research, start-ups, and cluster-building work better in the UK than in Germany?

The good thing about Cambridge is that it gives you a lot of freedom, and that's what's driving a lot of creativity today. Cambridge is home to some big success stories such as the antibody technology created by Sir Greg Winter. Over time this has helped create an entrepreneurial ecosystem here. I feel lucky because I'm also not sure whether I could have done this in Germany.

Is it because there's more freedom or more capitalistic spirit in the UK?

In the UK, there is less hierarchy, and the university system gives you more freedom at an earlier stage of your career.

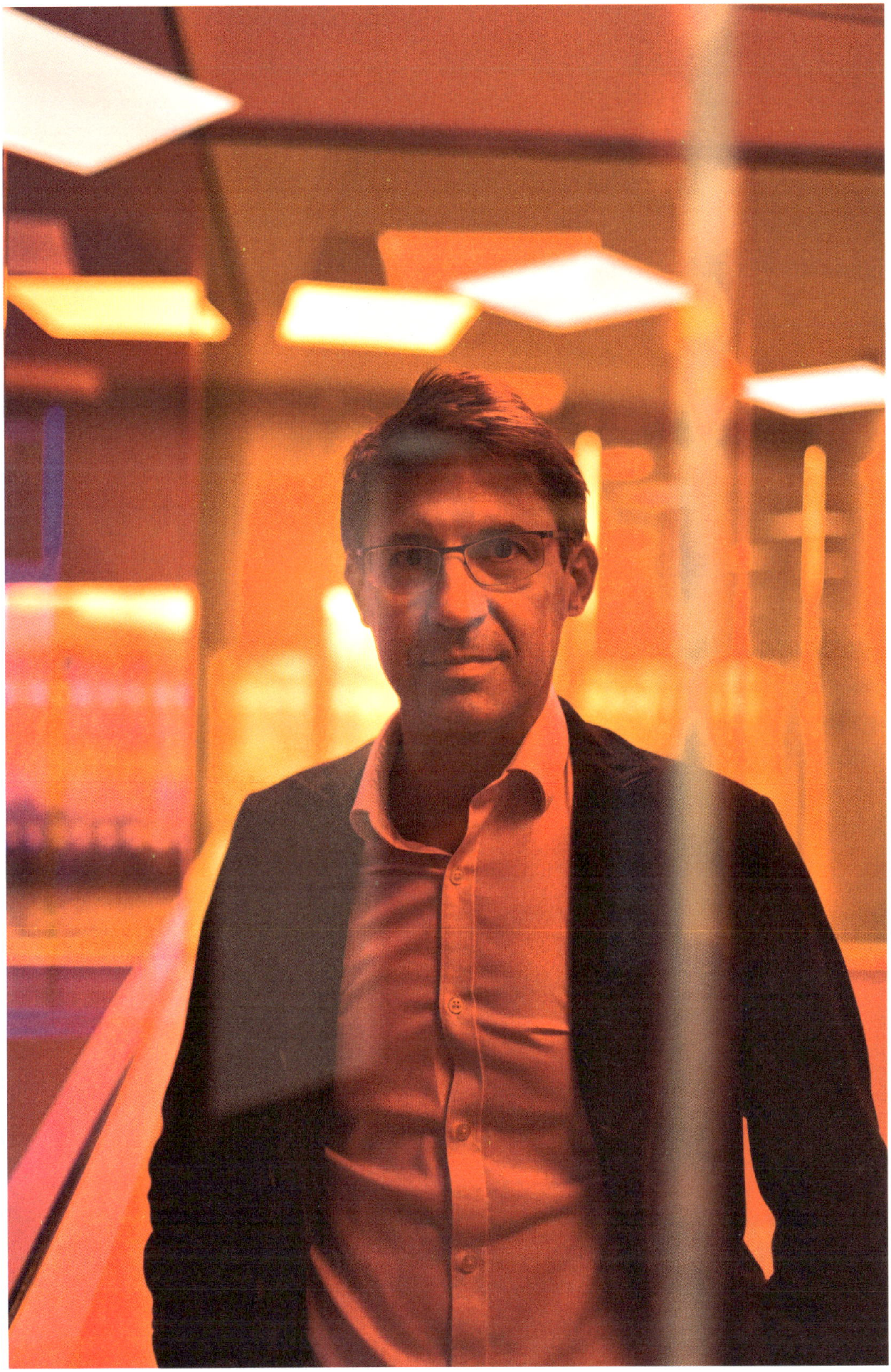

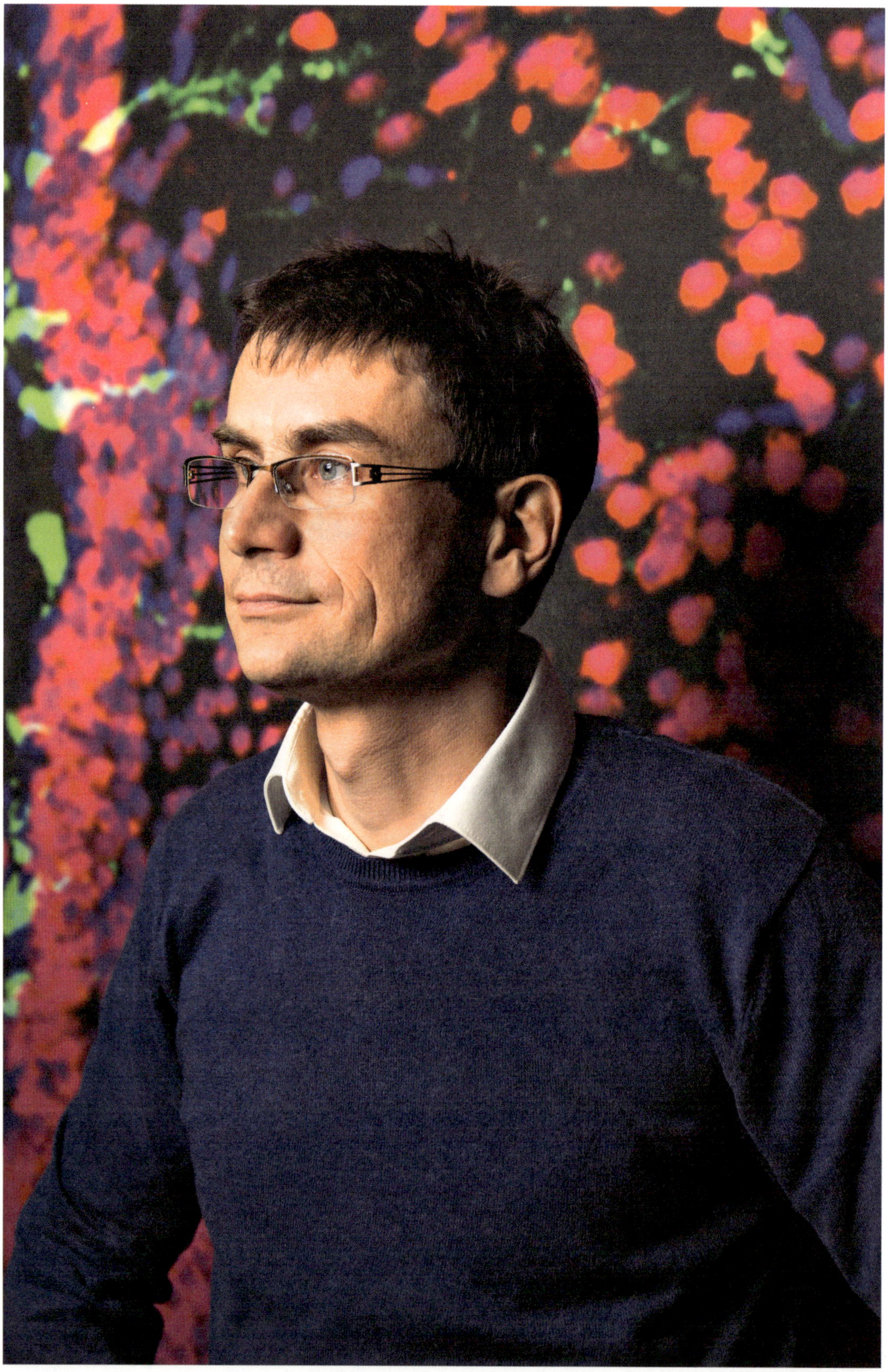

BENJAMIN M. FRIEDRICH

In Search of the Algorithm of Life

The salamander *Ambystoma mexicanum* (axolotl) has the incredible capacity to regenerate its limbs and organs. They grow back completely and are fully functional. But how do the injured body parts know which types of cells are required and in what order the cells need to be produced for them to grow back again? Scientists at the Technical University Dresden are attempting to unlock this secret of bionic regeneration.

Professor Friedrich, you have recently assumed the Heisenberg Professorship for Biological Algorithms at the cluster of excellence Physics of Life in Dresden. That's a particularly poetic name for a research cluster: it proclaims that what is at stake here are mechanisms that are of fundamental importance for the existence of living creatures.

The thing that makes our excellence cluster so special is that scientists from very different disciplines are conducting research here together on fundamental issues in biology, and particularly in developmental biology. I originally studied mathematics, but ever since I was a child I have been fascinated with how biological systems function and make life on our planet possible. For me as a theorist, it is extremely exciting to work with experimental biologists.

You are investigating the architecture of tissue and, using simple organisms, are exploring how lost limbs grow back. If you find the key to this mechanism, this could also help people extend their own powers of healing at some point in the distant future. What would it be like if our limbs and organs could grow back on their own? We wouldn't need any prostheses or transplants anymore.

First we want to thoroughly understand the mechanisms that dictate development in a simple organism—that is, the pattern formation during the development of an embryo, from the fertilized egg to the organism with its differentiated body plan. Even in simple organisms there are a lot of surprising phenomena and various mechanisms that proceed in parallel, mutually influencing one other and becoming very complex. Biology is working its way along here, bit by bit, and we mathematicians and physicists are attempting to systematize these mechanisms with mathematical models.

So you are searching for the algorithm of life?

That is our visionary goal, even though we tend to expect a toolbox of nature with repeatedly recurring, basic mechanisms, rather than one single algorithm that explains everything. But the recipe for success here is reductionism. First of all we want to understand, for example, how the body axis in a simple organism is determined during early embryonal development. Where is the head positioned, where does the tail go? Or, for example, in the case of the axolotl, we ask what happens when an amputated limb grows back. While this question is also still too complex, we zoom in on a subquestion: what are the first steps in the first days of regeneration, during which the tissue first has to differentiate and body axes are determined, so that every cell in the tissue knows where it is? Is it on the anterior or posterior side, more in the direction of the head or more in the direction of the tail, further inside or further outside?

As a layman you initially picture it as a pile of stem cells.

In embryonal development this is actually the case, at least in the beginning, but then the totipotent and pluripotent stem cells differentiate into the terminal cell types, which build up our bodies.

Do the genes organize this differentiation?

Yes, but the exciting thing is that a liver cell has exactly the same genes as a nerve or a muscle cell. What is decisive is that in every cell a number of genes are switched on and others are switched off. Markers sitting on the DNA specify which genes are even read off.

What switches these genes on and off? Enzymes? Transmitter substances?

Exactly. You can basically picture it as follows: the fertilized egg cell has divided a number of times, and the DNA doesn't yet have any markers. The unmarked DNA is like a cookbook without any bookmarks—everything seems possible. But then pattern building processes start, which, for example, specify a body axis in an initially undifferentiated pile of cells. A fundamental mechanism involved here is known as *morphogen gradients*, that is, different concentrations of special signal molecules throughout the pile of cells: for example, the concentration is high at the head end and low at the tail end. Morphogens are signal molecules that control the pattern formation. Every cell can independently read off the concentration of these signal molecules at their position. As a result the cell knows whether it is positioned closer to the head or the tail, and it places epigenetic markers on its DNA. As a consequence a number of genes are expressed more strongly, and others less so. As a result the cell's possibility space is restricted and what types of cells it can differentiate into are preselected. For the cells that are close to the center of the head, this is a different set of possibilities compared to the cells in the tail region.

But what is the coordinating entity that sends the respective signal molecule in one direction or the other?

There are a number of possibilities. It can be influenced from outside. For example, the egg cell notes the position where it was penetrated by the spermatozoon.

So there needs to be an anchor. Every neighboring cell knows how it has been marked and whether the stamp is somewhat stronger or somewhat weaker. Is that the correct image?

Exactly. If a signal molecule has been stamped at a specific point on the egg cell such as at the point where the spermatozoon penetrated, and then the egg cell divides one or two times, only one daughter cell inherits this increased concentration of signal molecule. As a result, this "marked" daughter cell receives this signal molecule at the very beginning and can change into a different cell type and produce signal molecules itself, which spread.

How many cells does it take for the differentiation to begin? Sixteen, or 65,536 cells?

I think around a few hundred cells. The progeny of this one daughter cell, specially marked at the beginning, emit signal molecules that are distributed throughout the entire embryonal tissue, with the highest concentration near the source. The concentration of the signal molecules is a measure of the distance from the source.

And then we already have an axis.

Exactly. A cell normally finds itself in the tissue and can communicate with its neighboring cells, but it doesn't know its position within the tissue. The morphogen gradients, through their concentration gradient, define the body axis and provide the cells in the tissue with information about their position.

Now this is where you come into play as a mathematician who can recognize and calculate patterns?

On the one hand, mathematics comes into play because naturally we want to quantify these patterns, which the experimenters have observed. Biology has been experiencing a transition from a qualitative, descriptive science to a quantitative science for the past few years. On the other hand, we can also use mathematical models to provide crystal-clear formulations of possible mechanisms. A famous example from Wolpert describes cell differentiation as a response to a morphogen gradient with the aid of threshold values. When the concentration of signal molecules is lower than an initial threshold value, the cells differentiate into cell type A, when they are in an intermediate range, into cell type B, and when the concentration is high, into cell type C. As a result we already receive a predifferentiation into three different compartments. If you translate such an idea into mathematical equations you can already see whether this mechanism will function at all, whether one has considered all the details or whether further parameters can be determined, for example, through a comparison with experimental data. A further question is: is this mechanism at all robust? Every biological system is subject to strong disturbances and fluctuations. For example, in an embryo cell there is never always exactly one hundred mRNA transcripts of a gene, as gene activation and transcription are subject to random processes. Sometimes it is eighty, sometimes it is 120, but the development must still produce exactly the same result and proceed robustly. Ideally we have two or three mathematical models that correspond to different working hypotheses. The goal is to make quantitative predictions and propose experiments to check this, which we can use to test our mathematical models. Ideally we can then actually show that only one model from a series of different models corresponds to the new experiments, and then pursue the mechanisms represented in this model. The reality is always a lot more complicated and detailed, but at least we have an initial basic principle.

How does the morphogen concentration determine the cell type—does the morphogen act on the DNA in the cells?

Ultimately, yes, but there are a lot of intermediate steps. The morphogen is a molecule that bonds to a special receptor on the surface of the cell. Then a chemical reaction takes place. The receptor activates another messenger substance, but now within the cell, which perhaps activates a further messenger substance that penetrates the nucleus. Then, for example, there are transcription factors that attach to the genome and mark specific genes to be read, like a cookbook that is opened to a particular page. But there are also markers that remain connected to the DNA as a permanent bookmark, marking gene sections that belong to the essential repertoire of this cell and all its progeny.

In effect the cell receives the section of the human construction plan they are responsible for, and then this cell exclusively produces this special protein—for example, for building a liver?

Exactly. So you can say that of the ten thousands of human genes there are a few that every cell needs, that belong to the basic repertoire. But then there are special genes that only liver cells need.

How does the organ demarcate itself? When does the liver know that it should stop growing? When does a protein pattern reach its limit and stop expanding spatially? Who gives the signal here, and what happens at the boundary to the rest of the body?

Now it has happened: you ask a simple question, but we do not know the answer: we have already reached the limits of our knowledge. In a number of systems we know how the size of organs is controlled through patterns. But they are a lot more systems where we don't yet know that. This is currently being researched. As theorist we can conceive various mechanisms. For example, one mechanism would be as follows: for the sake of simplicity, I assume that tissue in a certain region only has one axis along which I record the concentration of a messenger substance. The messenger substance is produced in one corner, and the further we move away from the source, the lower the concentration becomes. And one mechanism could be as follows: as long as the concentration is above a certain threshold value, the tissue grows, and it stops growing when the concentration at the outer edge falls below this.

How can the different cells separate? What happens, for example, when a heap of spleen cells strays into a heap of liver cells? Can the liver also tolerate misdirected cells, or do they reject each other?

During embryonal development there is actually a phenomenon where different cell types repel each other—like oil and water. Something similar happens with cells that establish contact with neighboring cells via surface molecules. Cells of the same type have a preference for binding to one another, while different cells repel. This can actually result in cells segregating.

Do the cells recognize each other based on the protein composition at the cell surface?

Exactly. On the cell surface there are special surface proteins, as well as special sugar molecules.

If the other cell is not sweet enough, it is repelled?

You could say that. Cells always communicate with their neighbors and establish mechanical contact with the surface proteins. In addition there is also communication at the level of the tissue, the organ, and finally the organism as a whole. There are growth hormones that flood the whole body, and then there are morphogen gradients in the individual tissues and organs that you can calculate. There are numerous mechanisms, when cells stop dividing and an organ stops growing: for example, the concentration of the signal molecule falls below a certain threshold value, or the concentration no longer changes over time. There are also tissues or organs located in a cavity, and at some point they reach limits as mechanical tensions affect the growth.

So with the mathematics it is about describing patterns and their emergence. But what role does physics play?

Physics can play a role at two levels. First of all we have the physical interaction of molecules, which, for example, generate adhesion. When the signal molecules spread within a tissue and diffuse, it is a physical process. Secondly, physics has a storehouse of models for describing inanimate matter. We can then attempt to explain the dynamism that we observe in biological systems, at least approximately, using a model that was previously developed for inanimate matter.

Could one also imagine thermodynamic models?

Thermodynamics, or more generally statistical physics, allows us to describe the interaction of large numbers of particles mathematically. One can imagine the particles as gas particles—

That collide and exchange energy?

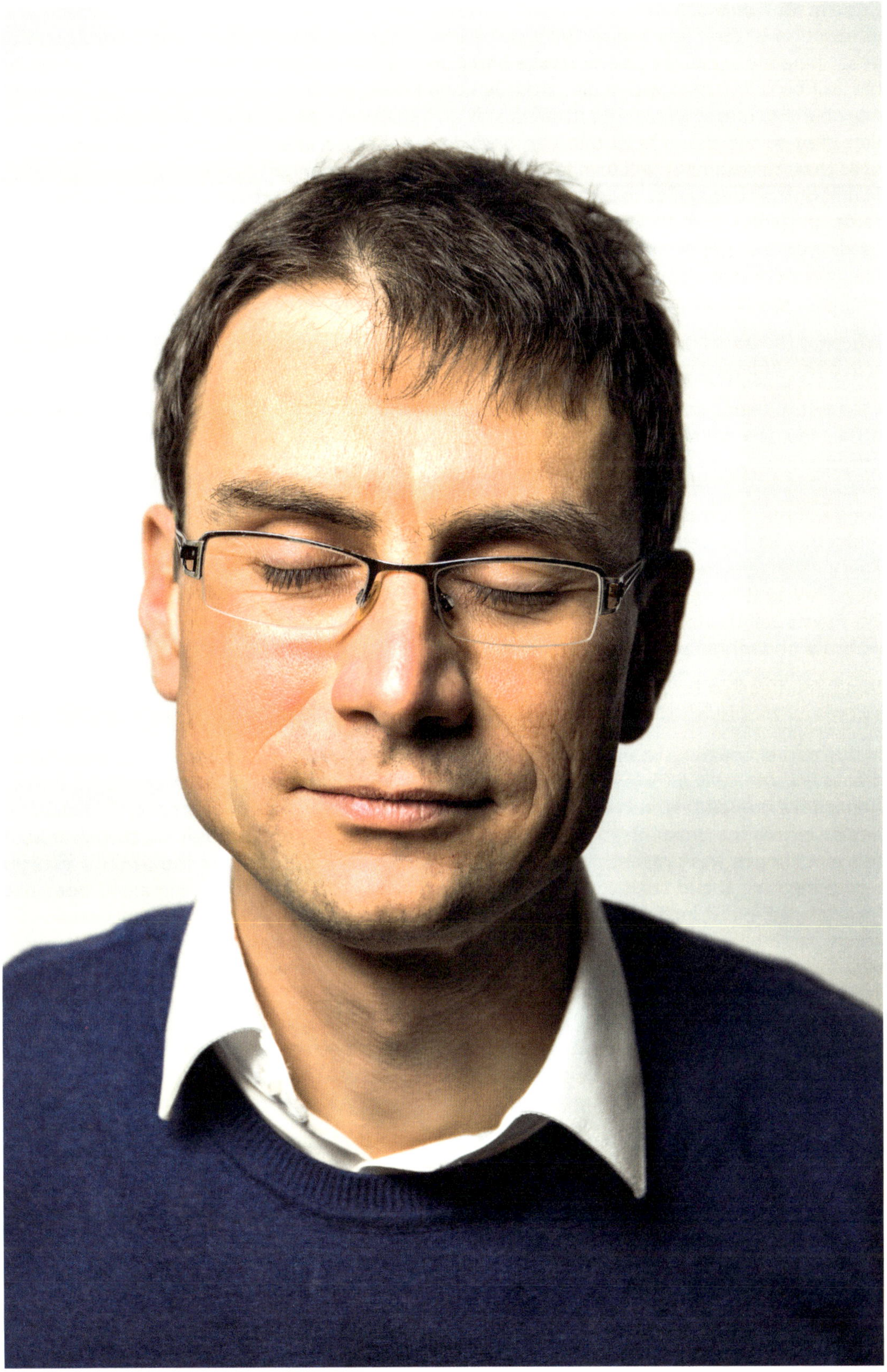

Exactly. But one can also use some of these models in modified form to describe the interaction of cells in a tissue. With statistical physics we have a tool that is very good at describing the dynamics on an average basis. Just as it is not possible to predict which number will be rolled next with a die, although one knows that the number six will appear with a probability of one in six. The dynamics of an individual cell is often random, like throwing dice. But at the tissue level one can provide a definitive answer as to the percentage of cells that, for example, will turn into type A. A further important tool is nonlinear dynamics from control theory. For example, how do I need to build a feedback loop so that a system remains stable? How should I build a thermostat with temperature sensors to control the heating so that the temperature remains constant? These questions are addressed in control theory, and we employ this modeling language to build mathematical models that, for example, describe how cells communicate with each other.

Where is the approach path for the biologists, for turning the genes on and off so that the skin color changes or an organ develops differently? Are the morphogens the lever?

I have only sketched very rough, basic ideas. In the moment when, for example, you turn off a gene, the cells lack a protein.

And thus possibly an important signal?

For example, it could mean that suddenly this morphogen is no longer there as a signal molecule. So the cells no longer know what their relative position is along the body axis. There have been neat experiments with the flatworm where a whole worm has been turned into a tail by turning off one single signaling protein. It was possible to transform the whole worm into tail tissue. Or in another experiment employing the opposite signaling molecule where a phenotype was produced that grew heads everywhere.

Don't you need to turn off all the other genes such as the head gene and the torso gene, so that all that is left is the tail? What do you turn on and what do you turn off?

In our earlier roughly simplified picture the body axis is generated by the morphogen gradients and the cells differentiate into three cell types, A, B, and C at different concentrations. If we completely turn off this morphogen, everything has to become A, for example, because here the concentration is at its lowest. Everything becomes tail. Or the reverse, if this morphogen was present everywhere at a high concentration, then the whole organism would become C and start forming heads everywhere. Naturally one gene alone does not form a head, but it instructs the cells to run through a highly specific genetic program. You can picture it like this: this morphogen activates one or several especially important genes, which results in the expression of further genes. That is like a second subprogram that is started exactly where the head should grow. When we have our gradients and a compartmentalization into A, B, and C, the cells know, through their positioning, that they are type A, for example. They can then begin to build another pattern and divide into further cell subtypes. And there is not just one but several body axes. The tissue, which is initially completely undifferentiated, continues to divide into ever smaller areas where the cells become increasingly finely differentiated. In a later phase, even cells that at the beginning were only listeners and merely reacted to the signal concentration gradients in the tissue can suddenly begin to emit signal molecules themselves and divide this compartment even further.

How far does this subdivision go? Do we eventually reach "n equals one"? In other words, is there eventually a cell that is as unique as a snowflake?

This is actually the case for roundworms. This goes down to individual cells. The roundworm has around one thousand cells and each individual cell has a specific function that is determined by the genetic program. With complex organisms such as our own, there is more flexibility and the cell type does not need to be precisely set. But there are precisely defined structures such as the liver that have the right proportion of cell types in the right geometrical configuration. Naturally this needs to be precisely controlled.

Mutations are the read errors that occur during this process? Or does nature allow a little freedom in this strict plan?

A read error would just mean you have not followed the recipe correctly and built the wrong protein. As a rule this is not so grave. There are also quality controls for proteins, and if they don't meet them, they are immediately dismantled. Rather, the problem is when real typing errors creep into the genome through the copying process. Every time the cell divides the genome has to be duplicated, and with a certain probability this results in write errors, which accumulate. Environmental influences and radiation can also result in genome mutations. That is why it is very important that complex organisms have special programs in the cells that check whether the genome is still intact. This can even result in the suicide of the cell when it has accumulated too many mutations. These cells effectively sacrifice themselves for the organism as a whole.

Does cancer occur when the cells no longer want to sacrifice themselves?

Yes, in the case of cancer a number of mutations must come together, and in particular, this cellular suicide program must be disrupted. As a rule the communication with neighboring cells or with morphogen signals is also disrupted. Maybe the cancer cell receives the signal to stop growing or dividing, but it simply ignores it. That is equivalent to antisocial behavior at the cell level. It is a cell collective that must work together. This demands a certain altruism of the cells in the tissue. Without this altruism you end up with cancer.

Who is responsible for quality assurance, and who recognizes the false cells? Is there a higher intelligence that has the building plan in its head?

Naturally, to an extent the immune system as external authority can identify mutated cells. But to some extent the cells can recognize this themselves.

So they feel the surface structure and know that there is something wrong with it?

The cells themselves always have an autonomous control program running. Even during cell division, instead of the processes simply running blindly, there are checkpoints where the cells check themselves to ensure, for example, that they have really divided the chromosomes equally among the two daughter cells. If this is not the case, the cell division is interrupted. This is just one example of such a control program.

Fifteen years ago mankind succeeded in completely decoding the human genome. We know the entire building plan and can say with some accuracy which gene is responsible for what. But this alone is of no use to us. You decode how the individual cells receive their orders instructing them what to build. But your vision is to turn genes on and off in order to generate positive changes, that is, to prevent negative ones. How can we develop new strategies here for the benefit of mankind?

In my opinion these approaches have a great deal of potential, but we must be honest: we are not talking about ten years here, but several decades. We are conducting basic research here. We first want to understand how healthy organisms actually develop and spontaneously build patterns, before we can move on to the next stage and address the different clinical pictures. The complexity of the human organism is so great that it is worthwhile to start with the simple organisms. Many of the mechanisms that were discovered there could actually be accounted for in larger organisms, but in a much more complicated form. I find it absolutely fascinating that humans have many of the same genes as a flatworm or even a number of protozoans. Nature uses the same building blocks and proteins time and again. We can actually—and this is our hope—learn a number of things about human developmental biology when we look at flatworms, axolotls, and fruit flies. This needs time and the corresponding perseverance. But it is still the right road.

You conduct research on these low-complexity organisms for precisely one reason: to discover the mechanism of their self-healing powers. This is highly exciting—even though this is still in the distant future for humans. But have you made progress in decoding the algorithm of life?

I believe that in a number of model systems we have actually been able to uncover fundamental mechanisms. In the case of the flatworms we have a mathematical model of how a head-tail concentration gradient automatically adjusts to the size of the worm, and can also spontaneously regenerate when one conducts a partial amputation.

So you are at the beginning. You have not yet found the key with the low complexity organisms, but your understanding is growing.

It is like a puzzle. Sometimes it takes two or three years to find a certain piece of the puzzle, but we are continually finding additional pieces.

But you couldn't write down a formula for life for me?

In biology we have thousands of systems and over a thousand different phases of development. And hopefully we will be able to write down formulas for every phase in every organism. Then maybe we will be able to see a few common principles, but there will never be just one single formula.

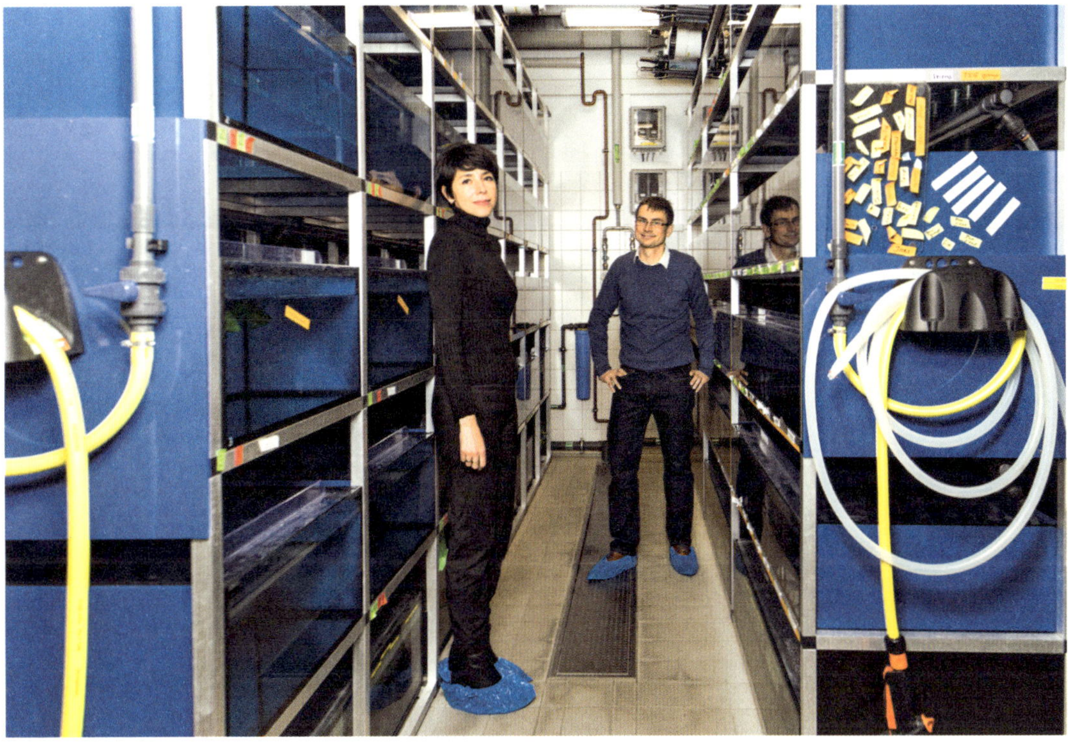

Tatiana Sandoval-Guzmán, leader of the research group Vertebrate Tissue Repair and Regeneration, at the Medical Faculty of the Technical University Dresden, and Benjamin M. Friedrich, Heisenberg Professor for Biological Algorithms at the cluster of excellence Physics of Life at the TU Dresden

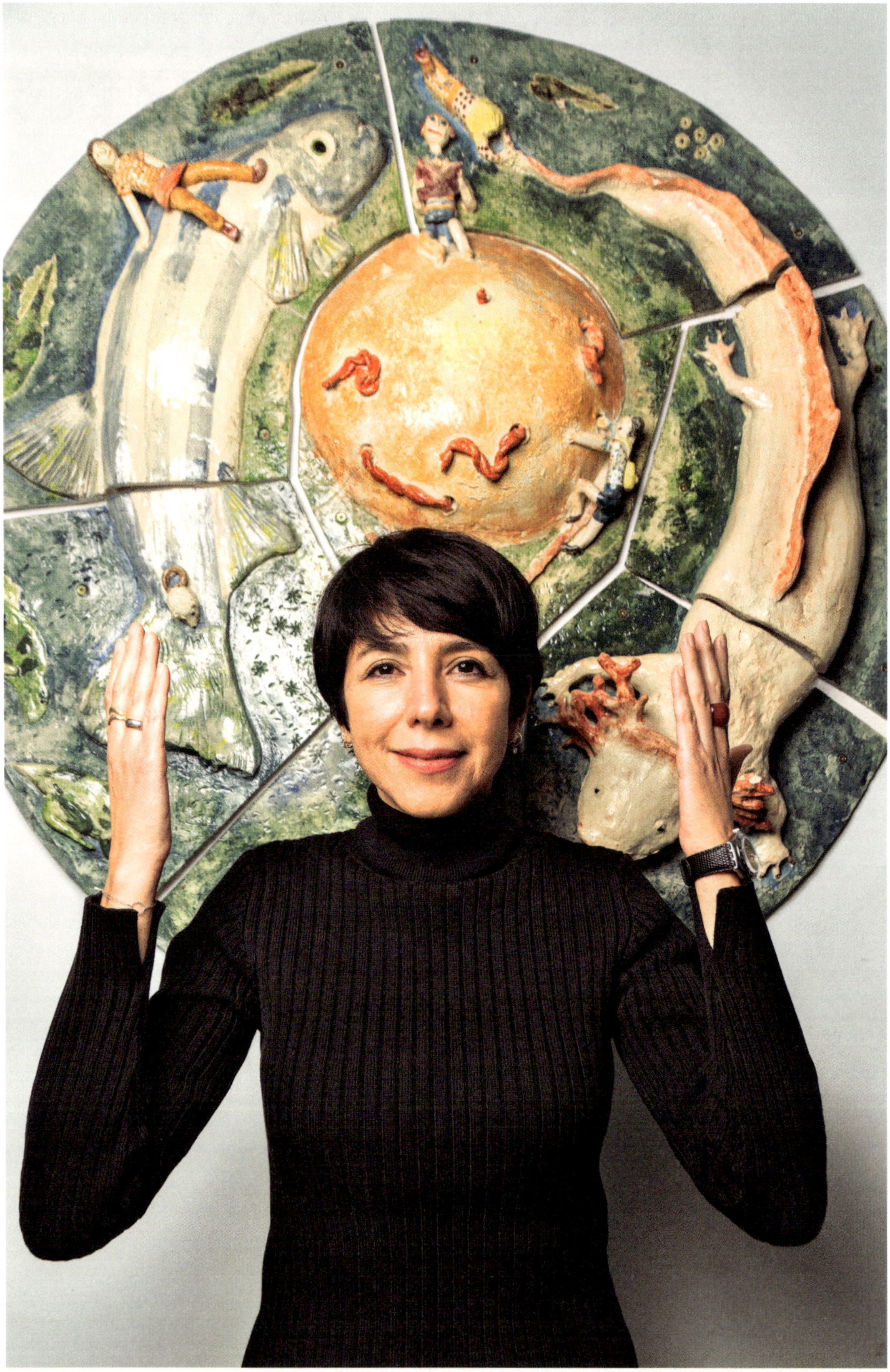

TATIANA SANDOVAL-GUZMÁN

The Mysterious Axolotl Holds the Key to a Long Life

Since humans can already regenerate their skin, fingertips, liver, and blood, the human body has great potential to expand its abilities. Animals such as axolotls are much more flexible and can regrow their own organs. Understanding how axolotls regenerate their limbs might be the key to finding the miraculous mechanism that is responsible for the self-healing power of the human body. Perhaps it will be possible in the distant future to grow human limbs in the laboratory.

After speaking with Professor Friedrich I have a much better understanding of what you are doing here. As the leader of the research group Vertebrate Tissue Repair and Regeneration, could you explain the model system to me?

Let's start with how salamanders regenerate limbs. This regeneration happens after an injury at any level of the arm. Although they can also regenerate the liver, organs such as the heart, and part of the brain, we use the limb as the model system because it's one of the most striking features observed in an animal model. It's quite impressive to be able to see this happening before our very eyes. One way to understand how it happens, we divide it in a plan that relates to Professor Friedrich's models.

Is there a theory about why certain animals can regenerate parts of their body and organs?

There are numerous theories, but we shouldn't forget that many animals can regenerate. It's just that their potential to regenerate varies. Humans can also regenerate the skin, the liver, and the blood. Women regenerate the endometrium every month. Humans can also regenerate the distal tip of the finger, so the potential is there. Regeneration is quite widespread in the animal kingdom. It's really like a survival trait that has been conserved, but some organisms can do it in a more impressive way. Our research is based on the observation that an injury of the limb in an axolotl can regenerate no matter at what level. This means that after a complete amputation it will need to regenerate the humerus, the elbow, the radius and ulna, and the whole hand, so it needs to have a plan of what exactly is missing. The way that cells are going to be organized has a lot to do to what Professor Friedrich has explained, which helps us understand how an animal can reform what is missing.

This is really a miracle. Even if you don't believe in God, this will make you believe that there is a creator. So it is actually possible to reprogram the cells of the human body?

Reprogramming therapies are already a reality in humans, and the potential is vast. What happens in the axolotl is that when there is an injury, some of the cells will go back in developmental time, becoming more like a progenitor cell.

They go back evolutionarily?

No, in terms of the developmental clock. During development, cells divide and eventually acquire a fate. These cells have enormous potential to become a tail or a head. They go from totipotent to pluripotent to multipotent cells. In the limb of an animal that is already formed, most of the cells already have a function: they are mature or differentiated. And so, when there is an amputation in the axolotl, some of the cells will regress developmentally, but not all the way.

Do they wake up and announce to the cells that they should grow again?

That's where the morphogens come in, the signals that tell them what to do and what to become. Then the cells are reprogrammed—this is what we want to understand.

So the system goes down, and it is updated by morphogens. When it starts again, it says, "Oh, we are missing a limb. We have to make a limb now." It's like that?

In a way, but not all of the cells will respond. Some cells are better able to do that. Or they say, "No, we don't participate. We remain here." Also in the different tissues, because the limb consists of bone, muscle, tendon, skin, et cetera. Some of these tissues will have stem cells just like we do. We also have a certain number of stem cells that renew our tissue as it wears out over time. And then during regeneration, some of these stem cells "wake up," as you said, and they can also contribute to making new tissue. So you have cells that already have a function, and they can go back in developmental time and then participate. You also have stem cells that are just sitting there, and when they receive the signal and get activated, they participate. There are other cells that are mature and transdifferentiated, or just change the fate of the cell. They don't go back in developmental time, but just acquire different characteristics.

How do you examine this? Do you cut off a limb and observe what happens?

As you saw in our axolotl colony, we use animals that lack pigment cells. This provides us with a natural window into the cellular processes, and we can make some of these cells express color or fluorescent protein. We follow this with a microscope in time, and then see how they interact with others, or what fate they acquire.

You can see the division process of the cell in a microscope?

Yes, you can make movies that capture the moment. But most importantly, you can follow the decisions that the cells make during the regenerative process.

Can you see if there is a differentiation in the different areas and organs?

Some of the cells will allow you to do so because of the shape they acquire. The shape will be so particular in the microscope that you can tell that a cell is becoming a muscle cell, for example. But not everything is evident to the eye. In those cases, we need to see the genes expressed to determine that a cell is becoming another type of cell.

Do you mark them?

One of the methods you can use is by looking at the genes that cells are using at a specific time. This pool of expressed genes tells you what the cell is up to. What you do is take the mRNA, for example. DNA is the genetic code that all cells contain. DNA gets translated into a molecule that is called *RNA*, making a copy of the code that is to be translated to proteins. Then you can see how many mRNAs from a certain gene exist in this cell at this time. This profile of translated genes gives you a better understanding of how cells change. Then you explore in more detail to see how the reprogramming or the change is happening. Not everything is evident to the eye.

What I didn't understand is how you can switch off a gene. With the axolotl we saw that it changes color and that it was manipulated. You reprogrammed an animal that was alive. You didn't go to the beginning of the process.

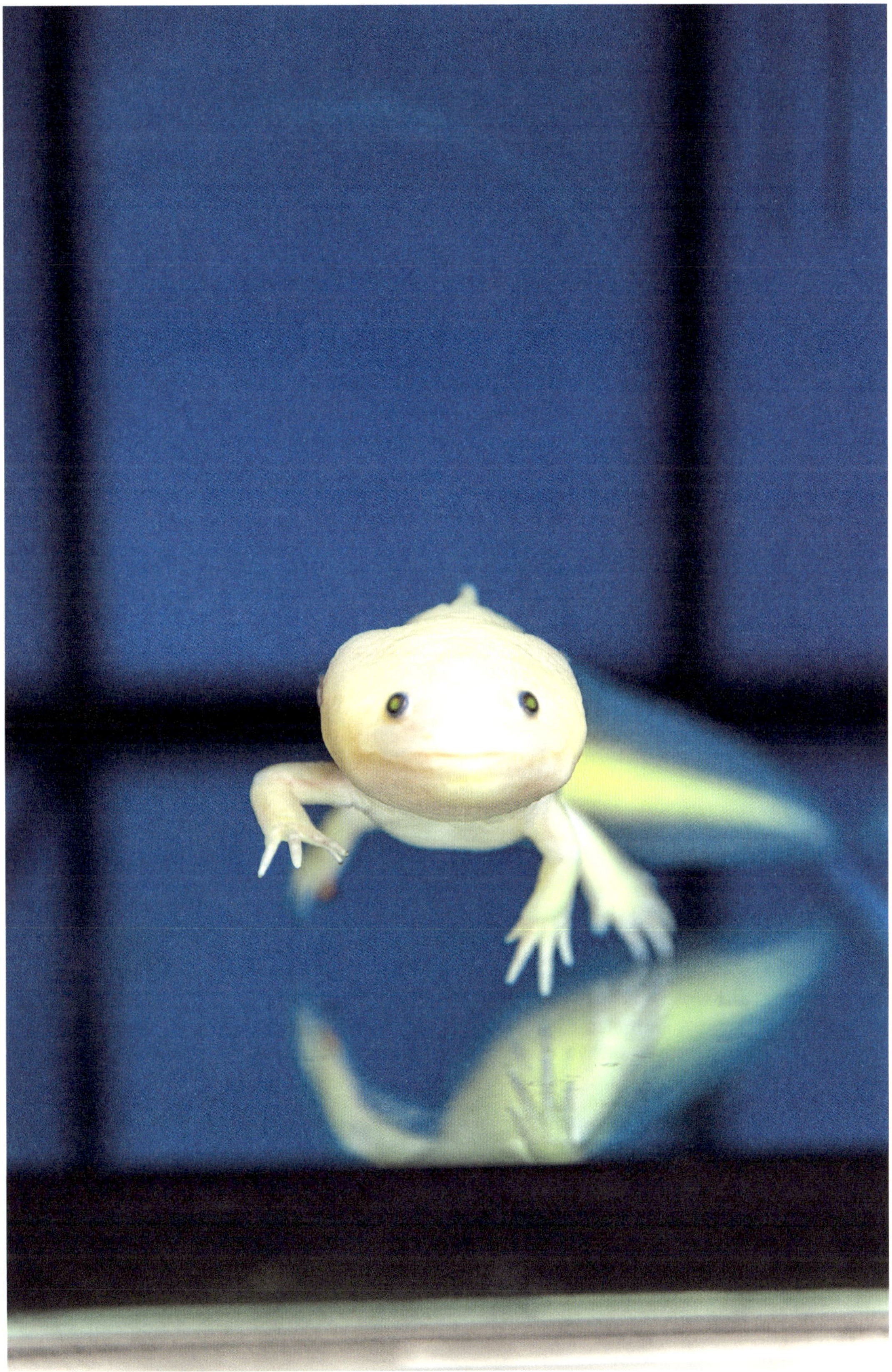

Well, there are different techniques to do it. We can start with one cell and introduce genetic material that will tell the cell to express a colored protein that can be easy to follow. Similarly, you can instruct the cells to make more or less of one protein. But we can also manipulate the genes in axolotls and many other animal models, when the animal is already grown. You can introduce genetic material into a specific part of the body to make the cells express more of this gene, or even express something that will inhibit the gene that you want. Or you can make it express a form of that gene that is not working. So you can tweak the system a bit.

But how do you switch it on or off?

There are many different ways. One way is to introduce a protein that degrades it very quickly. Instead of switching off the gene, I sequester or quickly destroy the protein so that it doesn't have a chance to have any activity. Or if there is a chemical compound that binds to it and prevents it from doing its job, there is no need to change the genetic material. I can basically inhibit the function without turning off the gene. Turning off the gene requires introducing, for example, a mutation into the gene that will remove a part that it needs to function. Trying to block something can be done at different levels. When the protein is already formed, you can inhibit the action. Or you can block the receptor of that protein. For protein it only works if you bind it to a receptor, and if you put something that sits on the receptor, it won't be able to activate itself. So there are many different possibilities. Of course, the relevance for humans is what can do the least harm with an effective function.

What was the biggest change you have made in one of these animals?

The changes are not big. Part of our work has been to understand the basics of what happens. Observing how things are happening to the cells, following while we look at them. We haven't got to the point where there are big transformations in the animal, and it not what we are looking for. You may not know that the genome of the axolotl was not available until just recently.

Isn't this very simple?

No, it's huge. It is ten times the size of the human genome.

But the sequence of the human genome is already known.

Right, but it's not correlated one-to-one. The location of the genes can also be different. We cannot easily locate the genes just because we know where they are in humans. Big manipulations are less informative about the process than small modifications. For example, we only examine the communication of cells in the limb, without changing the whole animal. It is even better to know how to do this—you know, if you change the whole animal, then what you see is more artificial. But if you just work in one area at a time with a chemical that would change a function, you're also able to remove it, so the effect is very concise and very fixed in time.

You can look inside axolotls without making an incision because they are transparent. What makes them so transparent?

These animals have a natural mutation that prevents the pigmented cells from migrating during development. That's why they look so white or transparent. You can't see through them when they have pigment cells.

So in addition to research on the limbs, you could also look at their organs.

They can regenerate the liver, parts of the heart, and parts of the brain. That's a bit harder to observe. In our lab we are currently addressing limb regeneration but with a focus on organs like the liver. What is interesting about the limbs is that they are made of many different tissues, and that its structure can be directly compared to the human fingertip process. For me, it's easier to start making comparisons of how the mechanisms are equal or different.

How close is the human genome to the axolotl? If the function of the cells in the limb is quite similar to human cells, you could make a lot of conclusions about the human body.

Exactly. We have the same tissue types in the limbs, such as muscles and tendons. In addition, we have nerves, of course, and axolotls do too. They also have blood vessels, a dermis, and an epidermis—although they don't have hair.

They have the same skin? I would like to touch an axolotl.

Let's say it shares some components, such as the epidermis and dermis. But the skin of mammals, such as mice and humans, is more complex. The other difference is that axolotls are aquatic. Their skin secretes a lot of mucus, which we don't have.

Does it have mucus like a fish?

More like frogs. But again, it's epidermis and dermis, separated by a basal lamina. The basic components are there.

If there's damage to the epidermis, how is it regenerated by the axolotl?

The cells surrounding the wound will just grow back and close it.

So it's like a human?

Yes, but it's much faster. In the axolotl it happens within hours, and in the human it can take much longer than that. And in contrast to humans, axolotls don't form scars.

If I cut my finger, it takes two weeks for it to heal.

Yes, for a wound to heal, but regeneration will take longer than that.

Are you examining the biochemical process or the cellular process?

In a way, both and more. Some biochemical changes take place that could reflect different metabolism or enzymatic degradation and change the function of cells. Ultimately many processes occur at a given time.

If you look inside the gene of the new cells, can you tell if the gene is different now or if it is switched off?

You can see or measure the function of a gene in many ways. We can see approximately how much each cell is making, which allows us to infer the function of that cell at that particular time. This complements the imaging. Of course, my scientific view is influenced by people that come from other fields, such as Professor Friedrich, who provide knowledge that I lack. This multidisciplinary approach helps things move forward. We are not always doing the same thing or interpreting the things in the same way, but combining our knowledge. Biology with mathematical modeling and biomaterials, for example, could move things along faster.

Can you explain the reprogramming in more detail?

The general aim of cell reprogramming is to change the identity of a cell. All the cells in an organism have the same gene map. These genes will be expressed in different ways depending on what a cell has been instructed to do. To change these instructions and thus change the identity of a cell, new regulatory factors are introduced. These can instruct the cell to make more molecules in particular, or introduce changes that make the DNA more

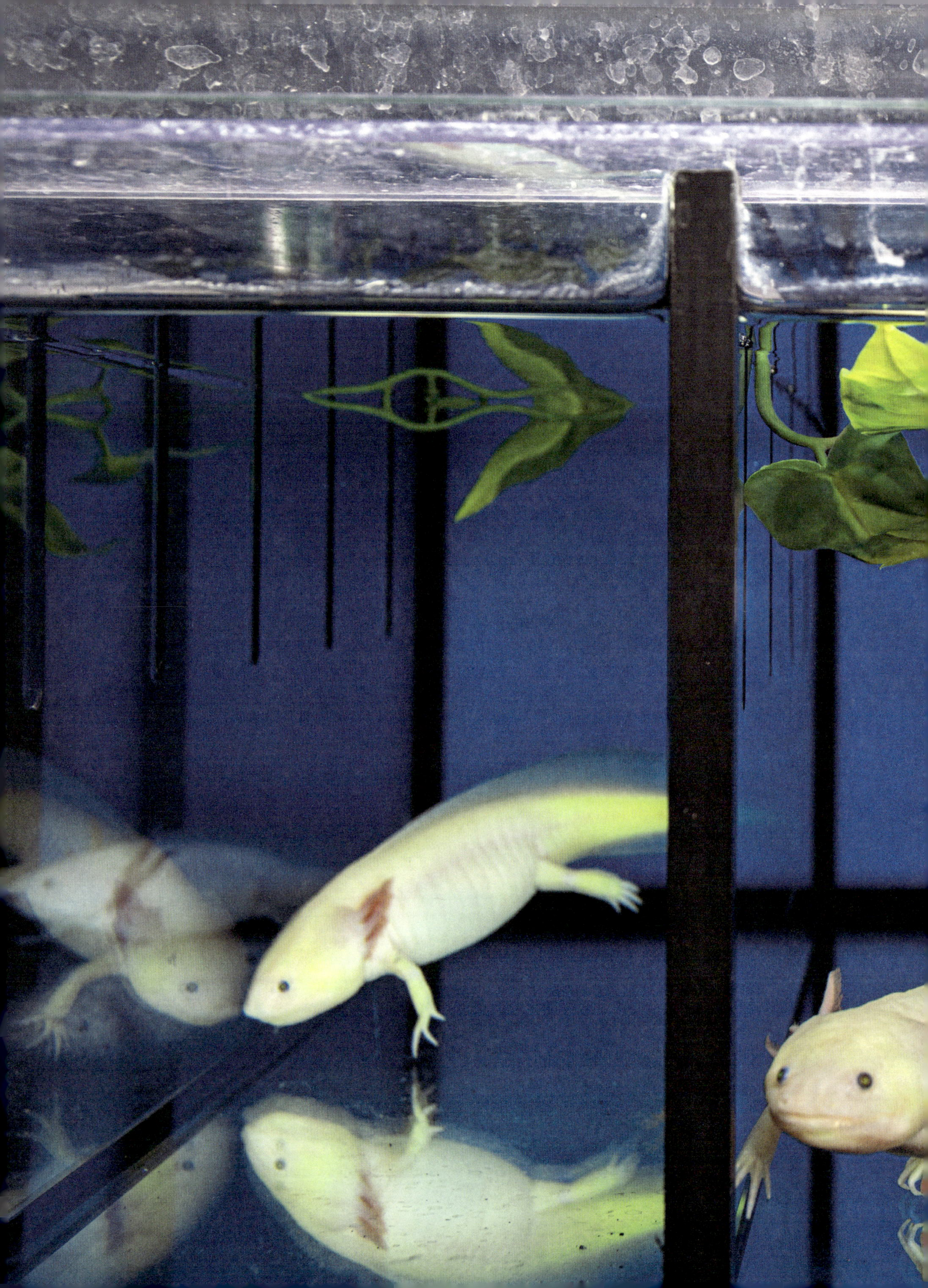

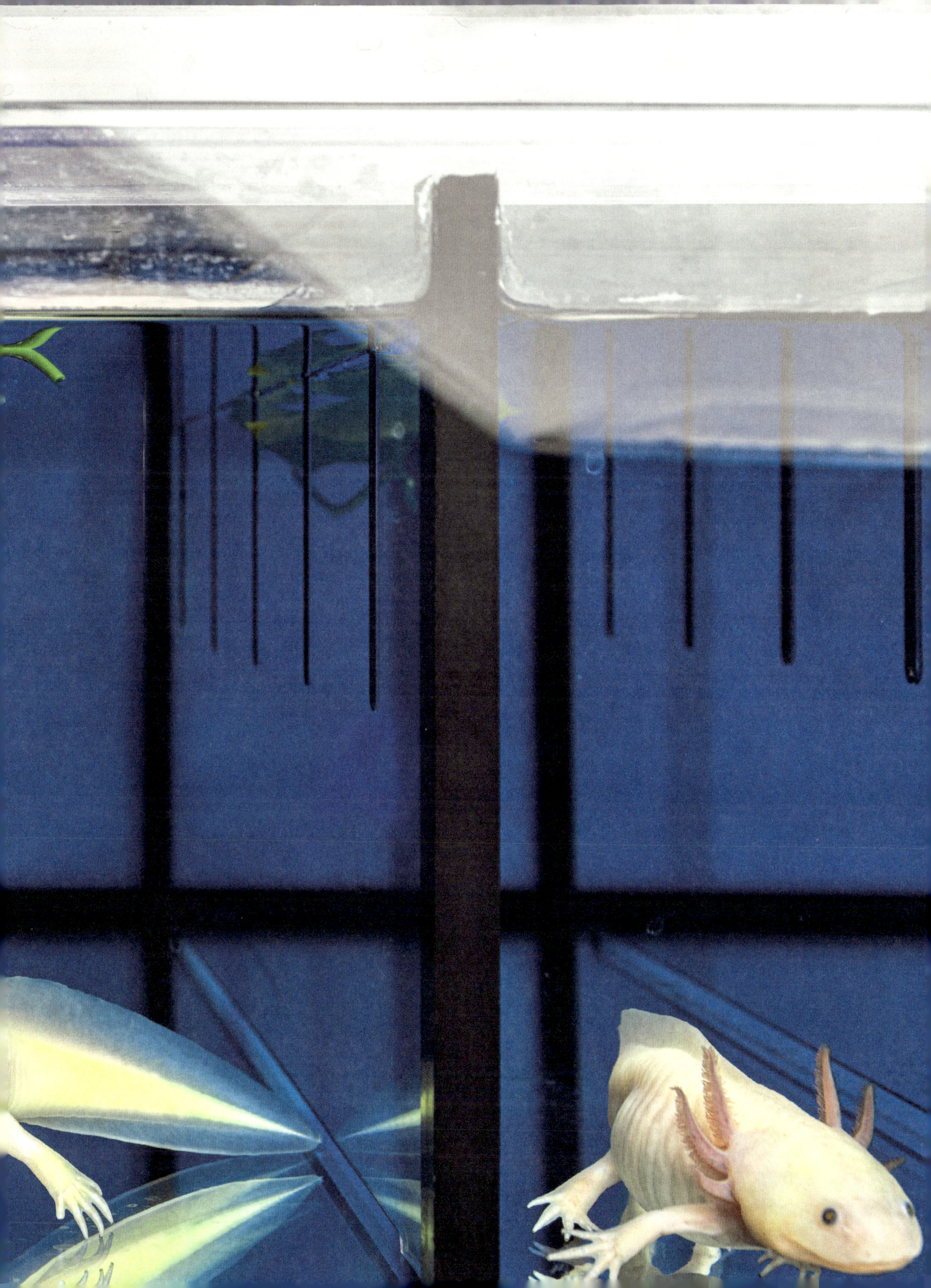

accessible to be read, et cetera. Normally, how the cell decides when these genes will be expressed or not is complex and relies not only on what the cell contains inside, but also on things that the cell is exposed to at a certain time. So if there is a lack of glucose, let's say, then the cell will respond in a certain way. This information will be conveyed to the nucleus through a cascade of different signaling factors. Another dimension to consider is time: how long should a gene be expressed? In some cases, it is important to not only turn on a gene, but also to control when it should be turned off. Nature does great job, but when we do it artificially, we should do it with sufficient knowledge.

Where did the artificial gene come from? Did you put it there?

You assemble it if you know the sequence. If you don't know the sequence, you can introduce a gene from other species.

Do you have the complete map of the axolotl already?

We have it, but it's not like you can just take a book and read it. It's a process that requires that parts of that map are named properly. Knowing the gene allows you to introduce specific modifications that will be useful in your research. This is a technique that has been used for a long time since the beginning of molecular biology. It's like inserting a new printed page into a book. That's how many biotechnologies work: you make yeast produce more penicillin by introducing fungus genes.

So this special artificial gene is then inserted into the DNA of the cell?

That is one way to modify the genome. There are various techniques to choose from.

And the cell starts to produce proteins that are necessary to recreate a limb or an organ. Can you describe the mechanism of how these new proteins are built up during regeneration? How does it work? Do you put an artificial chain in the right place so that it can be switched on or off? Is this the same as the process that you explained earlier?

We have to remember that what the cells go through during regeneration is highly regulated in a local and temporal manner. We are just beginning to understand piece by piece how cells respond to all kinds of signals at a given time and place. Using mathematical models, as explained by Professor Friedrich, helps us dissect and predict cell behaviors.

What is so special about this axolotl that it regenerates better than any other organism?

One thing that is different from humans is the immune system. Humans, and mammals in general, have a complex immune response to an injury. It's a trait that evolved to keep the organism safe. This immune response is also part of the trigger of fibrosis: instead of replacing the cells that are missing, the wound is just filled up with collagens. This causes a scar. It closes the wound, protecting the organism from external factors. It doesn't look perfect, and it's a different structure than before. Axolotls don't form scars. Another peculiarity is that axolotls retain juvenile features their whole life, and they do not transmute into a terrestrial state like other salamanders. We have recently described that axolotls continue to grow during their lifetime, although at a slower rate than in juvenile stages. What is the one feature that gives them the power to regenerate? Is probably the combination of many factors.

The main thing is that it's closed.

Right. The idea in regeneration is that the structure is similar to what it was before. One hypothesis is that the difference between axolotls and humans is the immune system. They have an immune system, but the response—

It has to be a better one than ours.

One could say that, but it could also be that the timing is different. The number of cells that respond is different, and the way they respond could also be different. I think we're just starting to understand how it works. We know that the immune system is necessary for regeneration and there is no regeneration without immune cells.

That's interesting. So how is it coordinated and connected?

One of the questions we have is how does the wound switch from a normal healing response to a regenerative response. There's a switch that would make a cell produce a lot of collagen or to start proliferating and become something else. There is a lot of effort on trying to understand how this transition from just closing the wound to making more cells and making the right cells at the right time at the right position occurs. And the immune system is certainly a candidate for the regulation of this switch.

Can human limbs be grown in the laboratory?

Probably not in my lifetime. But I think the important thing is that if we understand how the axolotls regenerate their limbs, we can start drawing the line tissue by tissue, so that we can understand a bit more about how the muscle is made, how the bone is made, et cetera. And we can see how humans do it and start connecting the dots in between the species. I think this is going to be quite beneficial in the meantime, while we keep dreaming about making a whole limb.

Perhaps you will develop a medicine that improves wound healing.

Yes. But then there is the problem of time. The axolotls take about twelve months to make an adult limb.

A whole year?

Yes. It takes them about twelve months from birth to adulthood, but they can regenerate a limb in three months.

Faster than growing!

Yes, the developmental process is quite well regulated. It has its own pace, but regeneration is different and faster. So it's quite a challenge for a human to make the enormous mass of material that we are missing. The speed is a challenge as well as the scaling. A combination of reprogramming and scaffolds could probably help the regrowth. That is not something that we can do now, but we should explore it. That's why I was saying that a multidisciplinary approach could provide more answers for these challenges. That is the scale of what we envision for regenerating and the speed that it needs to be done.

Have you looked at the evolution of the axolotl? Was there any evolutionary pressure for them to develop this regenerative ability in Mexico?

I don't think so. The natural habitat of the axolotls was an area that was so favorable, with sufficient food and no predators. It was so advantageous to them that they adapted by remaining aquatic their whole lives, exhibiting a local adaptation, and avoiding transmutation into a terrestrial phase. Axolotls are salamanders, and most salamanders have an aquatic phase, then undergo metamorphosis and become terrestrial.

If they had this good life in this little lake in Mexico, where it was always sunny and they had no enemies, why do they develop this ability?

It's a good question. There are other animals that can regenerate their whole body, which presents the question of whether organisms only keep traits that were or are purposeful.

Perhaps it was just an accident in the development?

Well, you could also say that accidents have led to biodiversity. But I don't know if I would use the word *accident* for an evolutionary occurrence.

So you are the young scientist from Mexico who will win the Nobel prize for Germany for finding the key to a long life. If you can regenerate everything, you can live for 150 years.

I can tell you what it felt like winning a prize. When I was doing my PhD, I worked on a group of neurons from the hypothalamus of postmenopausal women. These neurons become enlarged when women go through menopause, and they express different things. The lab where I worked during my PhD had worked on these cells and continued their work, and I just read a paper that one of the molecules that is secreted in the cells can be used as therapy for treating hot flashes in menopause. They cite our research, so my supervisor's findings were key in developing this therapy. Although it was not my discovery, I was part of it.

It was a team.

And it has already an impact on humans. For me, this was like winning the Nobel Prize, knowing that something I was part of has an impact.

Maybe it will take one hundred years for human bodies to get the ability to regenerate, but there are many ideas for therapy. We are talking about RNA, and now we have this new RNA from BioNTech. They want to fight cancer, starting with the ovaries and the prostate. They are very optimistic that they can do it by stimulating the immune system.

That's what I mean: with the knowledge that we have, we can start connecting the dots. It's not about waiting until we can regrow the whole limb. Drawing lines and moving from point to point can already be progress.

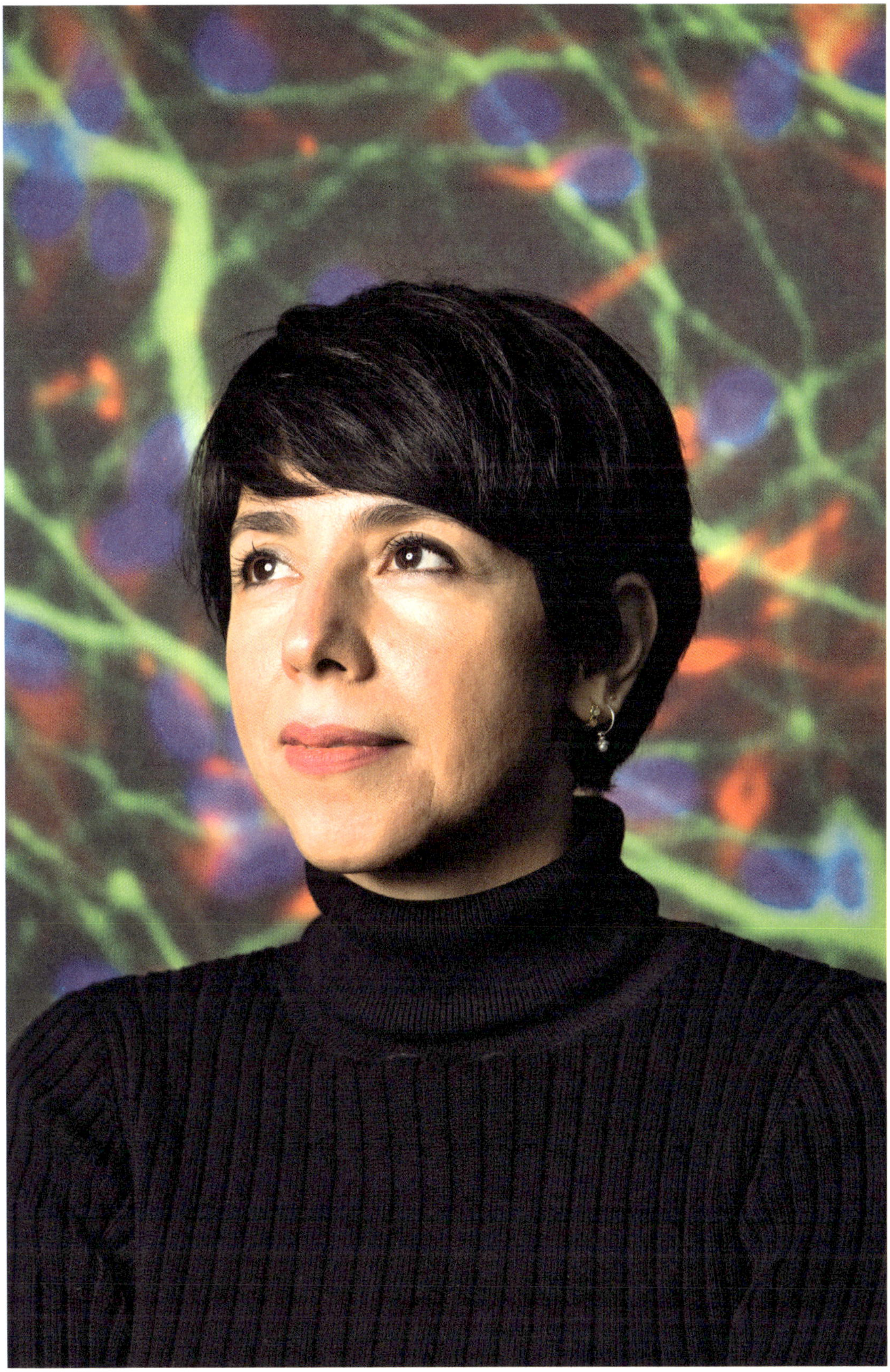

MARTIN BÖHM, HANS DIETL, ANDREAS GOPPELT,
BERNHARD GRAIMANN, AND GÜNGÖR KARA

Reflections on Bionic Regeneration

The experts from Ottobock are not in complete agreement on the extent to which the undreamed-of possibilities of life sciences could be utilized by orthopedic engineers in the future. A strong worldwide trend toward the convergence of these research disciplines and application scenarios can be observed.

Andreas Goppelt: I find this chapter on bionic regeneration fascinating, as it is in this field that the convergence of the most diverse branches of science can be seen particularly clearly. If different developments occur simultaneously within this vibrant field, we could soon see the next quantum leap in overall technological progress. Our goal at Ottobock has always been to combine the engineering sciences with certain potential solutions from the life sciences. Bob Langer and Anthony Atala are already uniting all the disciplines in their biomedical technology laboratories. This will be very relevant in the future. Christian Lüscher (see page 229) has shown in his research that nerves can be made to emit light. There are many other examples of this: for one, the genes of the deep-sea jellyfish or the glowworm can be deployed in human cells, causing the human cells to light up like a glowworm. There are homeobox genes that serve to differentiate an undifferentiated cell cluster through cell division in an inner organism along certain morphogen gradients, resulting in axes in bodies and organs. Benjamin M. Friedrich has described this well. It is interesting that higher organisms have lost this capacity for regeneration, which Tatiana Sandoval-Guzmán is researching. However, if a physician operates on embryos in the womb due to a valvular heart defect, they no longer have any scars on the heart after birth. Regeneration means perfect healing of wounds. Embryos can still do what humans have lost the ability to do over the course of evolution. Releasing this potential again for use in the future is an exciting challenge. I don't think that it will be possible to regrow a whole leg tomorrow or the day after. That's not what it's about at all. But the work of Mark Kotter, who can artificially produce stem cells on an industrial scale, is very interesting. This could become reality relatively soon for certain reconstructive tasks with a small cell volume. And this could prove to be very interesting for Ottobock.

Hans Dietl: Skin regeneration, residuum configuration, and bone lengthening could be very interesting for us. You don't have to immediately regrow the whole leg, but you could, for example, replicate the tissue from the sole of the foot using tissue engineering and thus cover the sections of the amputation residuum subject to the greatest loads to create a robust interface.

AG: Although you can print out skin with 3D printers and use it to cover tissue defects, this skin doesn't have any sweat glands and cannot provide any sensory feedback, which is very important for us. These are tasks that are being developed everywhere in laboratories around the world. These are relevant future perspectives for us.

Bernhard Graimann: If there were actually a quantum leap in medical progress, it would naturally be important for Ottobock. That is why we have to observe and understand it. As Hans Dietl has mentioned, there are already application fields for tissue engineering. It would be very interesting if we could improve the residuum with this. But I don't see Ottobock engaging in developing these foreign disciplines as we don't have our own expertise in these areas. What we should do is expand into the neighboring areas of orthopedic technologies where we can really move something with our competences.

AG: In the short term you are definitely right, Bernhard. But take a look at the laboratories of Bob Langer, Anthony Atala, Mark Kotter, and Hugh Herr. They are sites of powerful convergences and interdisciplinary innovations.

BG: In the history of innovations, many companies failed because they ventured too far outside their competences. That is not an approach to be recommended. It is a matter of making an entrepreneurial decision as to which research fields a company wants to venture into. This can only be developed from a company's core competences, otherwise it will overstretch its strategic focus. But it is definitely a good idea to work with partners, for example, on tissue engineering.

AG: That is only a question of time. This book provides a vision of a possible future, which will also open up new areas of growth for Ottobock. We don't have to be frightened of this.

BG: I think that the probability of failure is very high. But it's good that we sometimes disagree.

AG: If it were possible to successfully transplant diabetic islet cells, it could be really disruptive. Patient compliance contributes to so many diabetics being amputated: patients don't pursue their therapeutic possibilities as rigorously as they should. If this could be automated to restore the body's lost capabilities, it would be really disruptive.

HD: I am not a visionary—I am retired. If I could still do things the way I wanted to, I would naturally maintain contact with the laboratories and discuss with colleagues about small joint research projects and simply try out something like this. There should be brainstorming with Bob Langer on how to improve the coupling of the prosthesis with bioengineering methods, as this is still a sore spot for a lot of users—sometimes literally.

AG: Ottobock began as a handicraft and has had to continually reinvent itself. The company has always remained technology agnostic, and as a result has repeatedly taken on new challenges that were initially far from self-evident. If tissue engineering should become relevant—and I presume it will—it will play a role. For example, the prosthesis will be directly anchored in the bone using osseointegration or coupled to the implant through the skin. Effectively, the bone is extended with metal, which passes through the skin.

HD: And skin penetration sometimes causes problems because the skin barrier is not stable enough. Although there are frequently infections that have to be treated on the surface or with antibiotics, patients are prepared to accept this because the immediate advantages of this technology far outweigh the possible disadvantages: this method enables complete power transmission. There is no chafing or sweating any more. In the medium term this will become standard for higher amputations. Nevertheless, it is still a foreign body in the system, and the exciting question is how it can be improved.

AG: The pharmaceutical industry has now mastered organoid bodies, very small mini-organs that are used to conduct drug screening, and they are already being used commercially. It is also possible to cultivate skin. Things will be more difficult with greater tissue volumes, as the vessels need to grow in order to supply the tissue.

HD: For artificial skin to be durable it has to be able to regenerate, and it needs blood vessels for this. This is a very exciting topic that we should engage with—although, ultimately, there is reason to believe it won't work.

AG: We have many points of contact with biotechnology in the company. We are just beginning to see the very first delicate fruits of the convergence of these two major sciences—and that is a remarkable moment in our history.

Artificial Intelligence

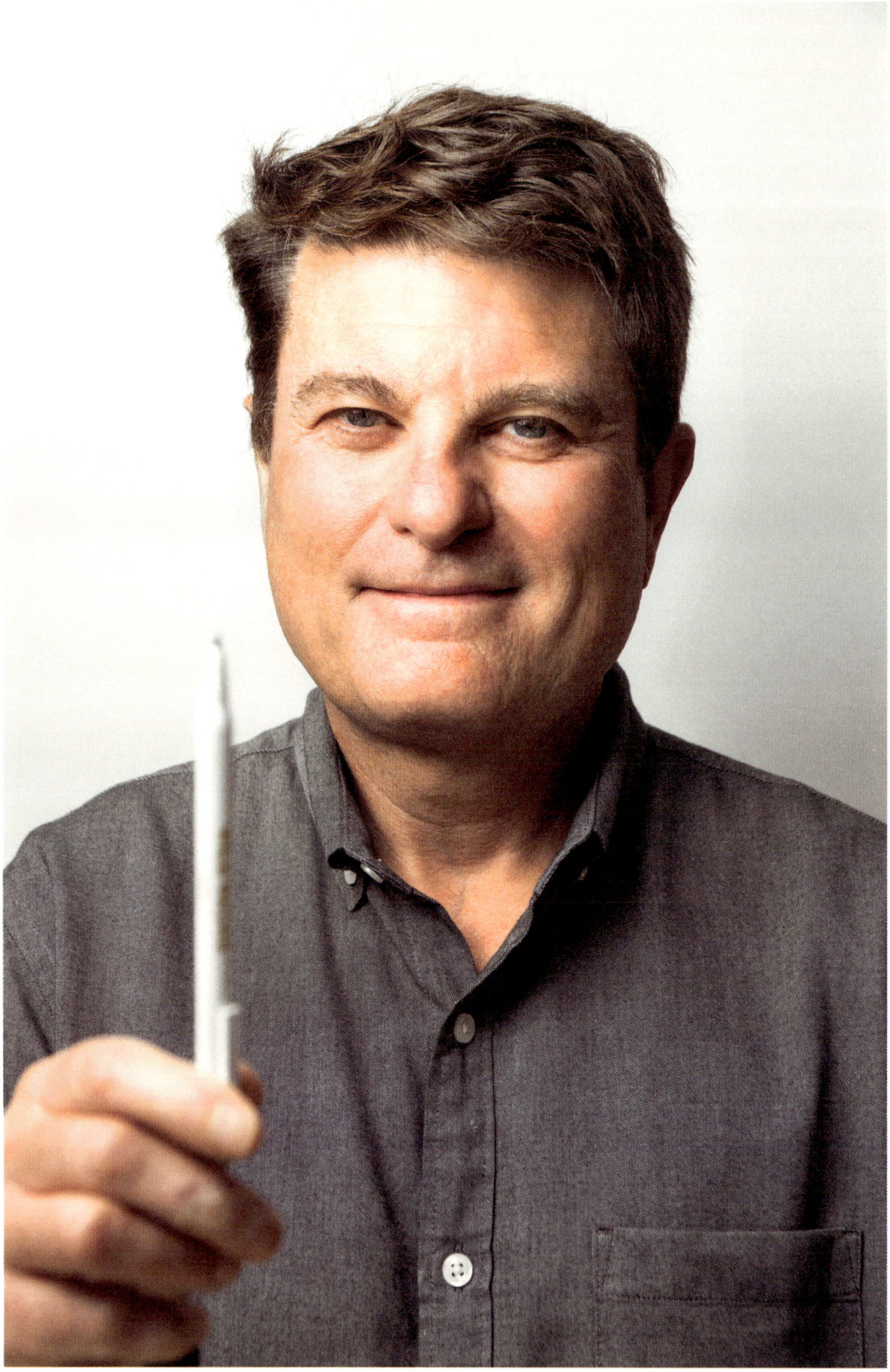

MARTIN FORD

"I Am Very Optimistic about Technology, But We Have to Choose between *The Matrix* or *Star Trek*"

The rapid rise of artificial intelligence is already making Hollywood's science-fiction visions a reality. The futurist and author whose focus is artificial intelligence and robotics describes the impact of these technologies on the job market, the economy, and society. He calls for humanity to make sense of this dramatic change so that instead of the dark version of *The Matrix*, our future resembles the bright utopia of *Star Trek* in which there is enough wealth, humanity's major problems have been solved by technology, and people don't have to spend all day doing boring, unrewarding jobs. This is the kind of thing that we should dream about—although it may be in the very distant future, it's the direction that society should be aiming for.

You were one of the first authors to describe the dramatic change that artificial intelligence brought to the workforce, and your view was quite pessimistic in 2009. Are you more optimistic today?

I think the same trends are in play now, and it's going to have a dramatic impact on work. People always say that I'm pessimistic, but I never quite viewed it that way. If we can come up with a new way to ensure that people have access to an income even if they're not working, then it's not pessimistic. It's actually a kind of utopia.

But won't technology take over the simple jobs? In your second book, you say that not only the proletariat loses jobs, but also the well-educated in finance and consulting—because algorithms are better bankers than humans.

It's not necessarily simple jobs that can be automated. It's predictable jobs, where you're essentially doing the same kinds of things again and again. Someday a machine learning algorithm will look at all the detailed data of all the things you've done, all the decisions you've made, the actions you've taken, and based on that it can learn to do a lot of what you're doing. An example of a job that's not simple at all, but may someday be highly threatened, is a radiologist who looks at scans to see if you have cancer. Machines may someday take all those jobs.

Radiology is a good example, because artificial intelligence is always effective when patterns are involved. Everybody has a different kind of lung cancer, yet doctors are taught to recognize and treat one example in textbooks. So if you could compare the patterns from all over the world, using data from patients in Tokyo, Vancouver, and Toronto who took this medication, had this radiation, and got this result, it would support the decisions doctors make, but do you think it will replace doctors?

I don't think doctors will be replaced in the foreseeable future. But I think what you say is basically right: AI is going to evolve to be sort of an automatic second opinion—maybe not too far in the future. For a doctor to diagnose or develop treatments without consulting the artificial intelligence will be considered malpractice. Doctors will be required to collaborate with AI. You will have to consult this second opinion and that will democratize medicine. You know, not everyone has access to the world-class cancer specialist, but in the future, even if you have a mediocre doctor, this doctor will be working with the AI that encapsulates the knowledge of the very best doctor.

There is already an app that can recognize if a picture of a black mole is a melanoma. I read that artificial intelligence is already better able to recognize melanomas than any doctor.

Right now it is quite limited, and you have to be very careful about the specifications of the image you're looking at. If there's some variation, with images coming from different places and so forth, it causes trouble, because the art of AI is very brittle. That's the word they use: *brittle*, meaning that it's not flexible, it can't handle small variations. The most famous case of that is the company DeepMind, which originally developed video games. They did a study that showed if you just shifted the frame by a few pixels, it would completely destroy everything, and you'd have to start from scratch and retrain. So these systems have no ability like a human being to make those adjustments. If it's just shifted a little bit, a human would just automatically adjust and probably wouldn't even notice. But it completely detracts the AI. So there's still a lot of work to be done to make systems that are more flexible.

There's a lot of hype about artificial intelligence, but do you believe that it will become more intelligent that humans and that the singularity will occur?

I don't know about the singularity. But I believe someday, far in the future, we may achieve human-level artificial intelligence. I interviewed all those people in my book *Architects of Intelligence: The Truth about AI from the People Building It*. Ray Kurzweil really thinks that we'll achieve that by 2029—which is less than seven years from now. I'd be very surprised if that happens. There are people who believe it can happen fairly soon, but the consensus is that it's probably at least fifty years in the future. There are many problems that need to be solved. First of all, you need lots of data—you'd need thousands, maybe hundreds of thousands of medical scans to make this work. It's not like a human doctor, who will learn from a textbook or with a few examples. But in every application where you have lots of data, where you're looking for patterns, AI is probably going to outperform people. There's a very broad range of specific tasks. So that's where I see the immediate impact being. But there definitely are initiatives working very hard to figure out how you can build machines that learn the way people do using much less data. Think of a baby learning to speak the first time, you teach it what a dog is, and what a cat is. You may only have to point to the dog once, and the kid learns. With artificial intelligence, you're working with thousands and thousands of data. So there is a lot of initiative in trying to understand how that works, and I think there is progress there. We're going to get gradually closer and closer to machines that are more human-like in general intelligence, but I think that's going to be a long path.

Ray Kurzweil has done some really great work creating a powerful narrative that is mobilizing billions of dollars, and all the big companies are following. Now that the hype is over, I think that people are focusing on the actual problems again, such as how AI can help us. It's not like those machines will overcome mankind and make us into slaves.

Not anytime soon. I mean, you still hear prominent people talking about that, like Elon Musk. But I think it's a good thing. We need to focus on the immediate things, the benefits as well as the risks. There are a lot of concerns and issues.

Can you give some examples?

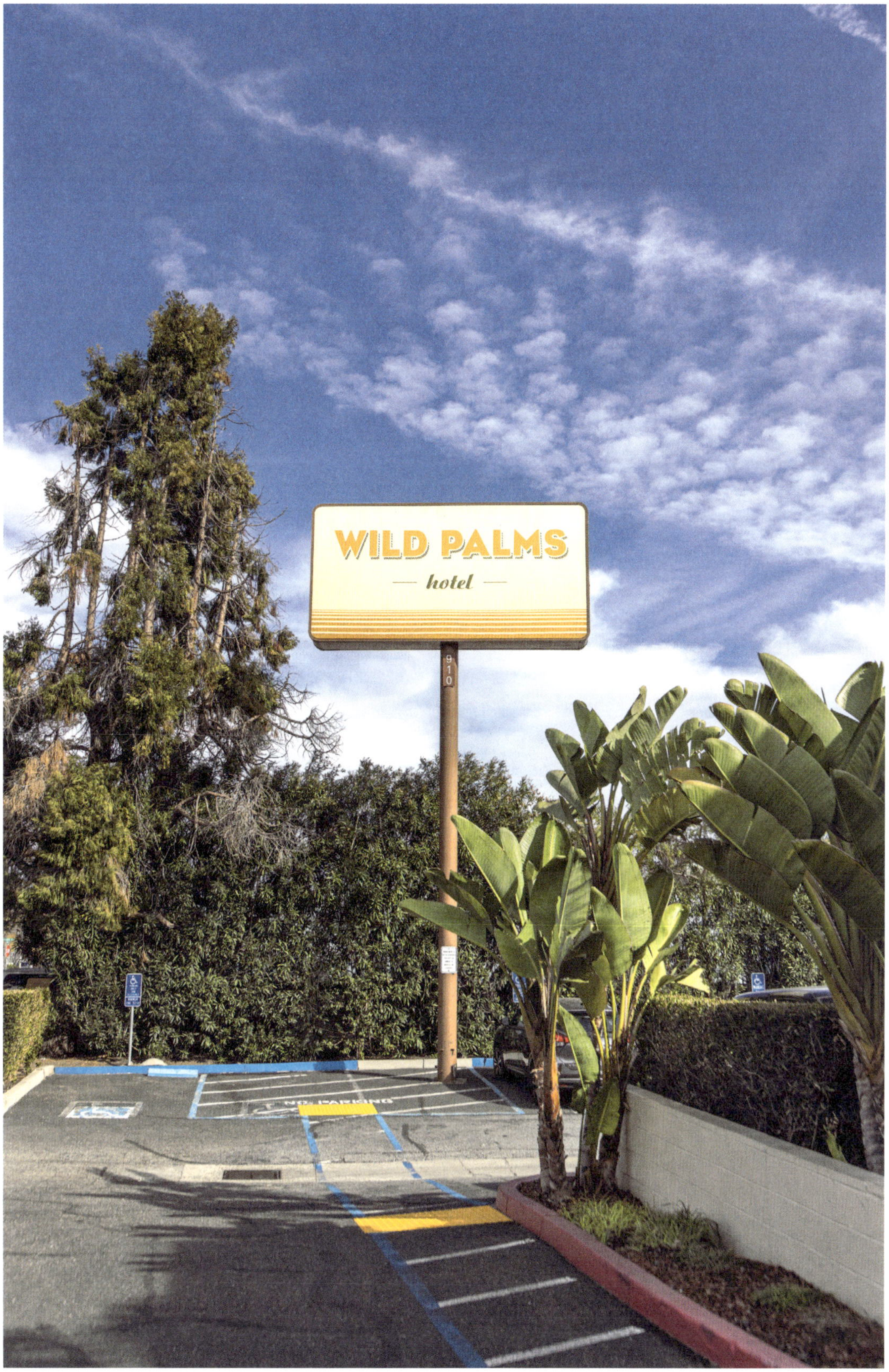
WILD PALMS
hotel
910
NO PARKING

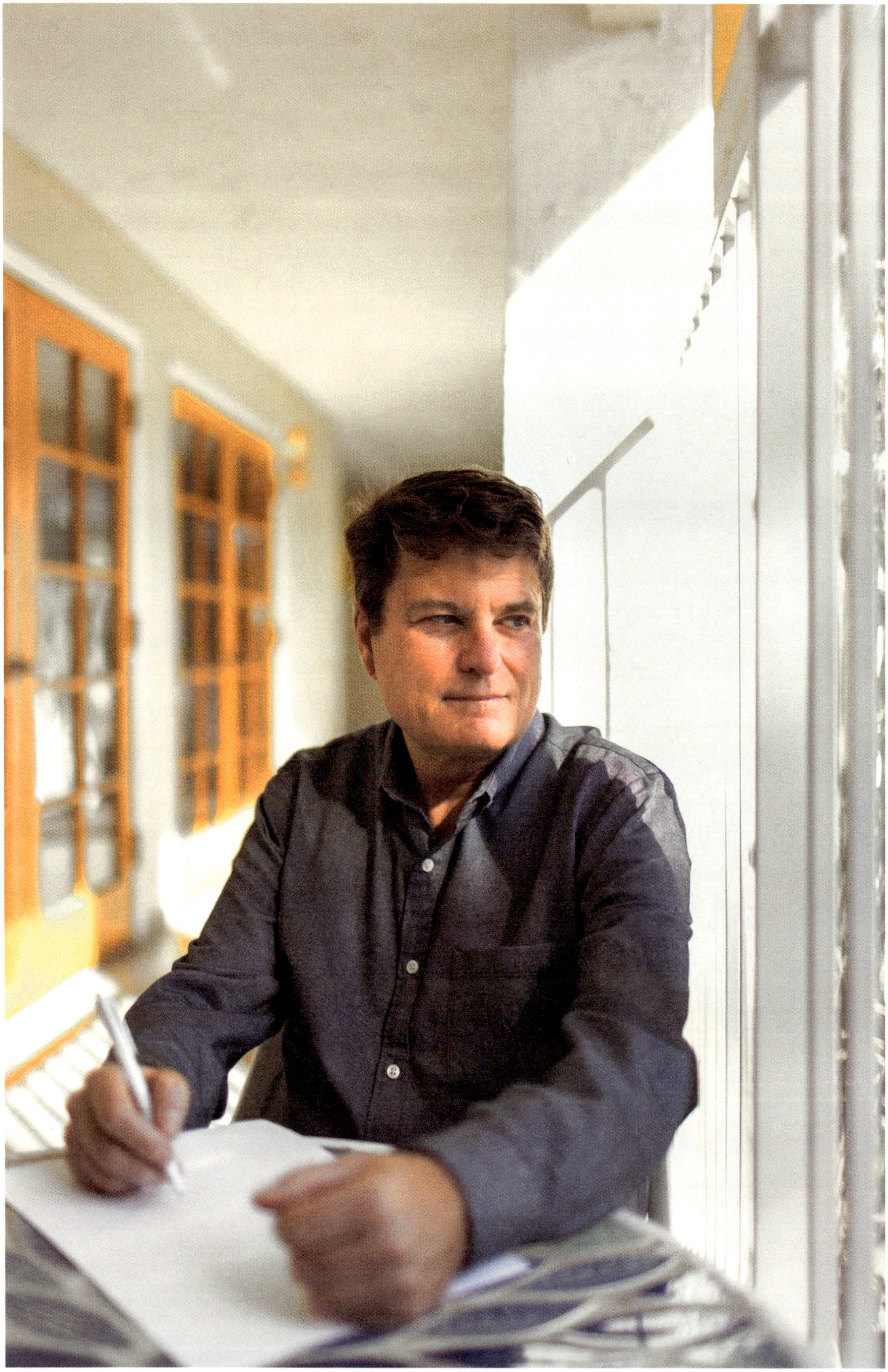

Hostile weaponization, autonomous weapons, and deep fakes, the ability to create these fabricated illusions that are indistinguishable from reality. As I said in *Rule of the Robots*, there have already been cases where people have used deep fakes of the CEO of a company. They play those over a phone call and have the finance department wire money to a bank account—they've stolen millions of dollars that way. So I think this is already becoming an issue, and you will see a lot more of those. As I said in the book, you could potentially append an election by having a politician say something very damaging just before an election, with no time to figure out if this is true or not. This is very scary. Democracy is already having lots of problems, and it could get worse. I worry about that. Of course, there are many issues around racial bias and gender bias in these algorithms. That is really coming very much to the forefront given the focus on those issues now. It's general security threats, the fact that AI is going to be used in cyberattacks and will also be used to defend systems. So there's going to be kind of an arms race happening there. Then all the issues around surveillance and particularly what's happening in China and how they are building an Orwellian society there, using technologies such as facial recognition and other forms of AI-based surveillance. And a potential for an AI race with China. We're really in a race with them in terms of intelligence.

They are investing billions and billions in artificial intelligence, and we don't know exactly what they are doing. Do you have any idea where it might go?

Well, there's no question that they are at the forefront of these technologies. In *Rule of the Robots*, I give a couple of perspectives on history. Some people think that that's been kind of exaggerated, and they really aren't at our level yet. But then there are others who think they're already pulling ahead, at least in terms of more practical applications. I mean, we have the people like those I interviewed—the most famous people tend to be here. They're not in China, and if they are there, we don't know them so much. But in terms of the people doing the more routine work, such as building practical applications and having access to enormous amounts of data, they really have advantages there. Just consider the size of their population and their focus on education. So there is an enormous number of people there who want to become AI researchers—and are very good at math. I think they're going to give us enormous competition in the future.

Germans are very concerned about the future of work. German factory workers have a lot of resentment toward machines, because they feel that it's a matter of who is the master and who is the slave. The workers are frightened that an anonymous power is taking away their jobs. There are already a lot of robots in the real world of Volkswagen and Amazon. If the human worker views the robot as a coworker who follows orders, they might appreciate new technology, but if it's the other way around, they will oppose it.

It's very hard to achieve that. Because if you look inside Amazon warehouses, there are a lot of people and a lot of robots. The people are working with the robots, but the people are also essentially working under algorithmic control. They're constantly watched, and the pace of their work is monitored—they truly are being treated like robots. It's like Amazon would really like to have this factory completely automated, but they can't do it yet because the robots can't do the things the people are doing. So they just plug the people into this big circuit. They are fulfilling this function that the machine can't do yet.

They are human robots.

Yes, but the reason they're there is that they have dexterity and visual perception that robots don't yet have. Mostly they put items on the shelf and retrieve items from the shelf to send to the customer. You're dealing with thousands of different types of objects, different weights, sizes, and textures, and robotic manipulation is not at that level yet. But that's going to change within a decade probably. Jeff Bezos said that he thought robotic grasping would basically match human level within about ten years. Within ten years those Amazon warehouses are going to employ a lot fewer people. What happens to those workers? New jobs will also be created, but the evidence suggests that the new jobs created might be very technical. You will really need a lot of more engineering-type jobs. Or they tend to be jobs that really require a lot of people interaction, because it's going to be a long time before machines can develop relationships with people. So not everyone who works in an Amazon warehouse will be able to transition into that kind of job. Some might but I think there's a real risk of a lot of people being left behind.

Could you summarize your last book as an examination of "following the money"? Can you tell us where most of the investments occur? It might help us understand what developments lie ahead. Is it autonomous driving, or medical applications, or military?

I don't know how the numbers compare between industries, but most of the money is going into relatively near-term practical application. Lots of start-ups use artificial intelligence in drug discovery, searching for molecules that have three-dimensional shapes that define their function. You can use artificial intelligence to search the universe of all the possible molecules out there and find the ones that might have medical applications. There are a lot of start-ups doing that. I think maybe a dozen or more are really focused in that area. That's kind of related to what DeepMind did with protein folding, which is a hugely important breakthrough. Applications like that are critically important. I mean, there definitely continues to be huge investment in self-driving cars, but I think those are going to take a bit longer, as I say in *Rule of the Robots*.

Will we ever reach level five?

I think we will, but it's probably a long time. In many ways driving a car approaches human-level intelligence. My guess is that there will be self-driving cars, but as a more limited application. Waymo has already got a carefully mapped system of virtual tracks, but I think it will be a long time before you can call a self-driving car in the way you call an Uber and just literally go anywhere you want. It's a real challenge. Tesla has been overpromising, and I think they're going to get in trouble. They're calling their system full self-driving when it clearly is not.

So you live in Silicon Valley, and you meet all these people. How do you approach them?

For *Architects of Intelligence*, most of the interviews were done over Skype. I live here and I'm part of it, but I wouldn't really view myself as kind of an insider. Many years ago I ran a small software company, but it wasn't big enough to really be part of the whole venture capital. When I interviewed those people, I had to go through the same process. It wasn't like I just called them up.

Who do you think will be the next big thing?

Demis Hassabis, the CEO of DeepMind, is a very impressive guy who may have some important ideas. He's really pushing that way. What OpenAI is doing is also very important. DeepMind and OpenAI are the two companies that are really pushing the limits on these technologies. They have somewhat different approaches: OpenAI is really all about scaling, building much larger systems, whereas DeepMind is more concerned with looking at the way the human brain works and incorporating that knowledge. I see a very interesting competition between many of these different approaches, and I don't really know which is going to be the one that produces the most progress. But I think it's fascinating that people have very different ideas about how to push this forward.

What are you working on at the moment?

I've just finished my last book, and I'm still doing some promotion. I do a lot of speaking engagements—at least I did before the pandemic. It's beginning to start up again, and I've gone to a couple of events.

Let's talk about the future of work. At some point we will need a machine tax. This discussion has been going on in Europe since the 1990s, when microprocessors were introduced in production and people were worried that automatization would take all the jobs. But instead, productivity increased, and industry remained competitive. So automatization helped a lot. You saw the same thing at Volvo in the 1970s, where they moved from assembly line production to teams of workers who were responsible for entire sections of a car. This resulted in much higher quality and workers who were more satisfied with their jobs. This is a good outcome of automatization.

That's a positive story. But I still think over the long run more of those jobs will disappear. I'm not sure what the numbers are in Germany, but in 1950 more than a third of the workforce in the United States was in factories. Now it's less than ten percent.

This is the result of the deindustrialization of American industry.

Well, yes, but the actual output of factories has increased dramatically. It's not just that all the factories have gone to China. A lot of manufacturing is still here, and in real terms more than in the 1950s. It's really automation that has driven down the number of employees. If you look at the graph, it's just a straight line. It doesn't suddenly go down in 2001, when China joined the WTO and all of that. It's just a very linear track. So, it's basically technology that primarily caused that and I think that will continue. As robots become more dexterous, they are going to take over more of those jobs in factories just like any Amazon warehouse.

So the future is more like the demise of Detroit?

Right, a lot of the factories have closed. What's happened in the US, of course, is that everyone has moved to the service sector. And those jobs are often not as good.

But these are the kinds of jobs that will be the first to be automated.

Well, up until this point the service jobs have been harder to automate. But now, with artificial intelligence becoming more capable, those jobs are also going to be threatened. Previously, this has happened on a sector-by-sector basis. The first one was agriculture. At least half the people used to work on farms, but now it's only one or two percent of the workforce. So the people who lost jobs moved to factories, and you had an industrial economy. Then the factories automated, or they offshored, and everyone moved to the service sector, which is where more than eighty percent of our workforce is now. But the technology is also going to come for that sector. What other sector is there? There's nothing. So I do think we're approaching the end game here, where you're going to begin to see this.

In your second book, you denied the possibility that new technology will also bring a lot of new jobs that we even don't know about yet.

Yes, they will bring new jobs that we don't know yet. But the question is: how many? There definitely will be some but will there be enough? Potentially you're talking about tens of millions of people losing traditional jobs. I'm very skeptical that there will be enough brand-new jobs to absorb all those people.

So what can we do to prevent this social drama?

Well, what I support is some kind of universal basic income, and I think there should be incentives built into that. Rather than just give everyone money, there should be opportunity

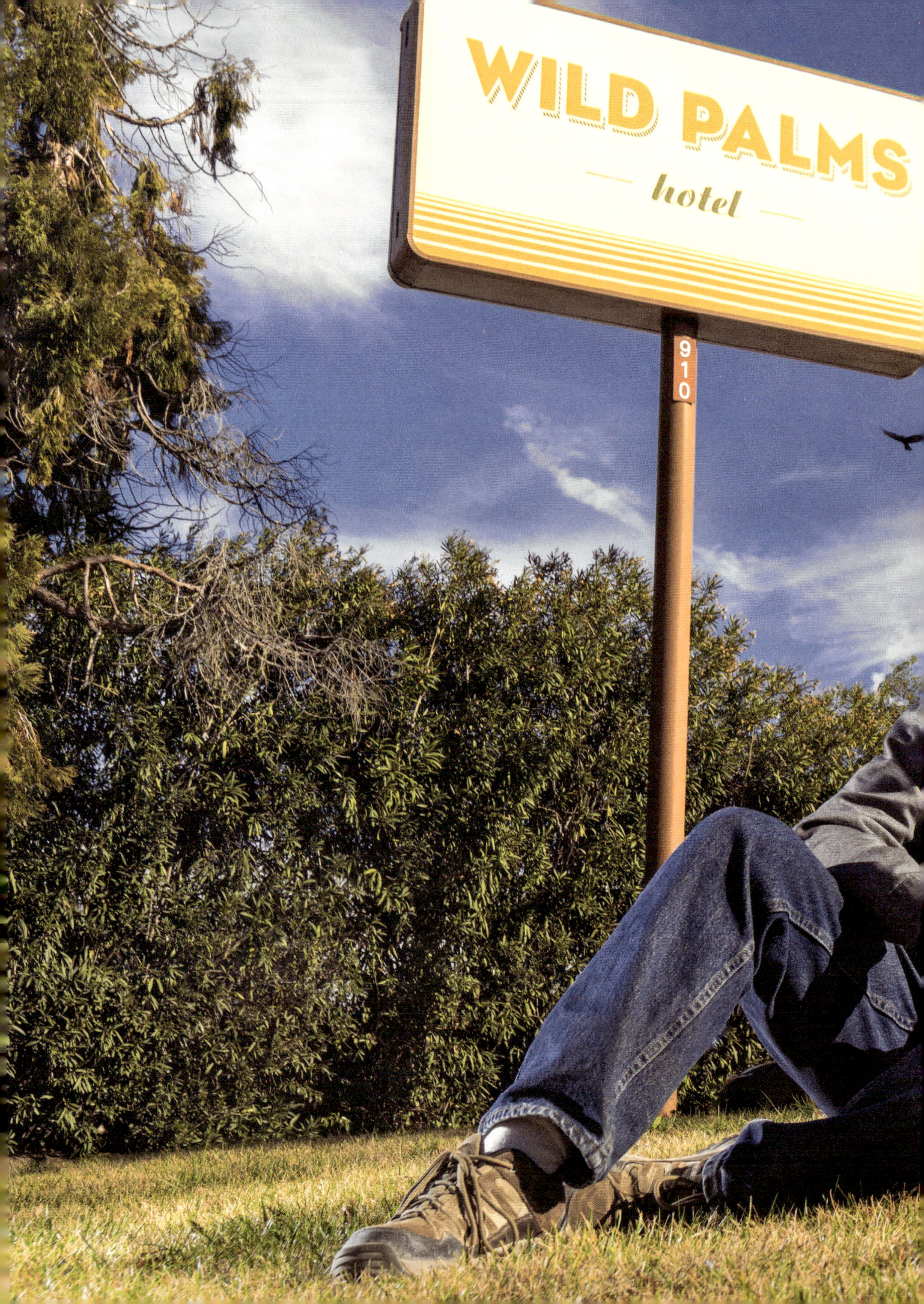
WILD PALMS
hotel
910

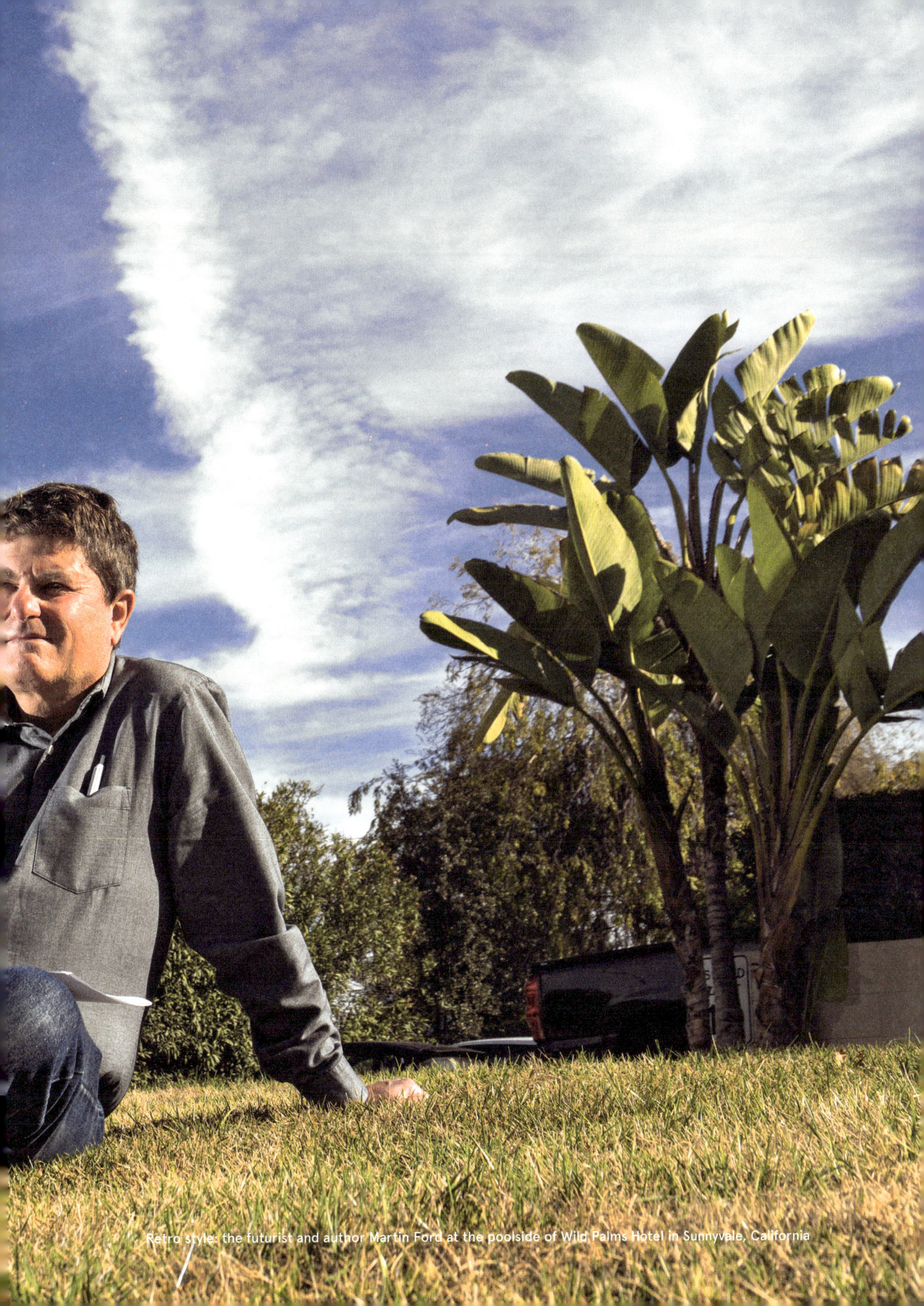

Retro style: the futurist and author Martin Ford at the poolside of Wild Palms Hotel in Sunnyvale, California

to get more money, go to school, study, and do something useful. Otherwise you create a society where people just play video games. In my most recent book, *Rule of the Robots*, I talked about this optimistic scenario of *Star Trek*, which is a utopian scenario. There is all this wealth, and technology has really solved our major problems. People don't have to spend all day doing boring jobs that are not rewarding. I think that's what we can dream of. Maybe we're a very long way from that, but that's the path we should put ourselves on. We have to at least be able to imagine a future where people derive an income from some sort of other job than a traditional one. But I contrast that with the pessimistic scenario, which is more like *The Matrix* movie—where the AI takes over and enslaves us. The film depicts a dystopian future in which humanity is unknowingly trapped inside a simulated reality that intelligent machines have created to distract humans while using their bodies as an energy source.

In The Matrix, *people are manipulated with drugs to augment their lives.*

We will have something like that—in a sense that already exists. Studies show that younger people who are not working and are not in school, especially young men, spend their time playing video games. But, you know, the video games are going to get much better. They will have virtual reality and artificial intelligence. They will become more and more addictive, maybe almost like a drug.

So the dream world of virtual reality is the answer?

You can imagine a future where people will feel they cannot get ahead in the real world, right? They cannot have a good life in the real world, no opportunity in the real world. But then there's this virtual world that is very attractive. So we could see a very big part of our population just checking out from reality and living in this virtual world where they play games or live in virtual environments, or maybe take drugs or whatever.

Is Mark Zuckerberg's metaverse the answer to that?

It could be very pessimistic. It's not like *Star Trek* at all, where the technology is working to make our lives better and more fulfilling, freeing us from drudgery so that we can do more meaningful things. Instead, the technology creates an environment where people choose to just leave the real world. That's what I worry about.

You know, I really believe that work is more than the opportunity to make your living. Work is also the center of the human construction of reality because your social life is still basically connected to your work. How will people have self-confidence or self-respect without work? How can they develop a personality without work?

Clearly human beings need something meaningful in their life. They need a sense of belonging; they need the sense of purpose. They need a sense that they are contributing something. And the thing that has always fulfilled that need has been work. But maybe in the future, there's a way to fulfill that need without traditional work, the way we think of it. Now, this idea of everyone having a job is relatively new. Before the Industrial Revolution people would sustain themselves on farms. So I think there's hope that we can have a new model in the future. That people will get the sense of meaning and fulfillment and accomplishment that they need, but maybe they will get it in different ways. Maybe by working in the community, not doing something that they earn a salary necessarily, but still, something that contributes or by pursuing education. You know, the answer might be different for different people, depending on what your orientation is.

So why should a young man in Silicon Valley go to university if there are no jobs?

I certainly hear that many people are very oriented toward technology. If you can be the person who studies and is a part of all of this creating this technology, then that's very exciting. It's just that that's not for everyone. Not everyone has the talent at mathematics, for example, that you need to work in artificial intelligence. But for those people who can, I think it's a very exciting opportunity.

Virtual universities on the internet make education accessible to people who live in slums in India or South America. Does it make sense to offer so much education if people will never find a job afterwards? The rat race is very unsatisfying.

I think education has value beyond just getting a job. It's not just about vocational education. It's about developing people and results in a more civil society and less crime. If people are more educated, even if they have more leisure time, they can utilize that time in a better way. The other thing is that there will be many entrepreneurial opportunities. People will start their own business, there'll be lots of opportunities in the future to do things to make money, and you'll be able to utilize these technologies to start a small business or trade or whatever. But in order to access those opportunities, you will have to be educated. You have to understand the technology enough to be able to utilize all these tools. So, I think I kind of imagine a future where maybe there's a basic income that provides at least a minimal level of income, but on top people earn more money beyond that by doing a small business, working, creating things. Of course, some people are very successful and build big companies just like today. But other people might work in a community doing useful things.

Asia and China have a very traditional approach. They believe in progress, and they embrace every improvement in technology. Young people there believe in a strong future, and they don't have these sentiments against new technologies. They want to build up a strong economy in China. I think their mentality is more like ours in the 1960s or 70s, when we believed in the future. They still believe in it. Isn't coping with artificial intelligence and new technology a cultural issue?

Yes, definitely. Like you say, in many Asian countries they're more optimistic. They're still at the rising phase on the curve. Of course, there are a lot of concerns about China's leadership in this technology, because it is an authoritarian country. They don't share our values for democracy.

Authoritarianism is not the reason why these technologies are successful; it is just that we don't approve of the way they use them. China is interested in improving the community and integrating more and more people in the middle class. They are very optimistic and don't have our doubts.

Yes, in the West, generally, we are in many different areas—not just worrying about technology, but we can't even discuss history anymore. We don't value our own history, we don't even value our own culture. We seem to have lost our self-confidence and our faith in ourselves in many ways, so it's quite scary.

So we will never win the race against China?

It's going to be a real challenge. It will make it very difficult for us to compete with China because they have a totally different outlook and enormous faith in their system.

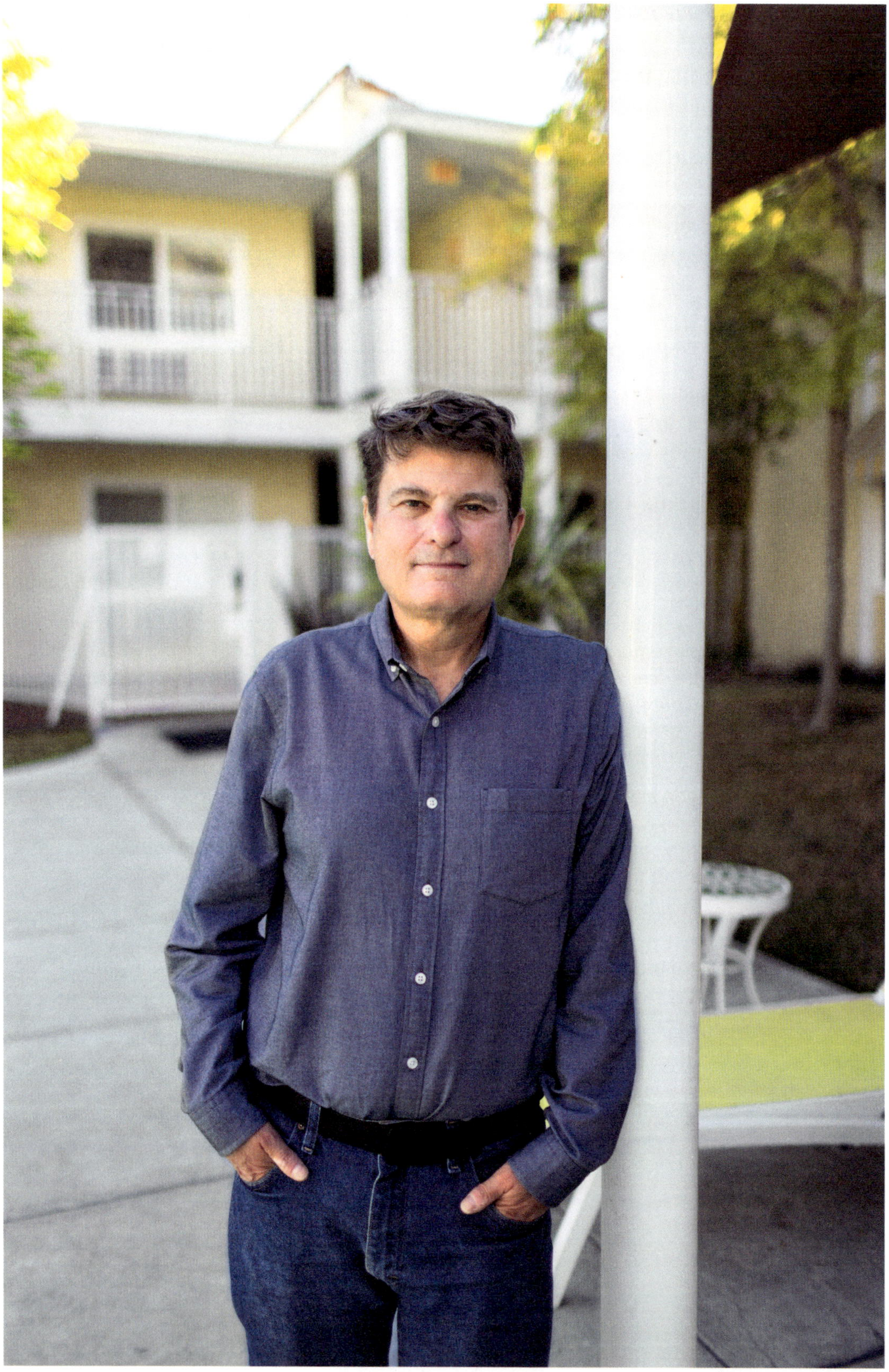

Do you give lectures at universities?

I was just in Boston. I really worry about the universities. We seem to be creating a future generation that's not going to be well equipped to run the country.

That's a very pessimistic end for our interview. Do you have any advice?

I am very optimistic about technology, but we have to choose between *The Matrix* or *Star Trek*. I believe AI is going to be an indispensable technology. If we want to solve climate change, we need innovation in many areas: not just clean energy, but new ways to make cement and steel, new approaches in agriculture to adapt to climate change, and to figure out how to live in a warmer world. AI is going to be critical for all of that.

Elon Musk's satellite constellations in low earth orbit will improve weather prediction, because how can we talk about climate change if we cannot predict the weather precisely? Climate is the most complex system on earth. We don't have enough data to figure this out.

Hopefully technology will help. I definitely support the technology. I think we should invest in it. We just need to be aware of the risks that come with it.

MARTIN BÖHM, HANS DIETL, ANDREAS GOPPELT,
BERNHARD GRAIMANN, AND GÜNGÖR KARA

Reflections on Artificial Intelligence

The five experts from Ottobock agree that artificial intelligence will play an increasingly strong role in the company's smart products and intelligent processes. AI and machine learning are employed in the analysis of customer data, and AI supports the mind control of prostheses. The increased deployment of artificial intelligence is opening up a large growth area with respect to extending the capabilities of the human body and capturing new markets.

Güngör Kara: We are already using initial aspects of artificial intelligence with the cranial helmets. These are individually tailored baby helmets known as *head orthoses* that correct the deformed heads of babies. Using facial recognition we create a unique face profile, the baby head is subsequently scanned, and then a digital model that resembles a shaped wire mesh is calculated by way of reciprocal comparison. The AI that we use for facial recognition is optimized using the additional scans, so that the algorithm is trained on the babies. That means that the AI gets better and better the more we scan. Babies are often restless and move during the scanning procedure. Through the facial recognition the scan finds anchor points in the face and can complement the digital wire mesh. We calculate a closed data mesh from the scan and use this for the design of the helmet, and from there we proceed to manufacturing. This corresponds to our vision for iFAB 4.0, by which we generate the meta-AI from the scan, the modeling, the printing, and the later adjustment. That becomes the baby's digital twin.

Andreas Goppelt: A further example is our hand prosthesis controller Myo Plus, where, in contrast with the past, it is not the person who has to learn from the prosthesis, but the reverse: the prosthesis learns from the person. This is currently the best available intuitive mind control. The user can explore the functions of the prosthesis control in a completely natural fashion while intuitively using remaining capabilities. The phantom is generally still present in the brain, and amputees can still imagine opening and closing their hand. We are not just employing AI for the sake of AI. The great strength of AI is in machine learning and pattern recognition. Eight electrodes are attached to the skin for the pattern recognition and integrated into the socket. When the remaining nerves and residual musculature in the residuum are activated, these muscle signals are transformed into myoelectric signals and directly translated into a movement of the hand prosthesis. If the user imagines making a fist, a network diagram indicates which electrodes are currently activated. The resulting pattern can be reproduced and can be machine learned. The AI assigns them to the corresponding movement sequences. As a result the prosthesis controller becomes more intelligent every day and can recalibrate itself. Every user can open or close a prosthetic hand after just a few minutes of practice. We have practiced this intuitive control with school classes, which functioned very well. We are now developing this for different, more complicated cases, step by step, for example, with upper arm amputations, and we will also employ this in the lower extremities for the control of active knee joints. However, here we have to take into account an important safety aspect. These are self-learning systems,

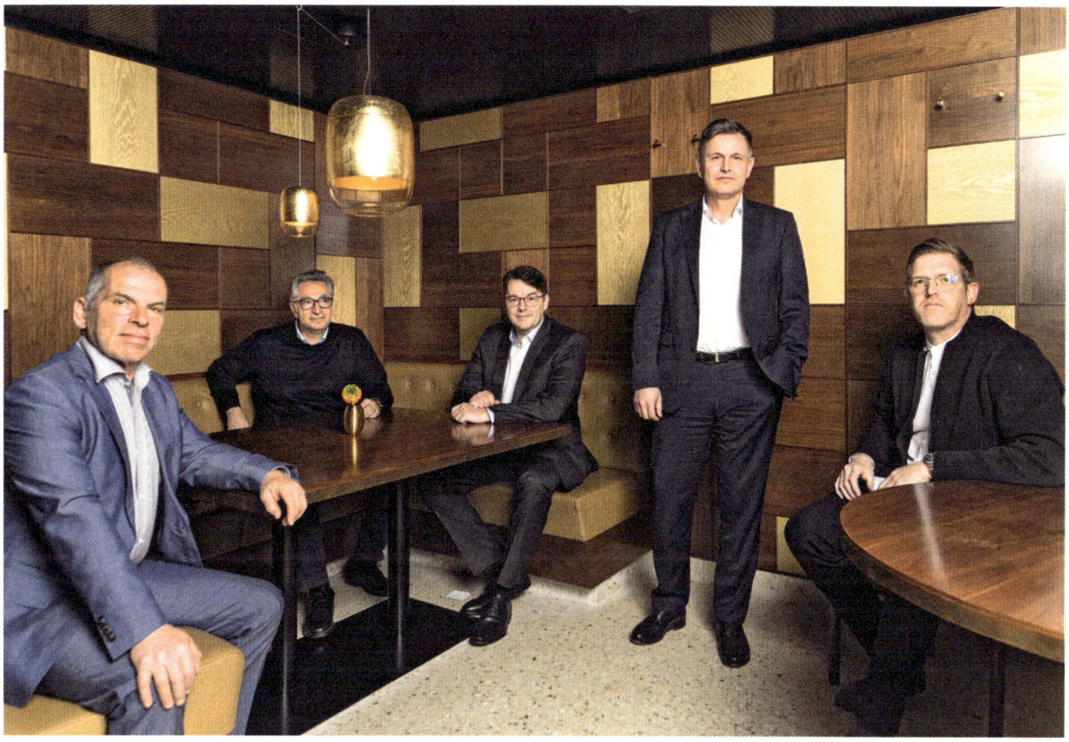

Hans Dietl, Güngör Kara, Bernhard Graimann, Andreas Goppelt, and Martin Böhm (from left to right)

which are a priori nondeterministic. An input A does not always result in the same output B. In the upper extremities this may not be so important as there is not much that can happen to the patient/user. But in the lower extremities users could lose their balance and fall if the prosthesis isn't functioning properly. The safety profile is naturally a lot higher here, and that is why we have begun with the intuitive prosthesis control in the upper extremities. But we will soon be ready to apply this to the lower extremities. And the advantage is clear to see: AI support generates significant user benefits. The patient no longer has to actively think about how he triggers a certain movement of the prosthesis via the contraction of a muscle and switches with the co-contraction of another muscle. That requires far too much attention. Through an intuitive control the brain can occupy itself with the pleasant things in life, instead of having to continually think about the prosthesis. A further advantage is that our mind-control technology has already been tested with many users and is already authorized for the health market. Here other founders such as Elon Musk are just starting out.

Martin Böhm: We also use AI and machine learning with our user data. We have data from over one million users with their individual user journeys. We use the standard items and algorithms of the Salesforce Einstein AI and are currently testing the Google Cloud Healthcare Data Engine. We anonymize the user data, combine them with the petabytes of the health-care service, and then analyze the anonymized data with Google BigQuery and then network the insights gained with our own data.

Bernhard Graimann: Much of what is now considered AI is not actually AI, but just completely classical data analysis using classical algorithms. But essentially, AI today and in the future can contribute to an augmentation of our skillsets. As engineers we develop solutions. And in the future AI will provide greater support in finding better solutions. In the future, with the aid of artificial intelligence, CAD systems will be a lot more intelligent; in other words, they will not just provide support as drawing programs with simple modeling simulations but will also be able to propose complete design solutions. The designers and engineers will benefit from this. But even here AI is just an intelligent support; it will not replace creative people.

GK: The compiling of an expert database is another good example of the diversity of artificial intelligence applications. It is not easy to find good orthopedic technicians. There are only a few experts who can build a really good socket. The externalization and transmission of this knowledge is not easy. That requires many years of expertise that one has to acquire oneself. Artificial intelligence can play out this expert knowledge at any location at the right time and to the right person. This opens up enormous growth potential for Ottobock, especially in the emerging economies, as the lack of qualified personnel in these countries is a huge barrier to market entry.

Human-Machine Interface

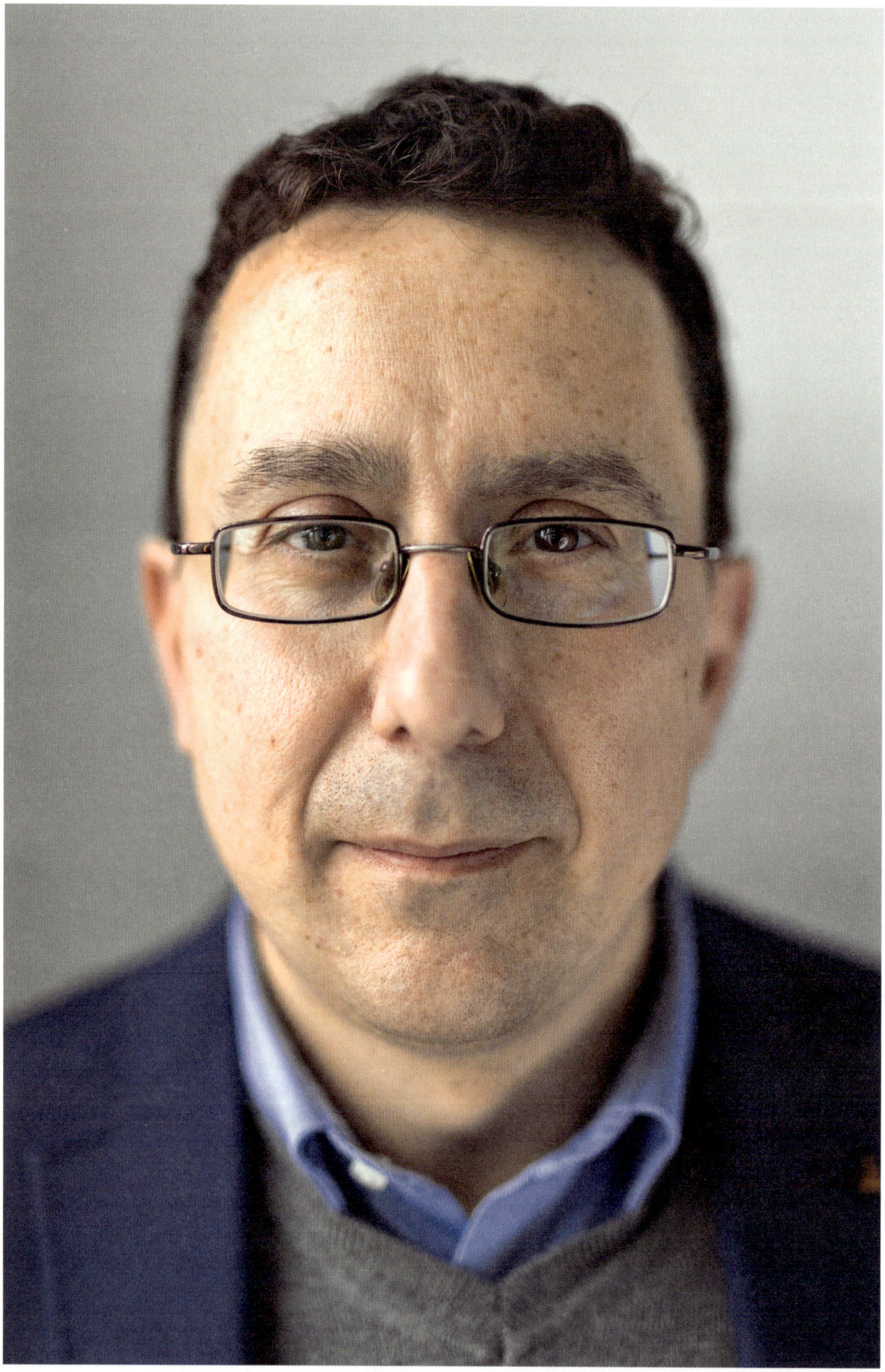

DARIO FARINA

Who Let the Bots Out?

While industrial robots are completely separated from humans by cages, wearable robots are used directly on the human body to enhance its abilities. To physically join them, the nervous systems of robots and humans must be fully connected. Designing this human-machine interface is the same as in prosthetics, where a robot is also connected to the brain. Since all muscles are connected to the spinal cord, every muscle basically functions as a small amplifier of the signal sent by the spinal cord. The professor and chair in neurorehabilitation engineering at Imperial College London and his team place noninvasive sensors on the skin that can measure a spike in the activity of the neurons in the spinal cord. The measurement data is evaluated using mathematical techniques that are based on artificial intelligence to hack the neural system and control it, with the goal of cracking the neural code of all human movements, including predicting what humans are thinking about doing and which kind of forces they are exerting. It's a very strong vision because it's the only example of neuron cells that can be assessed with wearables such as smartwatches.

You recently published a paper with Oscar Aszmann. What is it about?

Nerve transfers are currently made in many different parts of the body, but we transferred all the nerves to a single spot that we call the *bio hub*. That's Oscar's surgical part. Then we collect the electric signals from that spot. Our job is to separate them into the neural formation of each nerve that goes there.

What kind of sensors do you use?

We use electrodes, either invasive or noninvasive, but the vision is to have something implanted.

Not quantum sensors?

No, normal electrodes. To make a man-machine interface you have to imagine all the nerves in the arm with an array of electrodes. Electrical activity from the brain goes to the spinal cord. And then it's a mix of everything that was controlling the arm. We want to separate all this and reconstruct the neural activity for the individual nerves. The idea is to combine surgery and what is called a *neural interface*.

The rise of artificial intelligence and robotics in all industries promises a new paradigm of growth and wealth and a better life. But for the success of artificial intelligence, the design of the man-machine interface is decisive since all studies show that factory and office workers are afraid of robots. If we improve the design of the man-machine interface, people might accept this new technology. Although robots will take many of the jobs, humans and robots could also work together, which is already happening in the automotive industry or at Amazon. It depends on the design of the man-machine interface: is the machine friendly? Am I still the master and it is the slave?

The range of man-machine interfaces is very broad, so perhaps we should restrict it to industrial robots such as what we call *autonomous robots*, which are detached and preprogrammed to do a task.

They are in a cage and basically surrounded by a yellow line, so they cannot attack me.

Exactly. Autonomous robots are completely separated from humans because they are dangerous in industrial environments. In man-machine interfacing, robots are connected to the human nervous system. They're no longer preprogrammed, but since they are connected to your brain, you can control them.

Do such robots already exist?

Yes, you may think of prosthetics, where there's a robot that is connected to the brain. It's the same for exoskeletons, which can also be used in industrial environments to help workers do a job. The nervous systems of robots and humans can be fully connected and they can also be physically connected. An assistive device, such as an exoskeleton, can even be worn—we call those *wearable robots*. There are new challenges involved in the physical and neural connection of robots to humans. There are safety challenges, because now instead of having a cage around the robot, we are wearing it. It is conceivable that augmentation robots in the future will have additional arms. The key point is how to interface with our nervous system, how to transition from something that we preprogrammed to something that reacts and knows what to do—which is the same challenge we have had for decades in prosthetics. The technology is relatively simple in prosthetics, but that same approach can be expanded in any kind of man-machine interface with a robotic system.

So companies like Ottobock have an advantage because they have spent so many years working on connecting a robot to the brain?

You're right. If you think of the problem of controlling a prosthetic hand, that is one of the most challenging human-machine interface tasks, because in terms of motion there is nothing more complex than the hand.

But the hand doesn't have to be here on my arm. It could be on Mars picking up precious rocks.

Virtual robots don't have bodies or hands. There is no specific need of robotics once you can interface with a human. If I can take all the neural information traveling in your body from the brain to the spinal cord, and if I know exactly how you would move, we don't need our bodies anymore. We can read that information and have avatars moving in a virtual world.

It makes no sense to send humans into space because they are a logistic nightmare with all their needs for nutrition, water, oxygen, and emotional support. We should send robots instead.

Exactly right. This is another situation in which a human-robotic or human-machine interface would be important because we can send robots that are completely autonomous or semiautonomous. But if we were able to have some control, for example, with teleoperation technology, a human operator could react when the robot sees a special rock. This could also be used in dangerous environments with radioactive waste. It would require a human operator who can see what they see and then manipulate. That brings us back to the prosthetic hands.

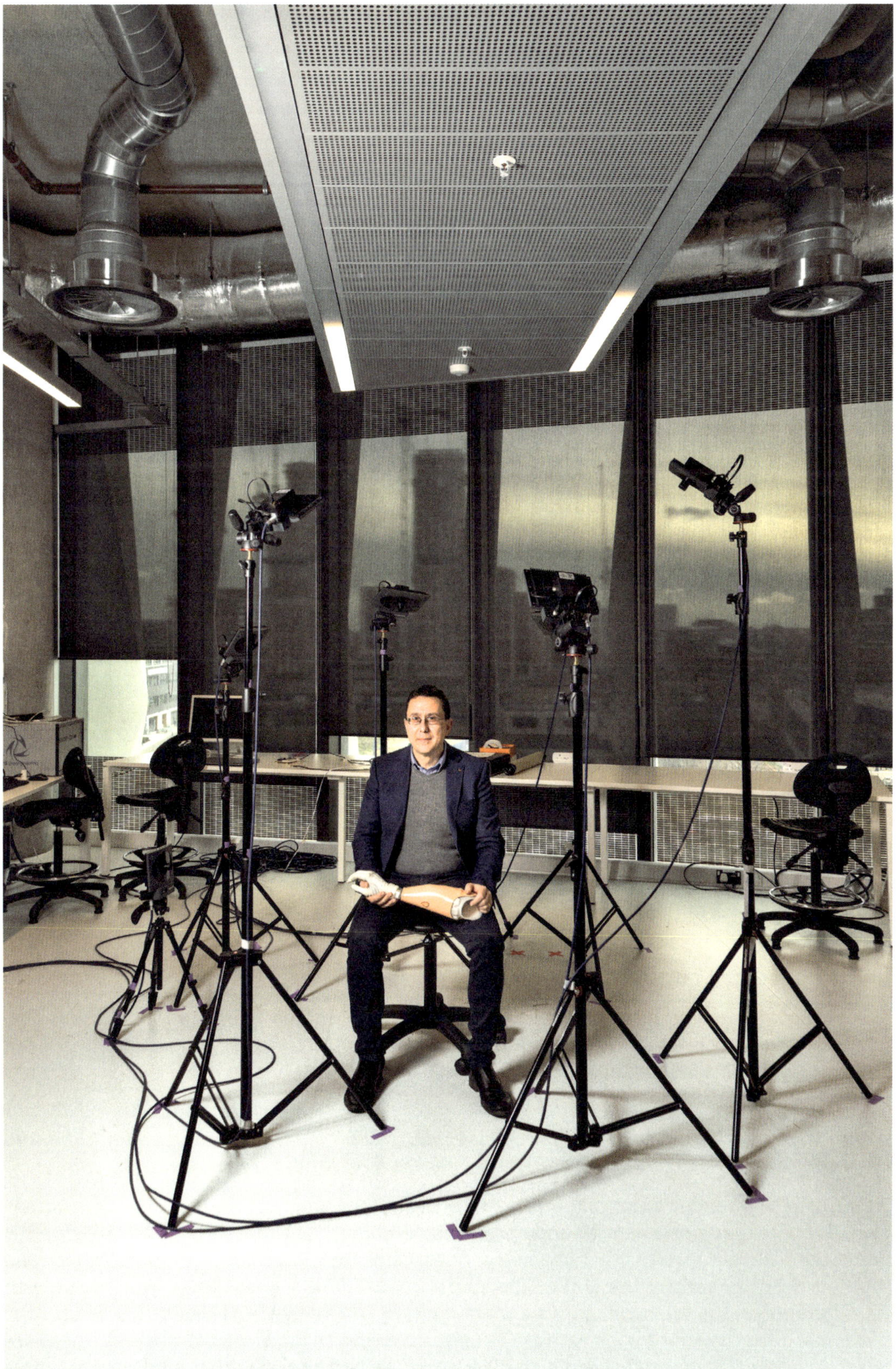

Instead of a joystick, the human will have a glove. This technology already exists.

Yes, but we can implant the electrodes that decode the data so that the robot can literally replicate the patient's movements. We can also send information in the other direction, so the patient can also have sensation. Because to make a movement, you need to have sensation. Patients can have feedback that tells them about the roughness of a surface, for example.

Sensory feedback is the holy grail of prosthetics. Hugh Herr is already on the brink of implementing it. He has some patients working with it, but it's slow.

It's much more difficult to encode that information.

Oscar told me it's nonsense to try to read the electronic signals in the cortex because it's just a multitude of information, and that the quality of information is only high enough when the brain signal reaches the muscle cell. That's what Hugh Herr says too.

The concept is to use as much as possible of what is still present. So if you have biological structures that are still present, it's better to use them instead of going to the brain. When Hugh Herr solders in muscles, using the natural stretch of the muscle, they already contain sensors that send information—and that information is perfectly natural. We don't need a lot of electronics to reproduce it.

I interviewed Matt Angle from Paradromics in Austin the other day, and he has been working on that for five or six years. For him, the human brain is just another computer. But it's not a computer, it's a biological machine.

It's a very distributed computer: instead of having a central processing system, the processing is distributed among billions of little neurons. Each neuron is very simple, but there are billions of them, and there are trillions of connections between them. It's like having a soccer field where there are strings connecting millions of little ping-pong balls. You would have to understand what is going on in every one of those connections.

Matt Angle wants to find one neuron with a needle. But there is no information there, because the neuron is only one hertz, and it's so slow. There's only information when it is connected to other neurons.

Exactly. We call that *population activity* in neuroscience. If you want to really understand neurons, you have to read them within a population that is large enough. It means that we would have to have millions of electrodes.

Where do you stand on this?

My point of view is that direct brain interfacing of the type that Elon Musk and other companies are working on is still very far away. At the moment we are able to read one thousand neurons. That's like looking at a drop of water in a lake and trying to determine if there are waves on the lakeshore.

Viola Priesemann, from the Max Planck Institute in Göttingen, is working on this. She told me that it's enough to have a few thousand nerves to establish a mathematical model.

That's true to a certain extent. There is correlation, meaning that there is shared information, but you need quite a lot to understand the community. The neurons that are directly connected to our muscles are in the spinal cord. Oscar may have told you that we can read the neurons that control the end of the spinal cord. There are less than two thousand of them to control the full hand, so it's a small number. We've been looking at the neurons in the spinal cord. There is a huge number of neurons in the cortical area at the end. There are fewer and fewer as you go down, which means that we can interpret that information better and understand what task each neuron is responsible for. This is very difficult cortically, but the neurons in the spinal cord and in the brain have some level of communality. Of course, you don't need to read every single neuron in the brain to understand, but you would have to read thousands of them. So, for example, in a patient with a spinal cord injury, you can implant just one hundred electrodes in the motor cortex and decode the hundred neurons or less. Even if you can teach that person to control a robotic arm, it doesn't mean that we have understood how the brain controls the robotic arm. It just means that our brain is very smart. And we can learn.

So you don't deny that Elon Musk might succeed?

No, I don't. I mean, there are already examples of human implants in the brain. I am a bit more skeptical about large-scale applications. I doubt that we will be able to do implants in the brain in the next twenty, fifty, or one hundred years. I believe that systems that read from the spinal cord can practically achieve something on a large scale faster than at the brain level. And even in the distant future I doubt that the technology will be so advanced for each of us to have a chip in the brain. Even if we have minimally invasive technology that is risk-free, it is difficult to imagine that large numbers of healthy individuals will have a brain implant.

What is your concept for the man-machine interface?

All of our muscles are connected to the spinal cord, so every muscle is basically a small amplifier of what the spinal cord is sending to them. We have mathematical techniques that are based on artificial intelligence. We place sensors on the skin, that is, in a noninvasive procedure. The sensors are arrays of very small electrodes. And we see a spike in the activity of the neurons in the spinal cord, exactly as if we had inserted these electrodes in the spinal cord. We do this mathematically. I think it is a very strong vision because it's the only case of neuron cells that we can assess with wearables such as a smartwatch. We are developing these things that treat our spinal cord. For healthy individuals, this implies that we can decode their motor intention as it is encoded in the spinal cord. Using wearable technology a person can write on an imaginary keyboard and we can decode what that person is writing. So this is similar to the approach in prosthetics. The difference is the idea of going back to the neuron source rather than remaining at the muscle source. This concept is also connected to the nerve transfers. We transfer nerves to go into muscles because muscles amplify the nerve activity. So we have biological amplifiers of nerve activity. We don't need a tool to read the nerves; we just put very small electrodes into muscles, and the muscle creates an electric field that we can read. Then we convert that electric field into the same activity coming from the nerve. The nerve transfer that has been done in prosthetics has the same concept: using our body as a screen.

It's all about augmentation.

It's also about augmentation. If you could see electric fields with your naked eye, you would see that my hands change color while I move them. These electric fields come from muscles but actually reflect the spinal cord activity. Because there is a spinal cord that determines this fist. So if we could see electric fields with the naked eye, our body would be a sort of a screen of the spinal cord activity.

So you intend to decode the human code?

Exactly. Spinal cord activity depends on what comes from the brain. Ultimately you can go backwards, not necessarily reading from the brain.

Can you hack this system?

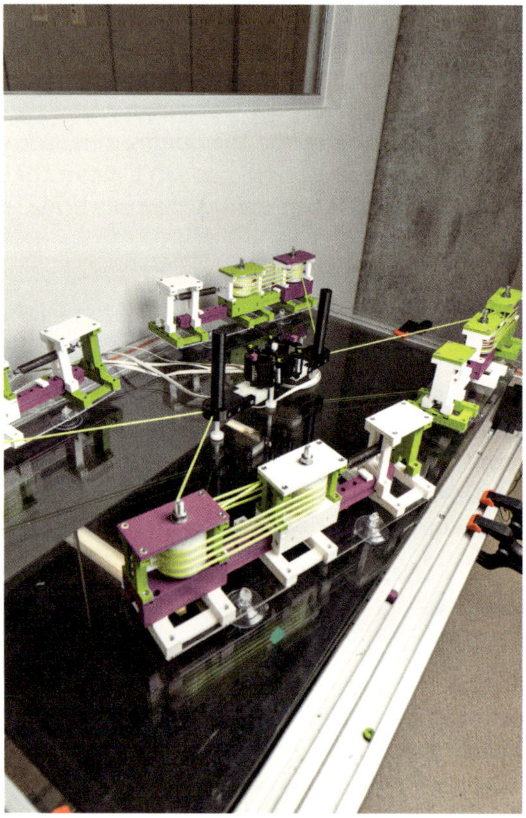

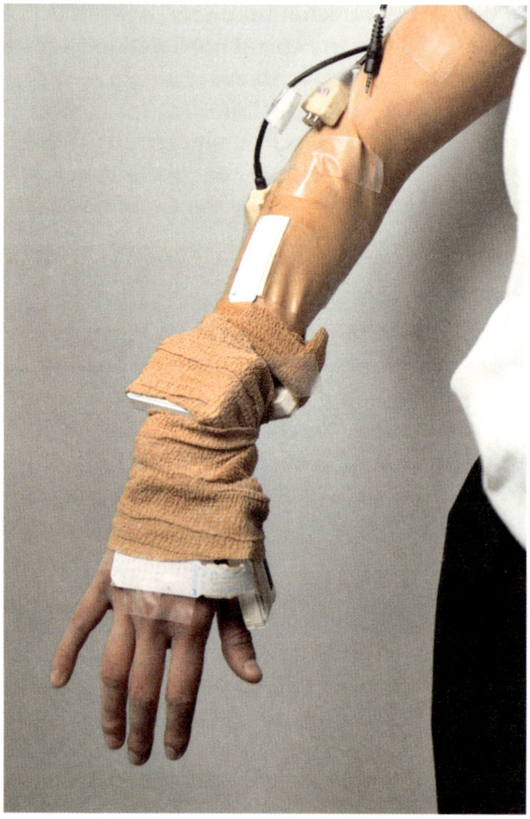

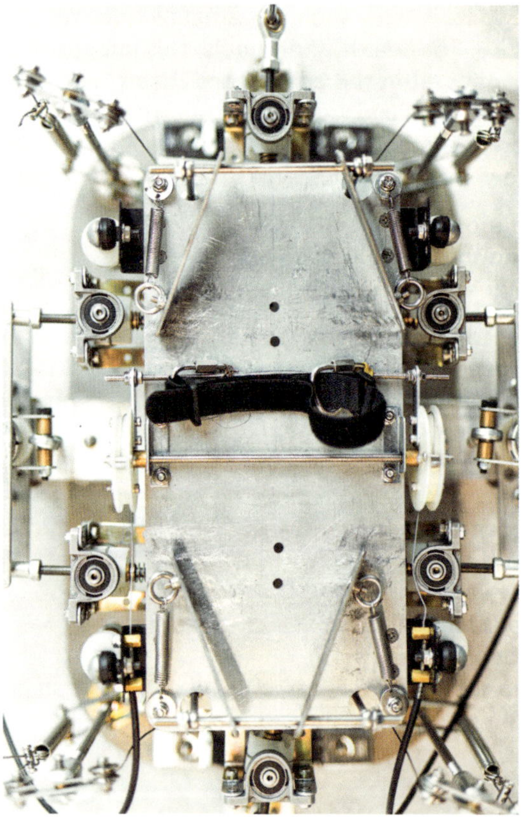

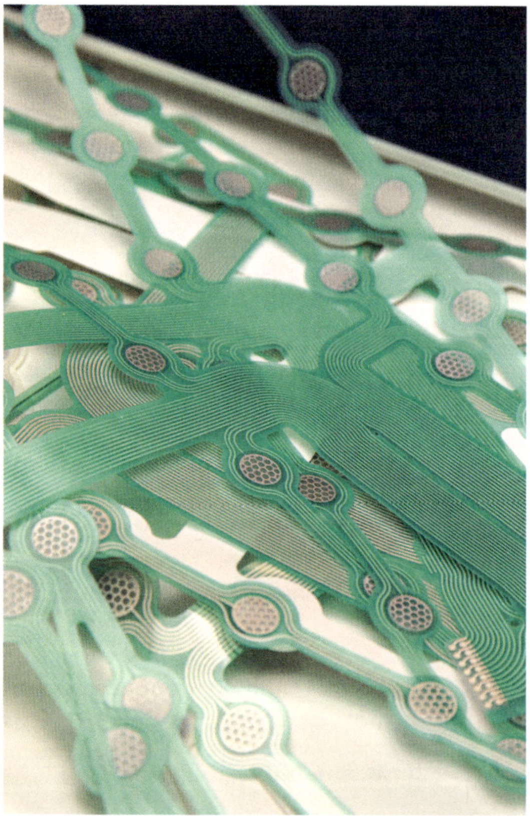

Yes. It's very interesting that you ask this question. Let's say that I read the movement of my hand, and I go back to the spinal cord. Then I look at all the spikes in the neurons. The question that we now ask our PhD students: we know which neuron is doing this and this, but is there any neural activity that we can control? Can we connect the neural activity to a sixth finger or a third arm? This is not a trivial question. We can decode this activity, we see all the neurons, and we see the movement of the hand. So we ask if we can explain the movement of the hand with the neuron. But then can we find any neural activity that the brain controls but is not doing anything in terms of the movement? If there is, can we enhance the body? Can we connect the neural activity to an additional arm, to an additional finger—or to a robot?

Is it just mathematics and coding, or do you do a lot of the work in your laboratory?

Our focus is mathematics, algorithms, artificial intelligence, and machine learning. We also do some electrode technologies, for example, electrodes that are simply inserted into the skin with needles. We also work on pure robotics, that is, designing robotic parts such as a robotic hand.

Are you working with Ottobock or any other companies?

We are collaborating with Ottobock on one project. We work quite a bit with Facebook now in terms of interfacing. There will be a paper coming out soon on a bracelet that decodes finger activity without any implanted sensor.

Are they marketing that now?

Well, they are doing a lot of research on this technology based on wearables to make a neural interface, and they have their own development teams.

Didn't they buy CTRL-labs, which developed a bracelet with electrodes?

CTRL-labs was a company we collaborated with earlier on before it was acquired by Facebook. The vision is to establish a neural interface with a wearable device. Now, wearables may be less fancy than implanting something in the brain, but if you can do something similar in a noninvasive way, it's certainly better. I think when we talk about disabilities, prosthetics, and orthotics, or about what we define as normal, augmentation is just a matter of the gray scale. At some point we'll be able to do prosthetics that are better than our own arms. They will augment humans, and that will do away with disabilities. It's just a continuum. It's just a matter of technology. Of course, it's easy to make a prosthetic hand that is stronger than our hands. The problem is maintaining all the characteristics of our hands, and then bringing selected characteristics above human performance. Then it's very different. But it's just a matter of the evolution of the technology. If you're able to do robotic hands attached to humans and make them better than our hands, then you open a bioethical dilemma. Would it be ethically appropriate to cut off everybody's hands and give them robotic hands just because they're just better?

That's the vision of the cyborg.

Exactly. The same applies for other aspects of augmentation. When you're working on medical problems, you have a clinical justification. When you have a patient with an amputation, you obviously do anything you can do. But there is the ethical issue when you are working on doing things better than before: enhancing memory, improving performance in car driving. Entire university departments focus on the ethics of biotechnology. There are also ethics related to privacy. If you have a brain implant, people can go into your implant and look at some of—

Mark Zuckerberg will do that, that's for sure!

I participated in an interesting neuroethics symposium on these topics. I think it is a legitimate concern for the neurotechnology field in general and not specific to a company. Of course, we are very far from that.

So what is it like, not only for prosthetics, but for other clients like Facebook? What is the goal for finding a way to measure the sensory feedback?

In terms of interfacing, the goal would be to decode the neural code of all our movements. It would entail having sensors in your body that can tell not only your movements, but also what you are thinking about doing and which kind of forces you're exerting. So if that is possible, it is also possible to make a complete replica of your body in a robotic body that replicates your movement and gives sensory feedback.

A digital twin.

Yes. That's a vision that is quite broad, and pieces of that vision can be implemented in various applications. In prosthetics, obviously, it will be immediately applicable. Those in exoskeletons will be applicable. And in other type of impairments. We are also looking, for example, at motor neural activity remaining in a spinal cord injury patient below the lesion. While these neurons are not sufficient for the patient to move, they enable us to understand what the patient wants to do. So we can read the neural activity of patients who are completely paralyzed. If that were possible, we could replicate the movement in a soft-texture glove—under the control of the patient.

And that will work?

It may. We have some proof of concept of this. Even if paralysis is complete, there is always some neural structure in the spinal cord below the injury under brain control. If you can read them, you can understand the patient's intention.

Can you translate them to the movement of the hand?

Yes, we have shown that you can translate them into movement in a virtual hand. The patient is paralyzed, and he's trying to move his index finger. The virtual hand does exactly what he thinks.

But not his real hand.

Once you have that information, you can move the virtual hand or the real hand by an exoskeleton, for example. You can also use this interface for the patient to write messages on the internet. So movement is not necessarily the holy grail to reach in some patient populations. You may have a tetraplegic patient who has an interface, like the one I described, and is able to control two virtual hands. He can write to his friends on a virtual keyboard, navigate the internet, and call for help if he needs a glass of water. You don't need to have a robot. The important thing is to be connected.

So we don't need physical robot hands when we can do it virtually?

You need them in some cases. The core of the man-machine interface is how much information you can pass on. If the patient's body is paralyzed, then all the information that moves in the body can no longer reach the external world because the body doesn't move. But if we can read the neural cells that tell us what the movements will be, we can transform those movements into something else. The important thing is that we have extracted that information. So man-machine interface is actually an information problem. The information may not be available because the patient cannot provide it, but I may want to read the information to augment your abilities. It's just a matter of information flow, and sensory feedback is also information. It's the same thing. Instead of reading I want to give you information.

But then you manipulate me.

Yes, of course. But I manipulate you even if I ask you to do an exercise. I mean, if I asked you to do a couple of squats here, I've changed your brain. Our nervous system is so flexible that if I teach you to play tennis, I'm substantially impacting your brain. In the same way, we can use this type of technology to exchange information to speed up the recovery of a stroke patient. This is called *plasticity of the system*. When we exercise, we learn a task. So our nervous system is constantly changing, it's plastic. And when we have a man-machine interface, obviously we impact on that. But our impact is not fundamentally different than our everyday interaction with the environment.

Are the muscle signals from the magnetic field, or are they from the electronic field?

We've done some work with magnetic fields, but it's a bit in the future in terms of real applications. Most of our work is in the electric field, which is a very old technique. The main thing is that we take these electric fields from the muscle and mathematically follow them back to the spinal cord. So we are transforming muscle interface, which is a classic interface like in prosthesis, into a neural interface using mathematics. The electric fields in the muscle are nothing else than the activity of neurons in the spinal cord that you can mathematically link in a relationship. So our core work has been to go from this level to the spinal cord level. We have techniques that enable us to literally see what comes out of the spinal cord.

Once you have the code, you can use it without the body to steer a hand or inform the brain about sensory feedback, which doesn't exist because it's only your algorithm. What is the goal of this?

Well, in terms of prosthetics, the goal would be to have better control. In current prosthetics, they take the global muscle activity and can basically say if you want to open or close the hand.

It's about six different movements of robot hand.

At the most. So, in principle, if you go back to the neurons, you can replicate individual finger movements. You basically have a mathematical microscope that tells you in a very precise way what the individual wants to do. That is what we are trying to do with Oscar. For example, combining nerve transfer and mathematical techniques. Once the nerves are transferred, you would have a main control that is still based on the global muscle activity. What we're trying to do together is to transfer the nerve and then follow the nerves back to the spinal cord. That's for prosthetics, but we can also do it on you.

What do you do with the data after you take it from my body?

If the data is from a healthy individual like you, we can enable you to control a robotic or virtual hand. If you have augmented-reality glasses, you can control the screen of your computer and write a message.

The future is all about augmentation. We will eventually live in augmented realities where you can choose another wall or floor color every day. You just need a table and a place to sit, and the rest—

Sure, but augmentation needs to be defined. It has to do with expanding our world by building a virtual world. Our work is concerned with what we call *motor augmentation*. For us, augmentation means "discovery." The basic scientific question was to discover if our brain can control more than our body in terms of movement. Now the question in this field is if we can read the brain or the message of the spinal cord and give the brain an additional limb. Would the brain be able to control that?

I did an interview with Professor Hiroshi Ishiguro in Japan two years ago, and he showed us a project in which a student with two arms was steering a third arm with his thoughts. It already exists.

Yes, but can you have full motion with your hand and control something else at the same time? It's different.

You could move me against my will?

Yes, I can make you move before you consciously know that you want to move. It sounds strange, but if I ask you to make a movement, it takes your brain almost one second to prepare the movement. With electrodes we can detect that you want to move before you actually move.

You can also tell me to move even when I don't want to move?

Yes, I can. It has been possible to electrically stimulate the brain for a long time, so it's not very difficult to move you. We put a magnetic field in your brain, and we can move you like a puppet. We can stimulate your muscles with electrodes to move you against your will. The movement can be elicited in the muscles by a standard electric field. You can do this in the brain or in the muscle. The aspect that is much more complex to understand is predictable movement, and to make you do something before your body can do it.

Or to prevent me from making a move.

Or to increase your reaction time. It is possible to predict what you would like to do because your movement is encoded.

And you can extend human powers with machines to give more power and force.

Yes. As I said before, we can do much more power. When we design a robot, we can put in a bigger motor to increase force. The difficulty is to maintain the other functions. I can make a hand that can squeeze anything, but it can only do that. It's very difficult to do a hand that squeezes everything and at the same time can play piano.

Or to give my real body ten times more power.

That can be done. For example, we can have a glove that has motors for each finger, and using electrodes we read what you want to do. When we know that you want to close the hand, the glove will give you more force than you normally have. We can read the signal that you want to do this and not this.

So this is not just theory anymore.

We are very close to achieving that. We just put electrodes on the muscles, as Ottobock has done for eighty years. But then we have a sort of a machine-learning microscope. Every muscle is made of a small group of fibers linked to one neuron, and there are hundreds of these neurons. We go back to these neurons to determine what they are doing.

Neurons always work in groups.

At the brain level it's a very big group, but at the spinal level it's not so big. Oscar counted the neurons in the spinal cord that are connected to the muscles of the hand. The hand has a lot of muscles, and there are less than two thousand final neurons in the spinal cord to control them. This is a small number from the engineering point of view, and we are close to being able to read each of them. By reading them, we will know exactly what the hand does. The motor path is a small amount of information in comparison with the sensory part: about ten percent versus ninety. So we focus on the motor path at the spinal cord level, not at the brain level. With Oscar's neurosurgery, we are able to read the neurons responsible

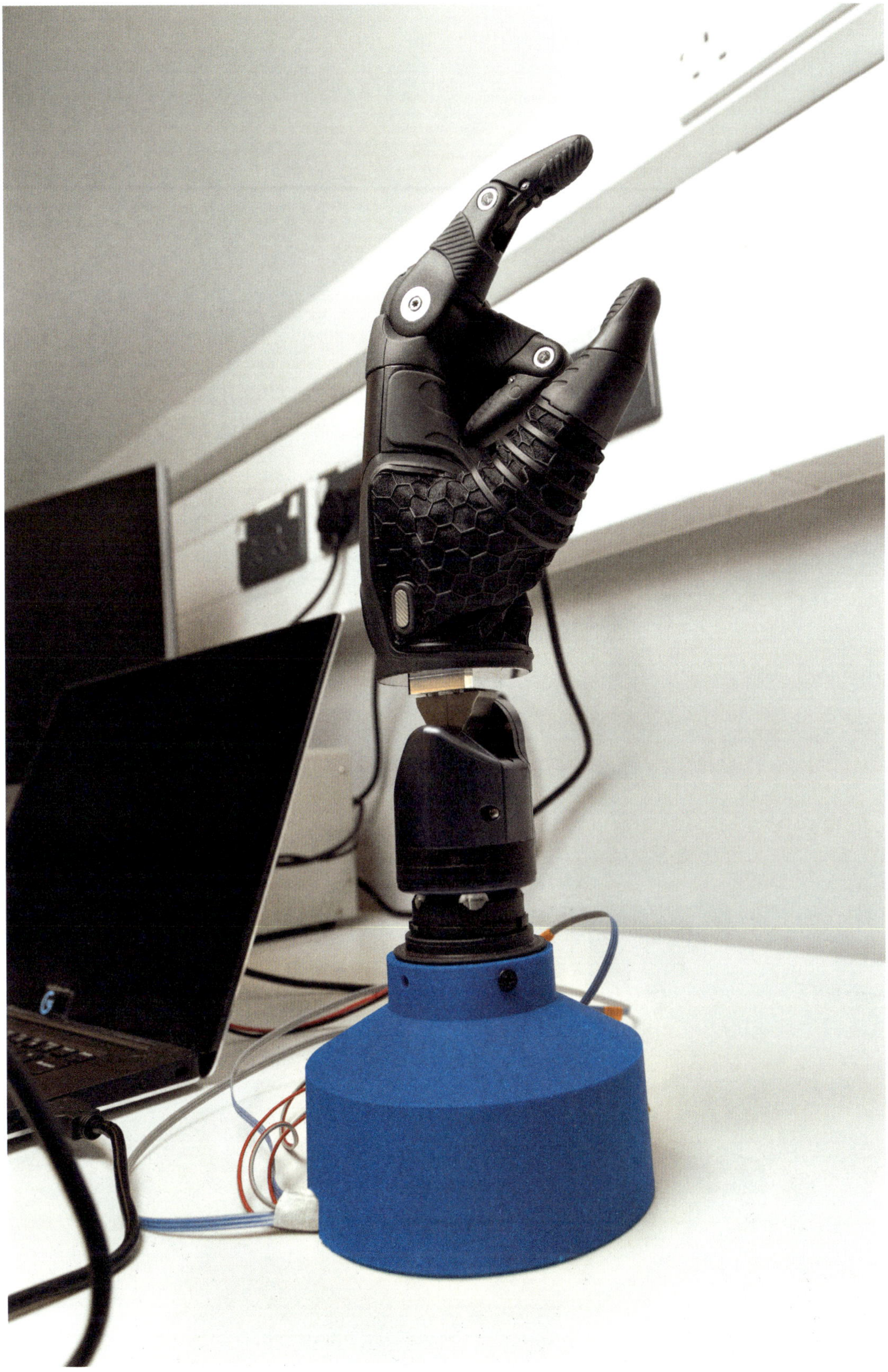

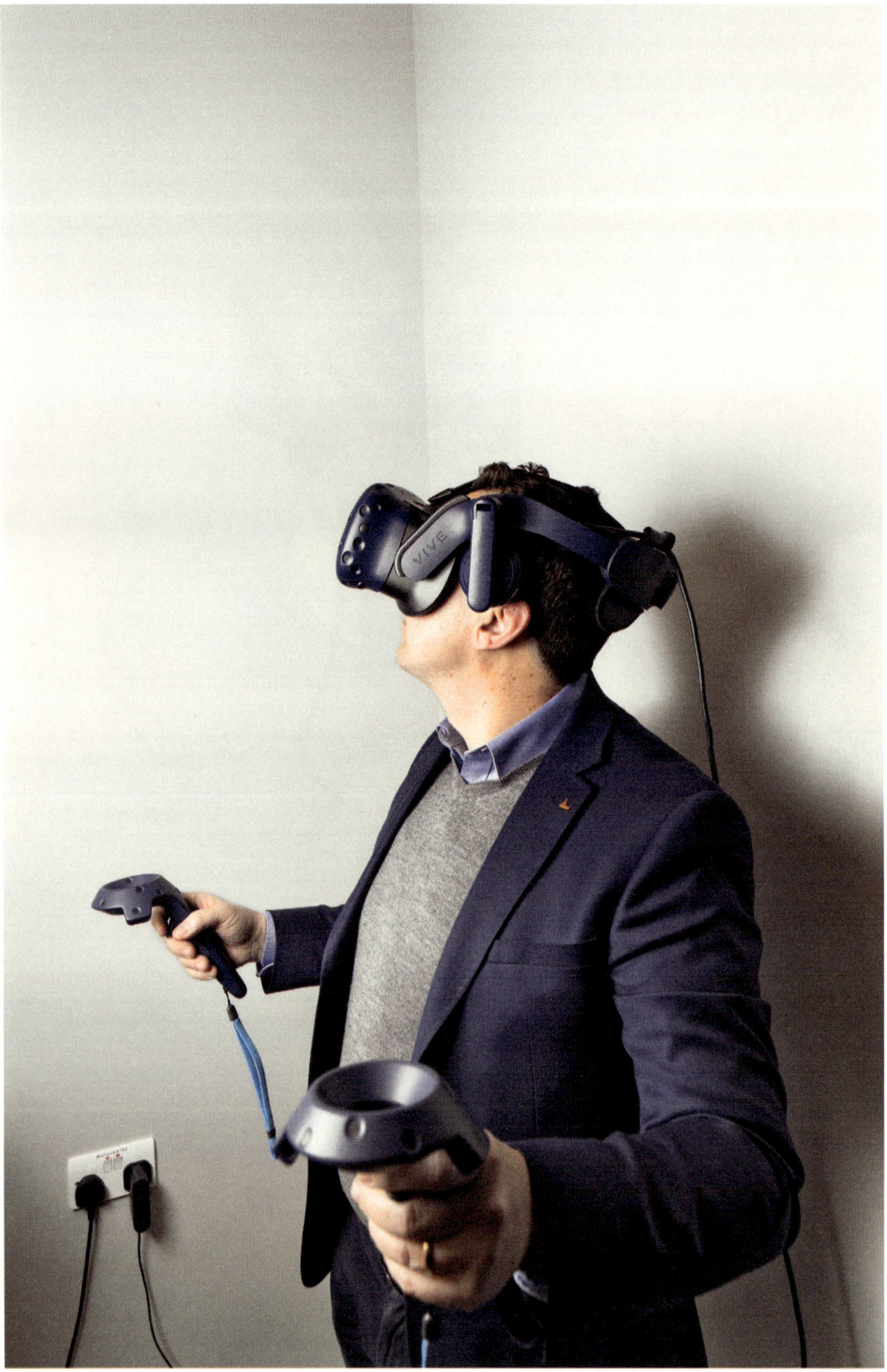
VIVE

for very delicate movement. I hope that we will soon be able to translate them. We have proof of concepts in the lab. But we still don't have it in a product. It will take a bit.

So you read the movement, put it in the computer, and the robotic hand replicates the same movement?

Yes. In the future, the movements will be even faster.

Can you change the movement?

You can map the movement differently. With prostheses we are always trying to replicate the natural hand, but that's not necessary. You could also have a different kind of hand. We could map the movements onto a crab hand or whatever. Once you have decoded the movement, you can transform it into something else. The movement is just what our anatomy is doing. But we can change the anatomy. We can make a prosthesis with more fingers or different finger anatomy.

Could you read the data of a really good musician and transform it to somebody who does not have these skills?

That's a good question. We can read the data of musicians and we know that they have a very particular way of controlling their muscles and their neural activity is very special. You know, there are some musician-specific prosthetics such as a drumstick that enables the user to have a very large number of drumbeats per second. Normally you have to limit it, because our body is limited. You can't have more than a certain number per second with your biological arms. From a robotic point of view, we can multiply them by ten. But that's an augmentation of the motor action that is not necessarily an augmentation of your neural capacity. From the neural capacity, you can just tell the system to do that, but you don't need to control it. So augmenting the body is mainly a robotic thing. But augmenting the neural, making your brain able to do more than what it does with your body, is another type. And it's a very difficult question that may have a negative answer. Because our brain has evolved for millions of years with our body, so why should it be able to control more? Maybe it can't. We're trying to answer the question.

That's the reason you went to the ethical conference because it is an ethical question.

Exactly. Understanding if your brain is able and if there will be consequences. Perhaps you can teach the brain to control a sixth finger, but maybe it's not a positive thing to do.

I think it would be okay, but other things are more difficult. Like can you really manipulate the movements of a remote person?

Yes, we had the example of teleoperating a duplicate of yourself. In that case, you would be responsible for actions that are not done by your own body.

Can you extend the capacity of the brain itself so that you think more or just directly connect to all the knowledge of the internet without an interface, like a computer?

That's really science fiction. The problem is that knowledge is nothing more than a specific pattern of connectivity in our brain.

It's a biochemical machine. It's not only ones and zeros, or a chip or a transistor. It's a very complex network of neurons.

To send out zeros and ones, a neuron gets input from tens of thousands of others. This input is really a sort of bombardment of neurotransmitters with biomedical substance—the biochemical substances around an error. And they change.

It is really an intelligent process to choose what information goes where.

That is what we are trying to replicate with artificial intelligence. It's a simple mathematical characterization of a neuron. In an artificial neuron, you also have many connections whose importance varies. When you change the importance of those connections, you change the behavior of the network.

You can calculate the probability of which signal will occur.

This is why artificial intelligent is becoming particularly relevant: we are now able to artificially create connectivity, which means it is merely something that we enter in the computer. Since we have much more computational power than in the past, we can have a huge connectivity. And the higher the connectivity, the better the systems behave like a real intelligent system.

But computers can only simulate and make a model of the one or two thousand neurons that we can already access.

Even the largest network that we can create artificially is still much smaller than the real architecture. And then there are some brain mechanisms that we don't understand fully. But we can expand the computation more and more.

Is your work more in biochemistry or statistics? How many people in your team are responsible for mathematics and how many for biochemistry?

We don't do much work on biochemical aspects. We work more on neural computation, meaning that we are more interested in the output of neurons, which consists of a string of zeros and ones. There are hundreds of them. Our group is very multidisciplinary. About a fifth of the people work in mechatronics, which is pure robotics; a fifth work in artificial intelligence, mathematics, and signal processing; and another part work on basic neuroscience, which is understanding the neural code. Others work from a clinical point of view. We also have people who go to the hospital and make measurements on patients, including amputees but also stroke patients and spinal cord injury patients. We work a lot with the tremor patients to try to understand how to suppress tremor. That is another interesting thing, because the classic neural technology approach to tremor and Parkinson disease is to implant an electrode in the brain, which is called *deep brain stimulation*. We are trying to find out if we can suppress tremor at the spinal cord level. Because tremor starts in the brain, goes down the spinal cord, and then from the spinal cord goes to the muscle. When the muscle reacts, the impulse goes back to the spinal cord. We analyzed the feedback loop and realized that at a certain point it becomes uncontrollable. So we are trying to break that loop with this interface. We read the neural cells in the spinal cord that are trembling. And we cut the resonance by having little stimulations that are out of phase with the tremor, which basically breaks the resonance. It reduces tremor with a very simple system. This is another application for neural interfaces. You need the human-machine interface to understand the neural activity of tremor, and you use it to control a stimulation in order to break it. We have a lot of different applications. Prosthetics is big but a relatively confined part of our activity. I have mentioned the spinal cord injury patient that we are working on, and we work a lot on stroke.

Does this project involve an exoskeleton?

Our main concern at the moment is to read whatever neural activity remained and to have stroke patients control a virtual environment that allows them to drive the neural activity in a specific way. We try to enhance plasticity by training stroke patients to control their neurons. We have a similar concept of interfacing. Again, we put the simple electrodes in the muscle. We look at the spinal cord, and then we connect the neurons to a video game.

What happens then?

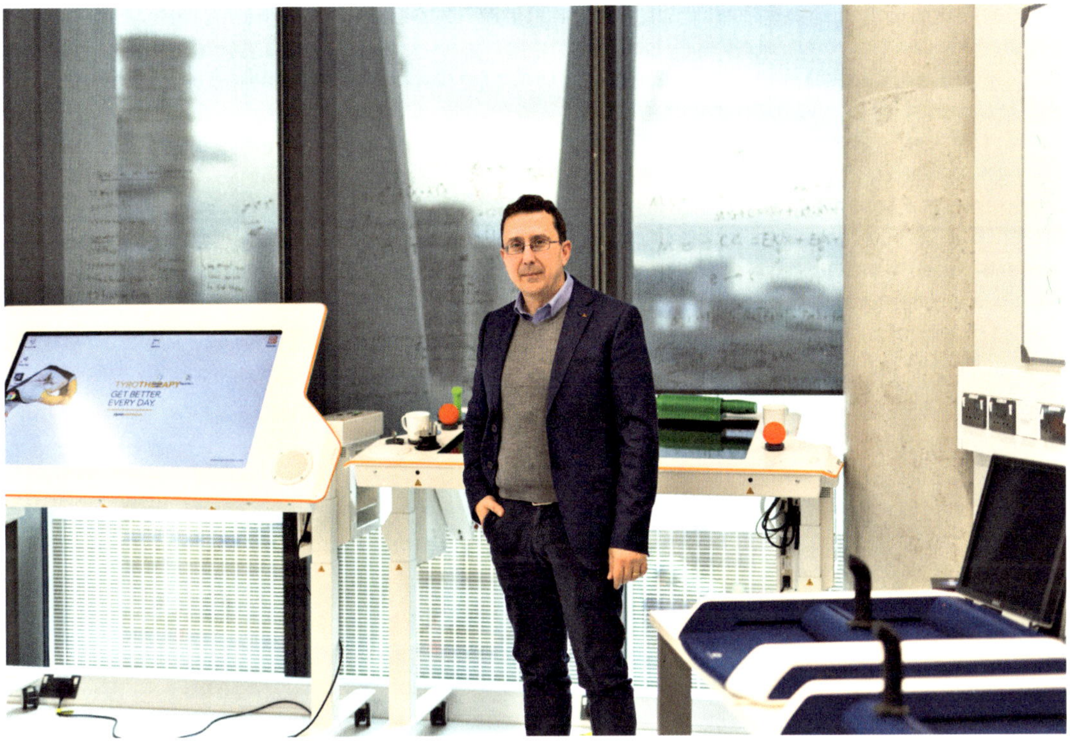

It's engaging because the patients learn to control the video game. Instead of a joystick, we used electrodes to read their spinal cord, and we tell them, for example, to move a space-ship and to avoid the asteroids. They have to control their neurons to move it up or down, which induces plasticity that can be very targeted, much more than if you do an exercise with these patients. Because you can target the individual neurons, and so they recover faster. We also read their intention to move. This was published by our group some years ago. We predict when stroke victims want to move. We give an electrical stimulation to the muscle that does the movement that they wanted to do.

Is it necessary to make the movement?

The reason we made the movement is not so they can move again, but because by moving there is sensory feedback. If you make the movement, the sensory feedback is natural. This was a concept that we fully translated a few years ago. We showed that the crucial thing was to give the sensory information with the same timing as in natural muscle activation. So if you are a bit late or a bit early, you don't have any plasticity. That's why you have to read the neuron information so well. We had a control group of patients that was having random sensory feedback. A group of patients was only having the physiotherapy. And then a group of patients was having this closed loop system. The rate of recovery was faster in this last group. So that's a very evident clinical application. And that's an interface application without implanting anything. We just put electrodes on the body—in that case, it was on the surface of the scalp, not in the cortex. So man-machine interface is really very broad. I think Ottobock has the possibility to expand a lot, especially in man-machine interface for motor impairments, but outside of prosthetics. If we would be able to translate even a simple man-machine interface for stroke rehabilitation, the numbers would be huge.

Fifteen times higher.

At least, and the number is growing because the population is getting older. So even a simple concept of man-machine interfacing for stroke would be a game changer. Other groups are working on it too, but the point is that it should be translated. It is one thing to write about it in a scientific journal, but it is a totally different thing to do it in a hospital as part of the routine clinical activity. Having some of the developments in universities is a very good way to accelerate projects. If a company develops something, it is a big risk. But if you invest a tenth of that in a university lab, and say, let's try to do this, the risk is much lower. If it doesn't work, it's fine. But it's a difficult discussion with companies, especially because they want to have a concrete plan for something that will return the investment within a certain period. On the other hand, I have to agree with many of the companies I talk to: universities are very slow. We do not work with a gun pointed to our head. It's a very different type of working.

What I am missing a little bit is that you don't think in terms of the industrial perspective, because all of this is coming now. It's already in the automotive industry and at Amazon, and we need a better understanding of robotics in the workforce. The machine that helps me load the dishwasher doesn't have to have a smiling child's face. It can be a gray box since it's only there to do a certain task. But in a lot of different situations you have to think about the interface.

This is a topic we didn't discuss: human-robot interaction, which is another big field. You brain reacts to having a robot around, and it reacts differently if the robot looks like a child, a gray box, or Bette Davis. The appearance of robots interferes with a task. So if the robot in your example loads the dishes, maybe at a certain point you will want him to unload the dishes. That's not easy for the robot's artificial intelligence because it has to understand what the situation is. So that's another type of interaction. It doesn't necessarily mean that our brain is connected to a robot.

The robot doesn't need artificial intelligence to load the dishwasher. It only has to follow my orders.

Yes, this is valid for the generation of robots, like the industrial robots we were discussing. But when robots are physically interacting with humans it's different. Let's say there are three robots in this room. It's different if they bring us coffee.

Then one should look like Bette Davis!

They should also avoid spilling the coffee on our laps. They should not be dangerous for us. That's social robotics, another big field, which is certainly coming. In industry we will work more and more closely with robots, but the robot needs to have artificial intelligence if there is a physical interaction. Otherwise, it will be dangerous.

Yes, but promise me that human beings will always be the master and the machine will always be the slave.

For the moment, yes. We are not even close to *The Terminator*.

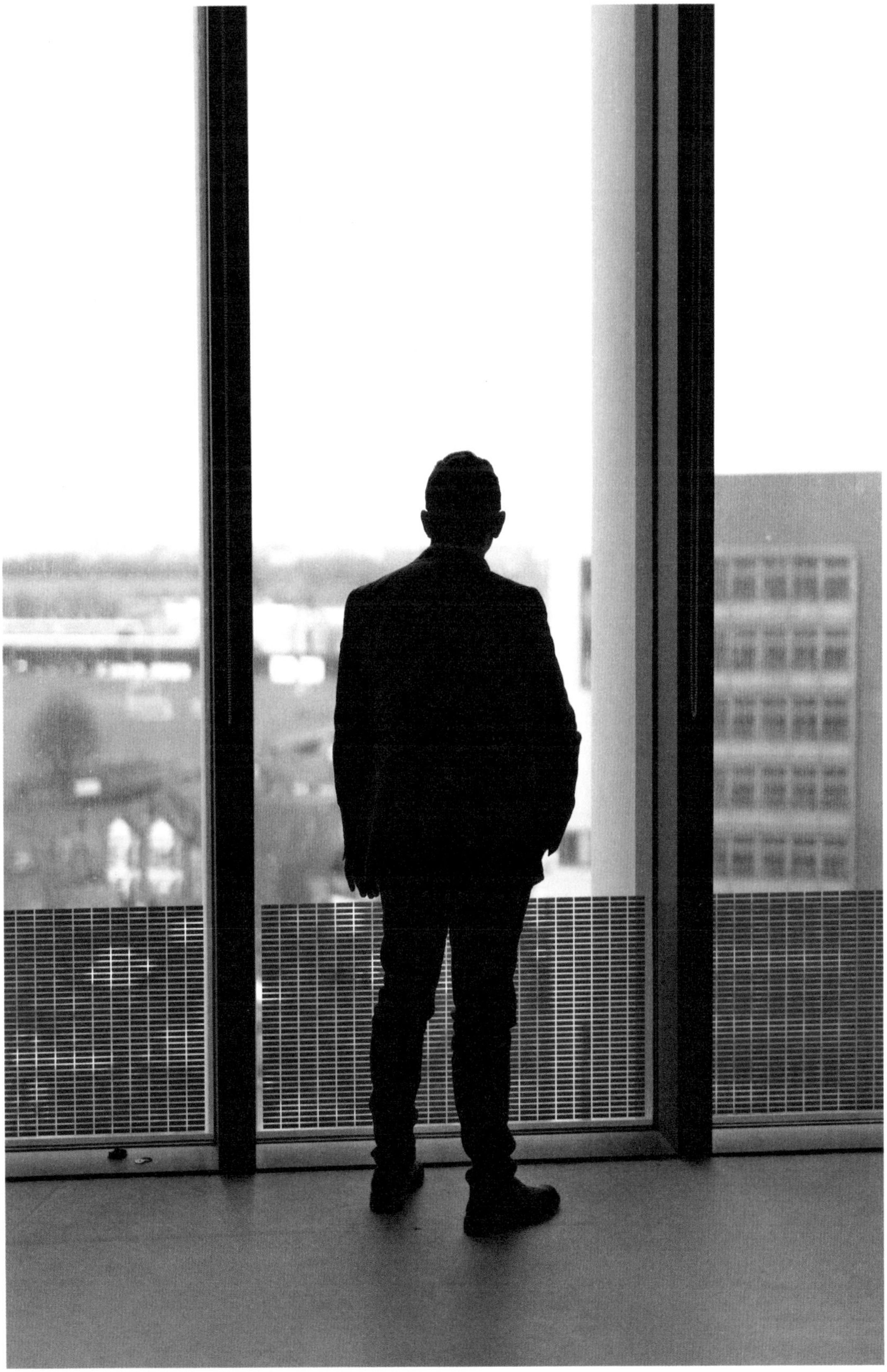

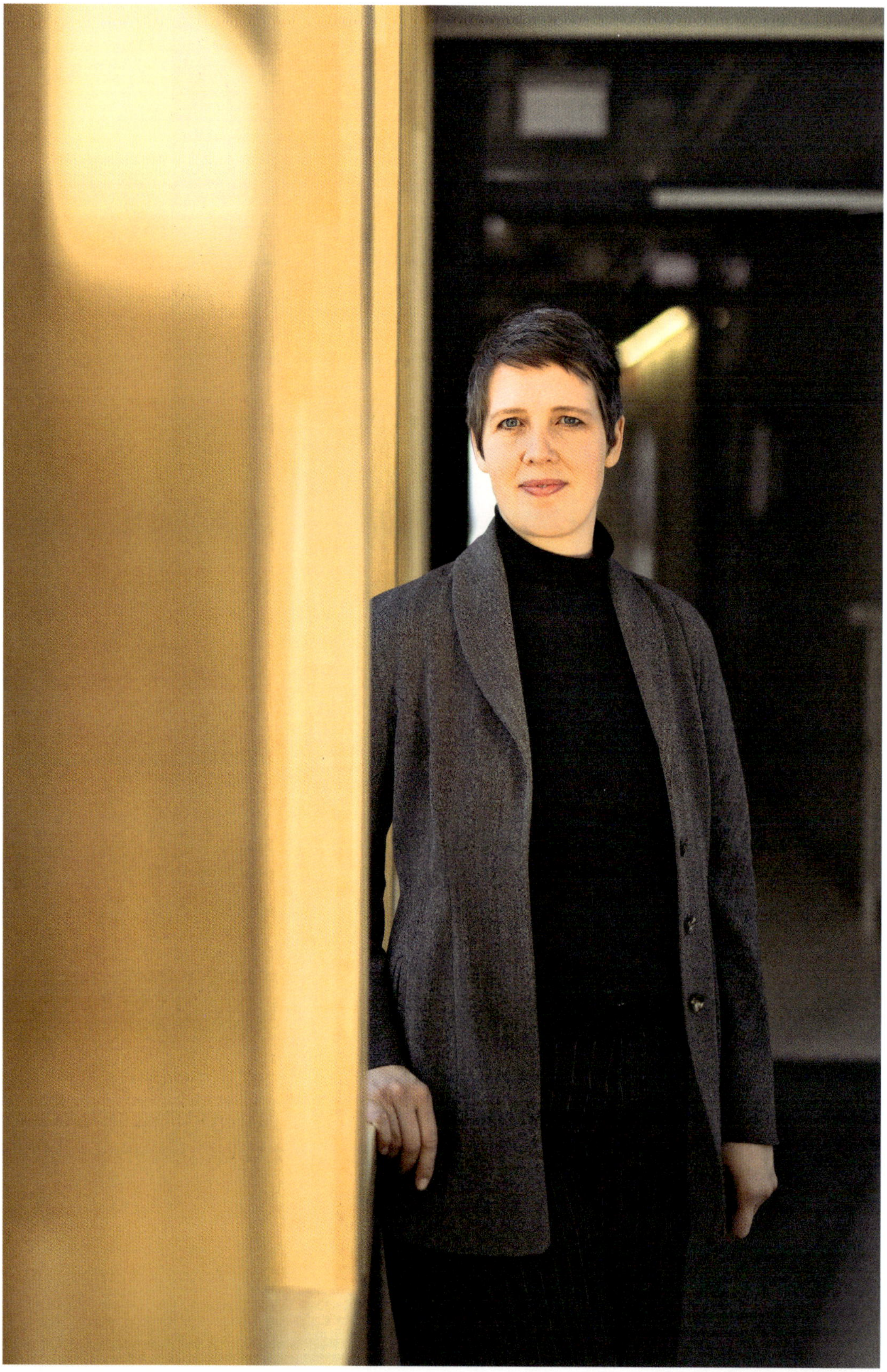

VIOLA PRIESEMANN

Human Consciousness Is Just a Question of Nerves

The human brain is a miracle machine composed of eighty billion self-organizing neurons that, through their mutual interaction as a network, have given birth to the processing of information, the control of the complex human body, the generation of thoughts, the experience of emotions, and the autocorrecting of human doubt. But is there actually such a thing as "free will," if everything is based on electrical signals and biochemical processes? Or is the whole perhaps more than just the sum of its parts?

Dr. Priesemann, as leader of the research group Neural Systems Theory at the Max Planck Institute for Dynamic and Self-Organization in Göttingen, you have become known to a broader public primarily for your mathematical modeling of the coronavirus spread. What does corona have to do with brain research?

There are a lot of similarities in terms of the mathematical description. In the brain we have a spread of activities in a neuronal network. The neuron is connected to many other neurons and can activate the other neurons and thus trigger chain reactions. With the spread of the infection in the COVID pandemic we are not looking at a neural but a social network. An infectious person has contact to a certain number of people and the virus spreads in this social network through chains of infections.

The mathematical modelers have been repeatedly criticized during the corona pandemic. You and several colleagues were personally pilloried by the newspaper BILD. *How often were your calculations wrong?*

There are two different research approaches in modeling. The first approach can be compared to forecasting the weather. That is a prediction based on case numbers. The second approach attempts, as in climate research, to understand the mechanisms. The determining factors are investigated by which we can make an infection wave smaller, larger, shorter, or longer—or even prevent a further wave. I have mainly conducted research into the fundamental mechanisms and principles of the pandemic. To this extent, it is not possible to say whether one was right or wrong, as these principles describe something fundamental. It wasn't our aim to make precise predictions for the next few days. We have developed scenarios in the area of medium-term predictions: if we start with a specific immunity and a certain R-value, and then were to remove all measures under these parameters, how would the hospital occupancy rate change in this scenario, or how long would it take to achieve herd immunity? Or, if we carry on as before, will the case numbers remain below hospital capacity? We have placed a special focus on the prevention paradox. However, it is only possible to forecast this effect to a certain degree: if the case numbers rise significantly, people automatically become more cautious. As a result, the effective R-value is reduced, and the whole course of the wave is weakened. Analysis has shown that both the voluntary restriction of social contacts and the willingness to vaccinate have increased with higher case numbers. This, in turn, influenced the course of the pandemic.

How can you calculate human behavior?

That is a nonlinear feedback loop, which makes it very interesting for physicists and mathematicians. We are just beginning to conduct research with sociologists and psychologists on how we can represent this human behavior in traditional SIR models. The SIR model assumes that an infected person (I) meets a susceptible—that is, a healthy but not yet resistant person (S)—and infects them at a specific rate. This leads to an increase in the number of infected people. That is a self-reinforcing mechanism—in other words, a chain reaction and a wave. And then at some point the immunity or resistance (R) increases sufficiently, which limits the case numbers and lowers them again. So in principle we have a clear understanding of how waves are generated and taper off again—when everyone's behavior remains unchanged. But human behavior naturally plays a central role. How do people perceive the pandemic? How do case numbers, intensive care bed occupancy, and news or fake news influence people's perception and thus their behavior? This feedback loop results in—and I say this very soberly as a physicist—a fantastically complex system.

A society is a social organism that functions more like the weather. There are an unbelievable number of variables that need to be considered. Intuitively I would actually expect such a complex system to elude such mathematical lawfulness. You have described the principles and now want to represent the different variables mathematically, in all their complexity. But aren't there limits to this?

In one respect the weather is actually a lot simpler than a social system: the molecules follow physical laws. And very importantly, a weather forecast will not influence the development of the weather. That is different in a pandemic: if we predict that case numbers will rise sharply, many people reduce their social contacts and thereby diminish the wave. The same principle acts also the other way around: when a relaxation is predicted, this also influences the wave. It is not possible to calculate the behavior of the society as a whole in all detail. In the models it is about simplification. But we know that this complex feedback loop exists. As physicists we now consider what actually makes up the central framework of this feedback loop. There are delays, retardations, there are tipping points—we have also found them with the contact tracing—there are exciting phenomena, depending on what this dynamic looks like. First of all, we pose the question of the properties of such a feedback loop's dynamic space. Then we consider the important parameters that can be adjusted and measured in society in a targeted fashion. There are what we call *stiff* and *sloppy* parameters. When the values of stiff parameters change, then this results in greater changes to a model's predictions. In contrast, sloppy parameters can be shifted on their own or in combination with other parameters without the model forecasts changing significantly. We calculate the effect of the stiff parameters in different scenarios. In that process we also examine how uncertain our prediction is. In the weather analogy, it is impossible to predict precisely that in two weeks there will be a certain number of centimeters of fresh snow on the mountain, but we can indicate an expected value with a variance, an uncertainty.

You have already mentioned the prevention paradox. This phenomenon occurred during the first lockdown. The science journalist Ranga Yogeshwar was the first to point out that solely due to the announcement of the lockdown people's mobility sank dramatically as a result of anticipatory compliance. People voluntarily adopted the expected changes in behavior in advance, before the measures even came into force. Some people drew the wrong conclusions and thought that the lockdown was not the reason for this decline, and thus had no effect at all. This is one of the reasons why trust in science and government has now declined dramatically.

The trust in politics and the media has sunk, but according to the science barometer trust in science has remained largely constant and, if anything, has increased somewhat.

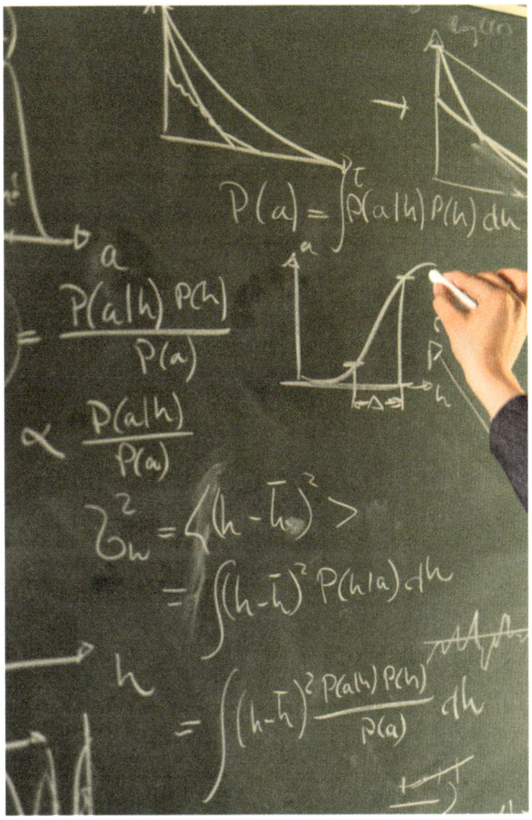

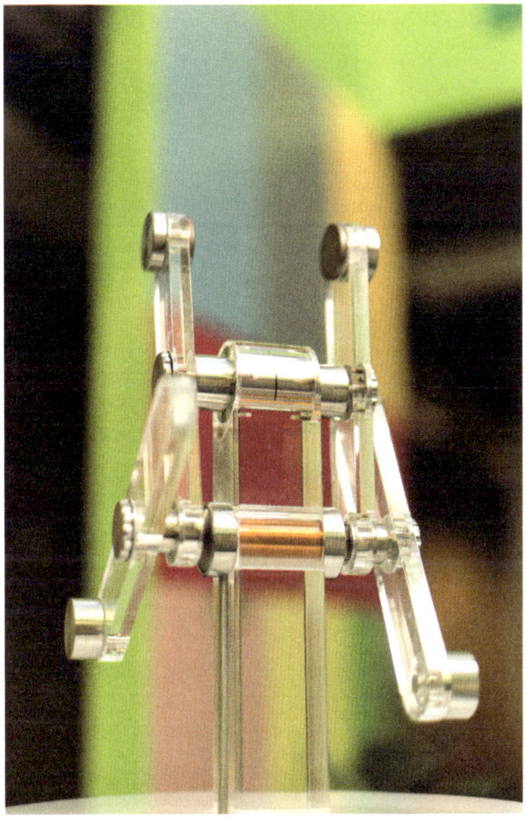

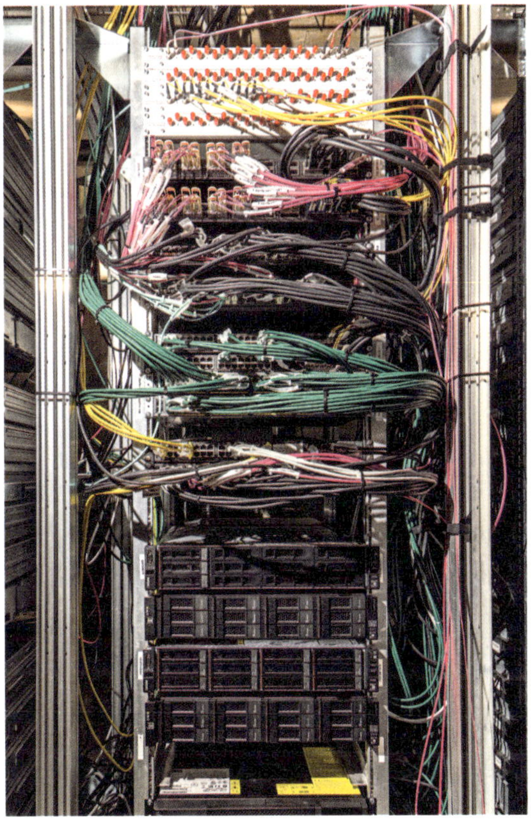

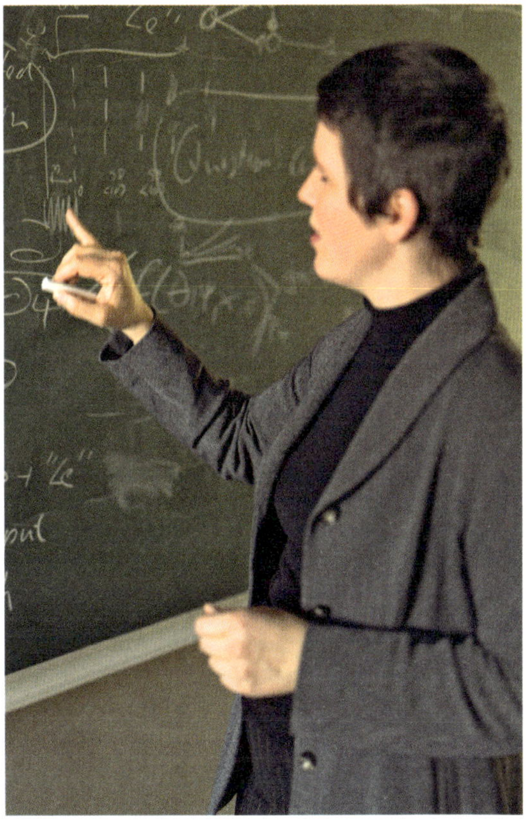

What variable is the "trust" in your equation?

That is a question that hopefully we will be able to investigate more closely in the future. As a physicist, I am very cautious about making statements outside my own discipline until we have done thorough research on it with sociologists.

But how can a complex system such as the global climate, a pandemic, or even the human brain be represented so clearly using mathematical models when there are so many uncertainties that cannot be taken into account? If you just look at a neuron—which operates at how many Hertz?

At several Hertz. That means, a few signals per second.

So a neuron conducts very slow computer operations at one Hertz and nevertheless achieves an enormous networking effect, which people are trying to copy in artificial intelligence using artificial neuronal networks. In your research you have focused on just a small group of neurons, and from there you attempt to generalize. At this snail's pace, how is it possible for neurons to conduct such incredibly fast and complex computing operations solely as a result of their networking?

I was at a workshop in London with the fantastic title: "How Dumb Agents Act Clever Together." Of course, the first question is: are the individual neurons really that stupid, or do they have a whole range of other capabilities? The second is the classic question that is the subject of research both at this institute and worldwide: how does emergence occur? How can individual molecules build a cell? How can networks of neurons ultimately formulate a consistent, coherent, and meaningful thought? Here we really are just beginning to develop a deeper understanding of neuronal networks, both in the supervised and unsupervised areas.

In which area?

There are two approaches to learning for neuronal networks: supervised and unsupervised learning. Supervised or monitored learning is classically employed in deep networks. One shows the network a large set of images—for example, of cats and dogs—and corrects it when it doesn't classify the picture correctly. The system learns as a result. Supervised learning trains patterns and connections on the basis of data with a target variable.

You have to show the dumb computer a million cat pictures before it can distinguish them from dogs. A small child only has to see three cats and can remember the difference forever.

That is the big question: is what you say really true? The small child has seen lots of other things such as dogs or other living creatures before seeing that cat and thus only has to learn a little more. In contrast, the network always begins with zero. But for this machine learning you first have to classify a million cat pictures as cat pictures and tag them accordingly in the metadata. Naturally Instagram and other social media are useful here, as they have already labeled billions of pictures with key words.

But I wouldn't call this artificial intelligence; this is simply the building of ontologies, which has been part and parcel of linguistics and computer science for a long time.

It is a classification that is now made automatically instead of manually. This learning in deep networks is currently a large field of research. But I am far more interested in the unsupervised learning approach, or autonomous learning. That means that instead of a supervisor for the human brain, the neurons independently adjust to one another. This autonomous learning would be very useful, because the brain is huge. It is probably the most complex organ that nature has produced. We have over eighty billion neurons, which is ten times the total global population. Every neuron is connected to thousands of other neurons. The brain is a significantly larger network than all the social networks combined. Here, autonomous learning would greatly facilitate the coordination between neurons.

The neurons connect to each another according to specific principles. When thoughts are generated, the neurons convert these employing electro-biomechanical processes.

Exactly. There are the biochemical connections between the neurons, and the strength of the connections changes through learning. We are researching how this connection strength changes autonomously. Autonomous means that the two connected neurons do not need to know what is happening in the rest of the brain to carry out their activity. One can neatly show that the entire network can stabilize itself when all the neurons follow a certain learning rule, that is, when they adjust their own synapses locally in a certain fashion.

It is a great misunderstanding to compare the brain with a computer and thus reduce it to electrical information.

We have a far greater variability. An essential difference is that the computer is very good at extracting the square root of pi. The human brain is not very good at doing this at all. However, in contrast, we can make far better decisions under uncertainty. We see just a section of a face somewhere and know that it is Uncle Herbert. And if I know that Uncle Herbert might appear in a certain context, the brain already expects him. That is part of predictive coding. Using internal models, the brain calculates what it could in all probability expect and compares this to what it is currently perceiving of the outside. If this corresponds to the internal model, the internal expectation, there is initially very little brain activity. However, if there is a surprise, the neurons are far more active.

The psychologist Ernst Pöppel has established that the brain is in fact very sluggish and only interested in information that it considers relevant. That is why the popular thesis of information overload is wrong. Because the human brain simply allows all the remaining useless information to whoosh by. He also says that the brain is constantly computing the augmentation of its environment and only consciously looks to see whether Viola Priesemann is still sitting at the table, or has already left, every three seconds.

There are a lot of good examples of this. If one of the bottles of water on our table were suddenly to disappear, we probably wouldn't notice. We assume that we are familiar with this room. We assume a continuity. We know that bottles don't simply disappear, which is why we wouldn't expect it. On the one side the brain generates an internal model, and on the other side it compares it with what it perceives from outside. When it registers a mistake or a discrepancy, that is signaled. What we see in the activity is in part just the deviation. But, in turn, to interpret the deviation, I have to know the internal model.

If only the discrepancies need to be calculated, the brain only needs a fraction of the computing power. That is also a lot more pleasant for the brain than looking to see whether the bottle is still there every thousandth of a second. Despite all its capabilities and complexity, the brain is actually a very sluggish, redundant system.

That is an elegant explanation, but it is not quite that clear: the model also has to be computed, and that also requires a certain effort. How efficiently the one or the other code functions in the brain is still an open question. But since Hermann Helmholtz there is a consensus that the majority of the information about the world is filtered out—in other words "thrown away"—and that the other part needs to be very efficiently encoded. Otherwise it is not possible to explain why—in spite of our retina's limited number of "pixels"—we see the world very clearly and not pixelated. I find it very exciting how the brain developed over millions of years in the face of this continual shortage of resources. It performs its tasks with twenty watts. Thus the brain consumes less energy than a conventional light bulb.

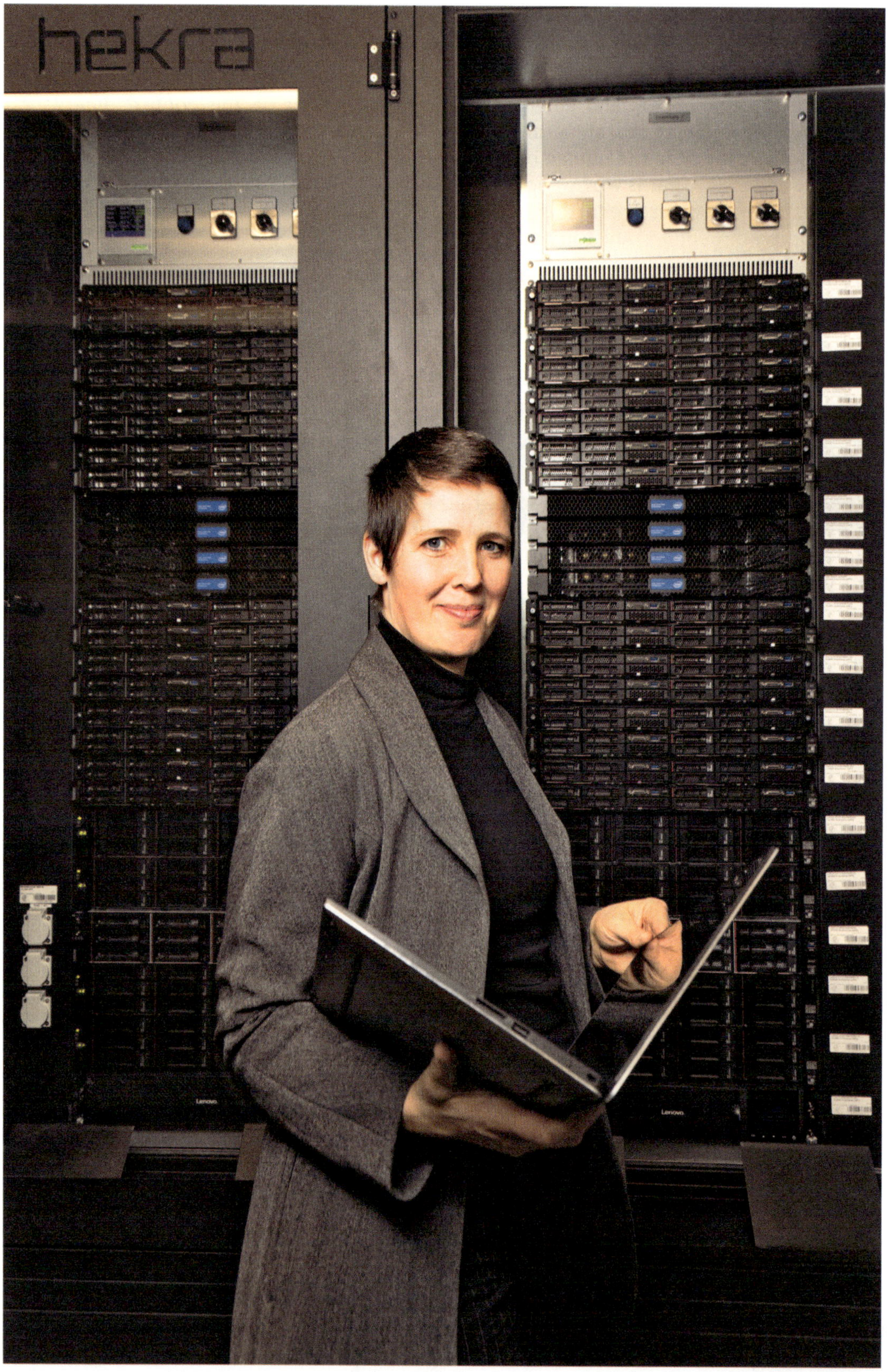
hekra

The real mystery is that our entire network of neurons has produced this miracle of human augmentation. That the brain has an idea of itself as person and can represent an idea of the world.

Yes, this self-organization and emergence is very fascinating. In this context emergence means the formation of new structures within a system as a result of the local interplay of its elements. As a whole the human brain is more than the sum of its parts due to the addition of networking. That is why we look at the activity of the neurons. Neurons have been found that can recognize recurring patterns or categories. They are called classification, grandmother, or Jennifer Aniston neurons—

Did you say "Jennifer Aniston neurons"?

Yes, there are actually certain neurons that only fire at the sight of a picture of Jennifer Aniston, and others only with a photo of Halle Berry. Let us imagine a scene with cows and birds. The brain will not pass on every individual pixel, but instead autonomously generates neurons that recognize a cow or a bird. The brain could be equipped with a type of autonomous encoder. The neuronal network can autonomously recognize patterns, which recur repeatedly, and distribute the representation of the pattern up among various neurons. To connect the one pattern with the word *cow* and the other pattern with the word *bird*, we need language and a supervisor to learn the concept word. But the real miracle is that the neurons succeed in filtering out these different patterns, in the complete absence of a supervisor. And they also distribute the patterns among themselves so that instead of all neurons coding for one object, like a cow, they specialize on different objects. This mechanism also enables objects that occur frequently or are important to be represented by more neurons. Accordingly, objects that occur less frequently are represented by fewer neurons, and in the best-case scenario the whole background noise is not processed at all.

What does autonomous learning mean in the case of our cow?

In contrast to supervised learning I don't need any label at all with autonomous or unsupervised learning. My goal is to deduce how the local learning rules are precisely configured between the neurons, so that the network develops optimally. We are still a long way off from this, but we have a number of helpful building blocks. If two neurons are connected, this connection can become stronger or weaker. According to the Hebbian Learning Rule, whether the connection becomes stronger or weaker solely depends on the activity of these connected neurons. It doesn't need any further information apart from that. If one neuron successfully activates the other, this connection is strengthened. As a result the chances are higher that this connection will result in a successful activation next time, and so patterns can form and be strengthened in the neuronal network.

And then the neurons sprout dendrites like little roots that connect to form a neuronal network?

Indeed, the neurons' dendrites pervade the network like a root system or a finely branched treetop. They connect with the axons—the neurons' branching, and sometimes meter-long "transmission cables"—and join to form a neuronal network. These numerous connecting branches of a neuron are called the *dendritic tree*, as they form this beautiful and unbelievably ramified tree structure. What is important is that these dendrites receive the signals from other neurons, so they are "receiver cables." The connections between axons and dendrites are the synapses. It is only where a synapse has been built that a signal can flow from one neuron to the next. The growth of the dendritic tree mainly takes place during development. After that, axons and dendrites grow very little. However, minute appendages grow on the dendrites like little buds or spines to form new connections.

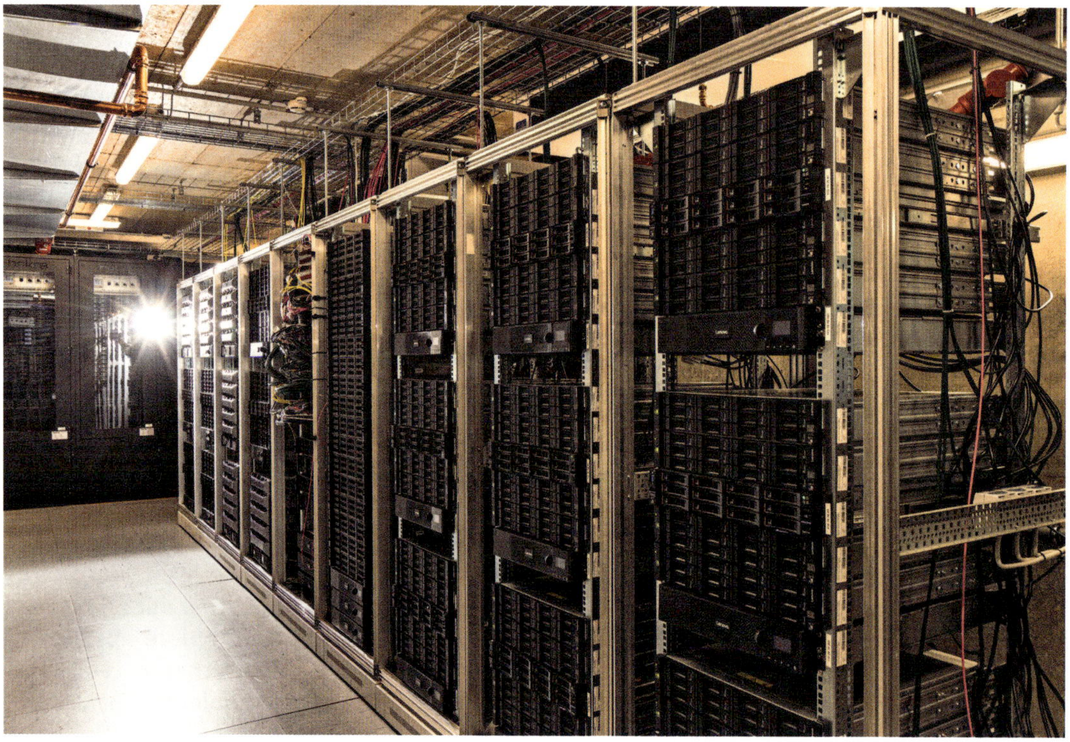

What role do the branches of this tree play?

Up until now these branches have often been ignored in models, and the neurons were just reduced to their signal transmission. But every branch of the neuron forms its own, almost enclosed compartment, in which certain reactions or processes can take place independently of other compartments. We have derived how the error compensation between the neurons can be calculated there at the local level. The result of the error calculation then determines how the local connection strengths change, so that the coordination works better next time.

The old saying that people only learn from their mistakes actually does apply to our brain.

There is even the theory that there are special neurons that are specialized in representing errors. However, it has not yet been possible to prove the existence of these error neurons in experiments. That has been a big gap for many years. However, we have shown in our research that you don't have to look at the network level with special neurons with would present a positive or negative deviation for this error calculation, but that far more efficient error calculations can be conducted per neuron in the local dendrites. As a result, the neurons also represent this uncertainty. Ultimately it is up to the neuron whether a signal is sent or not.

Who makes this decision?

That is the fascinating phenomena of self-organization: neurons receive signals from many other neurons, which makes their membrane potential fluctuate. When the potential is especially high or low, the neuron adjusts its local connection strength accordingly. With this adjustment of the connection strength the neurons specialize in specific information—for example, certain aspects of a cow. If the characteristics of a cow are sent to a neuron, the corresponding synapses pass on activations to the soma, the neuron control center. If a threshold value is exceeded at the soma, the neuron sends an action potential—in other words, a relatively strong, short signal—to all the neurons that it is connected to, and indirectly also keeps them from becoming active. This principle can generate neurons that become active when a person sees a cow—or a certain person.

Could you say that it is these error calculations that are the system's actual strength? When you say that neurons also represent the uncertainty, is this the origin of human doubt? If I am not certain about the cow, what is happening neurally?

You mean when someone says that is not a cow but a horse?

That would be an opposing thesis. Doubt just expresses an unease about whether it really is a cow.

That is a good question. A characteristic property of neuronal activity is its variability and flexibility. The network is in a continual state of flow. This variability also expresses the uncertainties, and this is an integral part of the inference about the ongoing in the external world. In addition, the brain is continually subject to change. Some processes and learning rules change the connections within a few milliseconds, some over hours, and there is the influence of neuromodulators. All this influences the functioning of the neuronal network. The neuromodulators can be produced by the body itself or introduced from the outside in the form of medicines and drugs.

It is possible to completely manipulate perception, representations, memory, and emotions with drugs and medicines. In his 1971 book, The Futurological Congress, *Stanislav Lem addressed this philosophical problem: what is hallucination, and what is reality? By taking an antidote, Ijon Tichy, the character in his novel, recognizes the brutal reality of a state that secretly feeds its citizens drugs to give the illusion of a beautiful world of luxury, although they are all sitting bedraggled in the mud. It is only during the course of the novel that the reader realizes that the "reality" could also be the actual hallucination that is triggered by the antidote. This question has been repeatedly raised, through to the film* The Matrix, *where, at the beginning of the first part, the hero Neo has to choose between the red and the blue pill. If he swallows the blue one, nothing changes, and he returns to the dream world created by the machines. However, if he takes the red pill, he awakens from his Matrix sleep and can see the "real world" with his own eyes. If that which we perceive as reality is just a product of these biochemical processes, and even a simple shifting of electrolytes can result in hallucinations, this is very sobering.*

We are a long way off from understanding the brain to the extent that we could generate such a stable and, above all, directed hallucination. Hallucinations only occur with a highly systematic shift in the neuronal characteristics. For example, if I intervene in the neurotransmitters or certain receptors, I shift the whole system in a certain direction. But at the same time compensation mechanisms are continually operating in the neuronal networks. That is probably one of the reasons why most hallucinations are so instable and can't be maintained for very long. But the exciting thing is that such phenomena occur. This is a strong indication that the neuronal network is in a transitional state between stability and instability—stable enough to function reliably the majority of the time, but close enough to instability to alter flexibly, and not get stuck in well-worn paths. As a result small changes in the network—a few milligrams of the active agent of some drugs—can result in strong changes in perception. But even without drugs there are continual changes. I have just learned something about *The Futurological Congress*, which means that the next thought will not be the same as before. The brain has to cope with all this variability. That is precisely why it is so important to decipher the fundamental organizational principles, that is, the question of how the brain organizes itself, how it can stabilize itself and learn autonomously. After all, we have eighty billion neurons up there that play well together without a conductor.

You have shown us a number of different learning rules that are clearly responsible for this self-organized marvel. You began your studies under Wolf Singer, who holds the deterministic view that all brain capacities, including the mental processes, are based on neuronal mechanisms and therefore free will is an illusion because every decision is preceded by neuronal processes.

Naturally I have been shaped by Wolf Singer as he had a great interdisciplinary department in Frankfurt. He also worked closely with my mentor, Theo Geisel, who is largely responsible for advancing the research here in Göttingen about chaos, self-organization, and complex systems. However, "free will" is relatively far away from our exploration of the function of neuronal networks. We still don't have a detailed explanation of how free will emerges in such models.

But you state that autonomous learning behavior—the whole creativity and wonderful capabilities of the human brain—is only possible thanks to its self-organization.

Exactly. Our goal is to completely understand these mechanisms. And the question as to how exactly free will—or the impression of it—can emerge from the interaction of neurons is completely open. In order to do this you have to clarify the term *free will* and translate it to the functioning of neuronal networks. How would one measure free will in an artificial neuronal network? Intuitively probably everyone has an idea what free will could mean. But it is not easy to measurably define this term. Again, this is an old philosophical problem. We have now found a number of approaches in brain research and physics as to how the conceptual systems of classification and pattern recognition could function, but it will take a long time before we can precisely explain how free will—or an impression of it—emerges. There is only one strong working hypothesis we are already quite certain of: human consciousness is just a question of nerves.

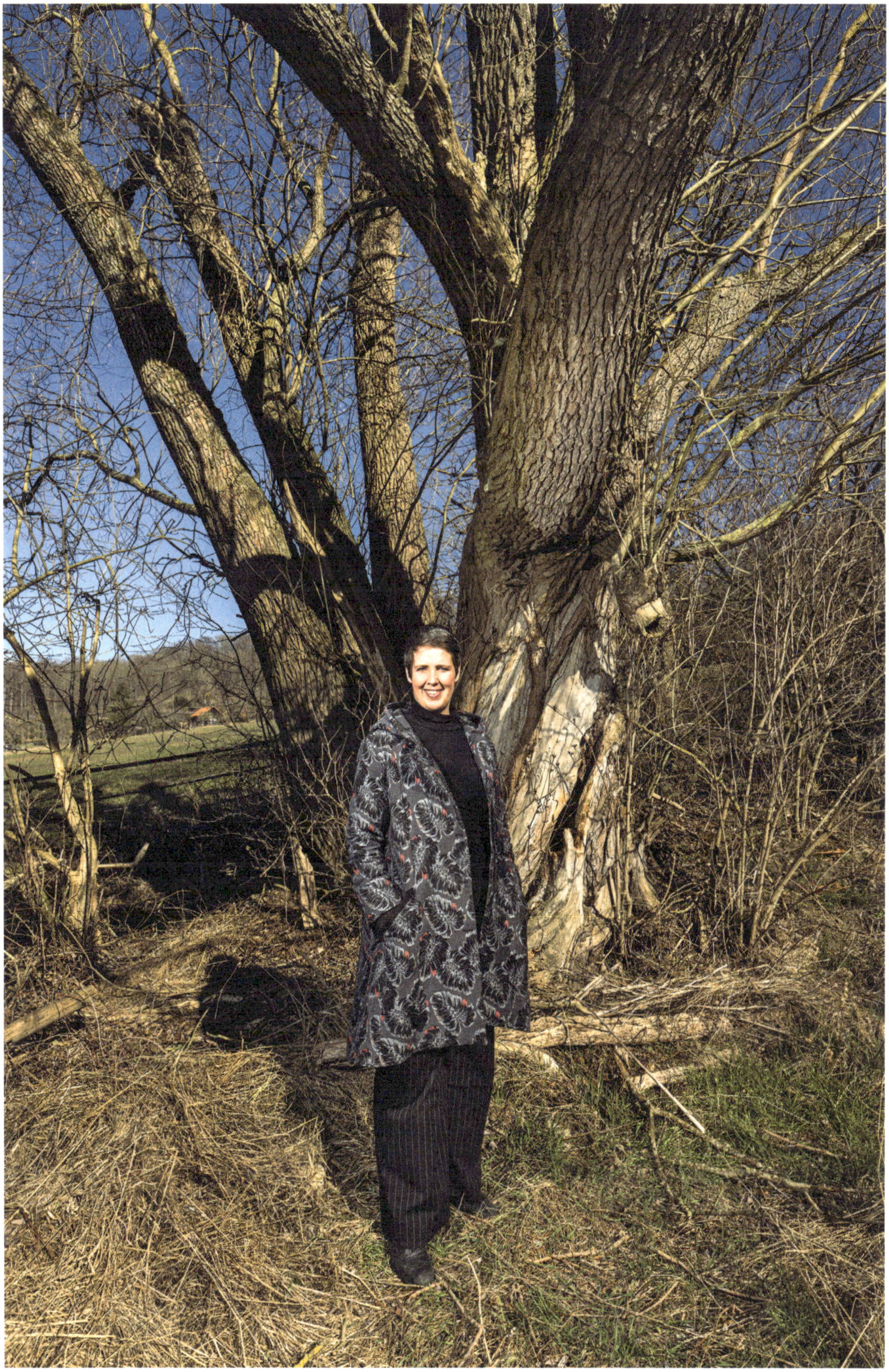

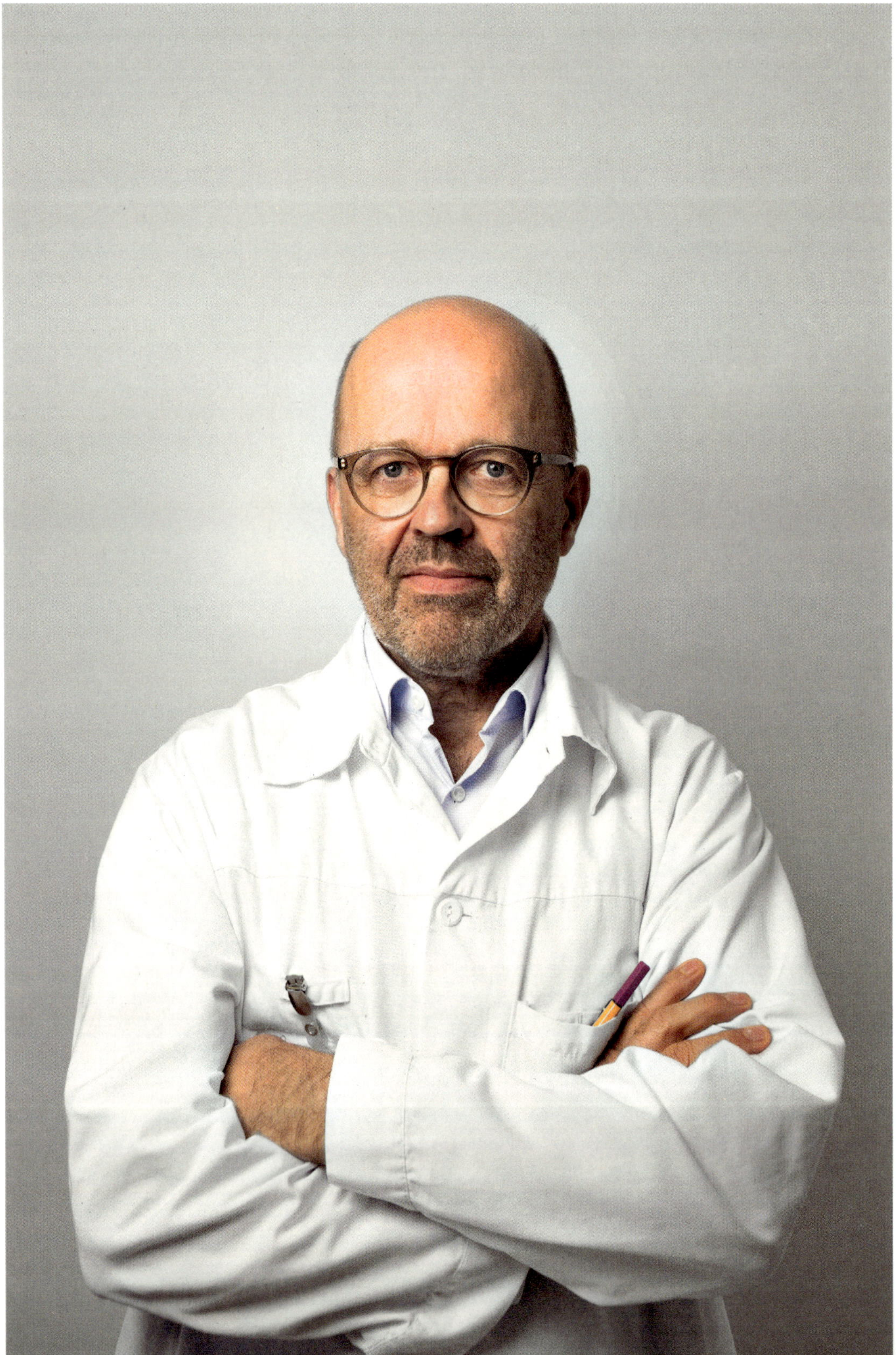

CHRISTIAN LÜSCHER

A Light Switch for the Brain

With his research group Addiction and Dopamine System, the professor for neurology and neuroscience at the University of Geneva conducts research into the cellular mechanisms of drug addiction. The rewarding effects of all addictive drugs are based on the release of dopamine in the brain. To build a circuitry model of addiction, the researchers in Geneva use sensor proteins in their experiments with mouse brains, which light up when the dopamine concentration and the cell activity increases. The research group also uses optogenetics, which relies on a genetic engineering trick that is used to express light-sensitive ion channels in the membrane of a cell. These neurons can be activated and controlled using light, which is introduced into the brain via a thin optical fiber.

Professor Lüscher, as a neurobiologist you head a research group at the University of Geneva where you conduct research into drug addiction and the causal relationship between drug-induced synaptic plasticity and adaptive behavior in mice. What does this involve?

We examine the cellular mechanisms that cause drug addiction. We believe that the core of the illnesses can be understood as pathological associative learning. Users know that drugs are dangerous, but at some point they lose control and take the drug even when the pleasure has been extinguished and they are socially isolated or economically ruined. The drug consumption has become compulsive; they are addicted. But what decides whether a person becomes addicted or not? Our research aims to understand and describe compulsive drug consumption. There are a wide variety of drugs that belong to different pharmacological substance classes: opiates such as heroin, certain stimulants such as cocaine, and drugs such as nicotine and alcohol. So the first question that arises for us is whether there is a specific addiction for each substance or whether addiction is one disease with a common core that applies to all substances. And here the evidence is clear: all addictive drugs produce the same initial effect in the brain. They increase the concentration of the neurotransmitter dopamine in the brain's reward system, in what is known as the *nucleus accumbens*. When a substance increases the level of dopamine in the nucleus accumbens, there is a risk of addiction.

Is there a risk of addiction because consumers want to restore this feeling of happiness?

I can't say whether it is a feeling of happiness or not. We know that the dopamine reinforces the behavior to take these drugs again. But meanwhile we also understand that some substances such as LSD do not cause a release of dopamine. LSD changes perception; tiny doses have a hallucinogenic effect, but LSD doesn't result in addiction.

When does the craving become an addiction?

Dopamine changes the neuronal transmission in the target neurons. It results in synaptic plasticity: the efficacy of glutamate, another neuotransmitter, between cell A and cell B is suddenly increased. This is when the addiction-like behavior begins.

So an electrical impulse arrives at the cell, which is translated into biochemical information in the synapse, which wanders to the other nerve endings—

They diffuse—exactly.

—and trigger the same electrical impulse at other nerve endings?

Not exactly the same. In the case of the presynaptic cell, it is called *action potential*. That is a large, very short, electrical phenomenon, which travels to the synapses. This results in the release of transmitters such as glutamate or GABA, which then diffuse. Each transmitter only has a very small effect on the postsynaptic cell. An action potential can only be produced when this effect accumulates across different synapses. But the sum is often too small. However, the transmission can be amplified when more transmitters are released or the postsynaptic cell adds more receptors to the synapse.

And the drug alters perception?

Initially it is a purely cellular phenomenon. The drug increases dopamine, which has a modulatory effect on this transmission. Dopamine doesn't alter the activity of the cell directly; instead it amplifies GABA or glutamate transmission.

So dopamine can fundamentally increase the activity potential?

It can contribute to the insertion of new receptors. The synapse has three partners: the presynaptic cell and the postsynaptic cell with the neurotransmitter glutamate, which transmits the information. Dopamine intervenes in the communication by making the postsynaptic cell much more sensitive. We first described this in 2011 as "drug evoked synaptic plasticity." The transmission of the information through the glutamatergic synapses changes, which then changes behavior. In the case of cocaine, for example, a very simple initial behavioral change is the increase in the locomotor effect. If you inject mice with cocaine for the first time, they become restless and run around in their cage. If you administer them the same dose of cocaine the next day, they display an even stronger urge to move than the first time. This is called *locomotor sensitization* and is considered an adaptive behavior that can occur at the beginning of addiction.

What role does optogenetics play in this study? How does it work exactly?

With optogenetics we can control the activity of the cells. With optogenetics you obtain a very high specificity: if you stimulate a cell in the traditional way using electrodes, all the cells in the vicinity are activated to the same extent. With optogenetics, by contrast, the light shines on all cells, but it only activates the ones that express the light-sensitive channel.

The neuron doesn't send it to all the other neurons in its network?

The networking doesn't change. But the original activation only occurs in the neurons that we have previously selected.

So what is the genetics *in the word* optogenetics*?*

The *genetics* consists in the fact that the selected cell type is defined by the genes that are expressed in a given cell. The expression profile can be analyzed with the aid of transcriptomics.

What does transcriptomics *mean?*

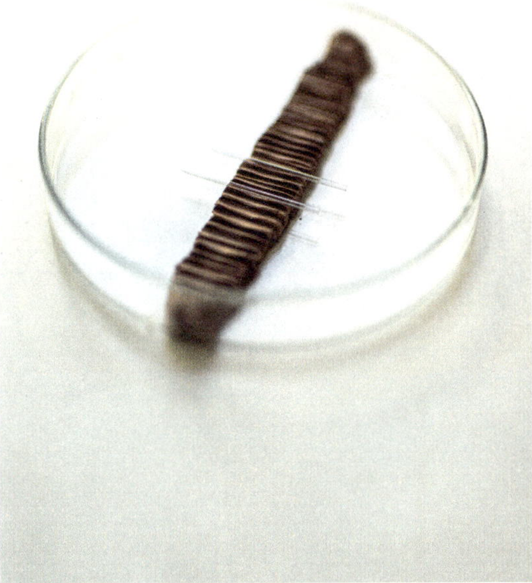

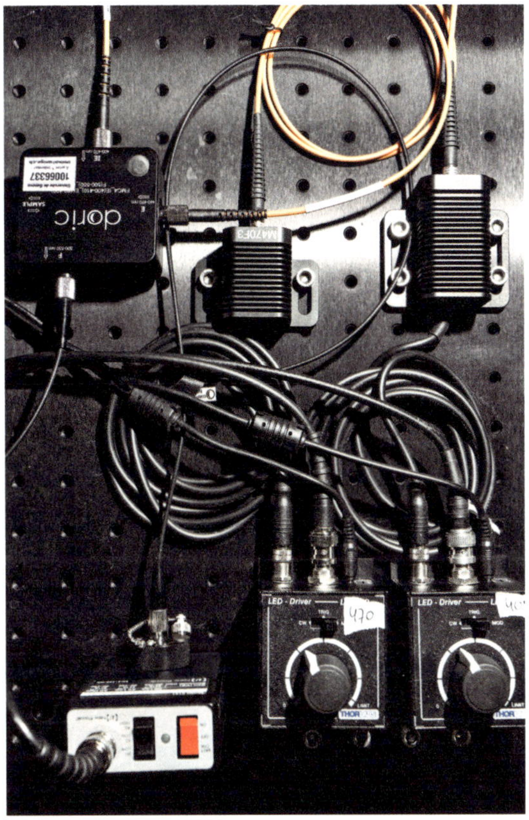
LED - Driver
470
LED - Driver

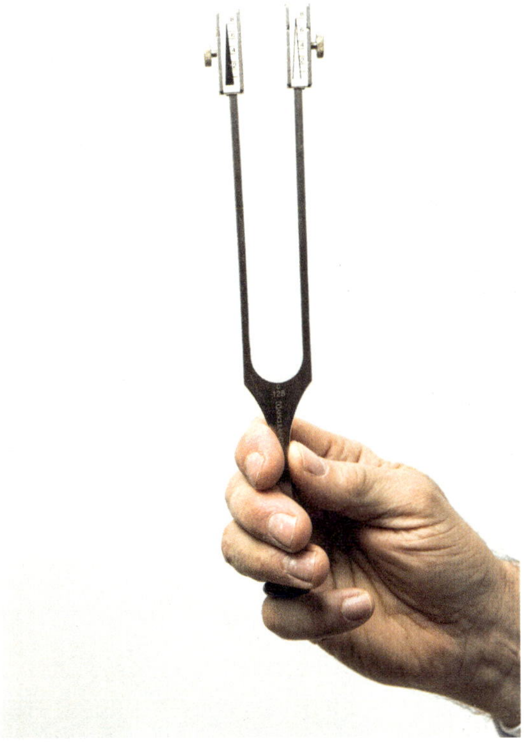

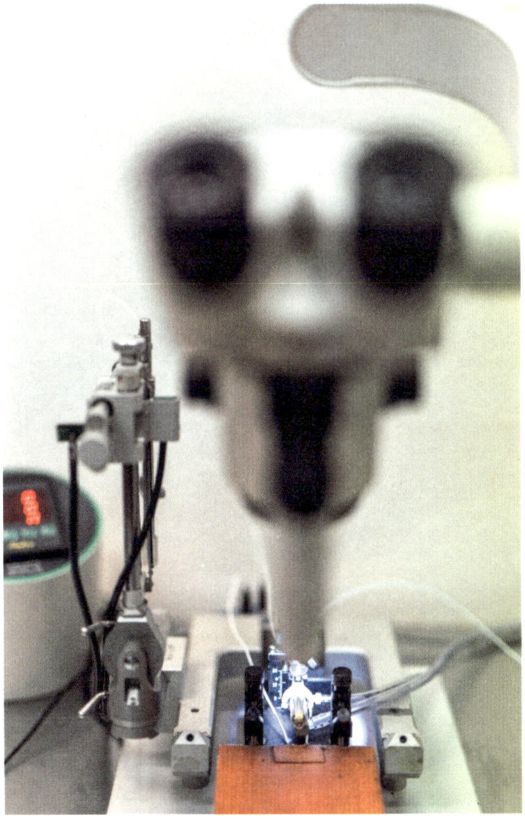

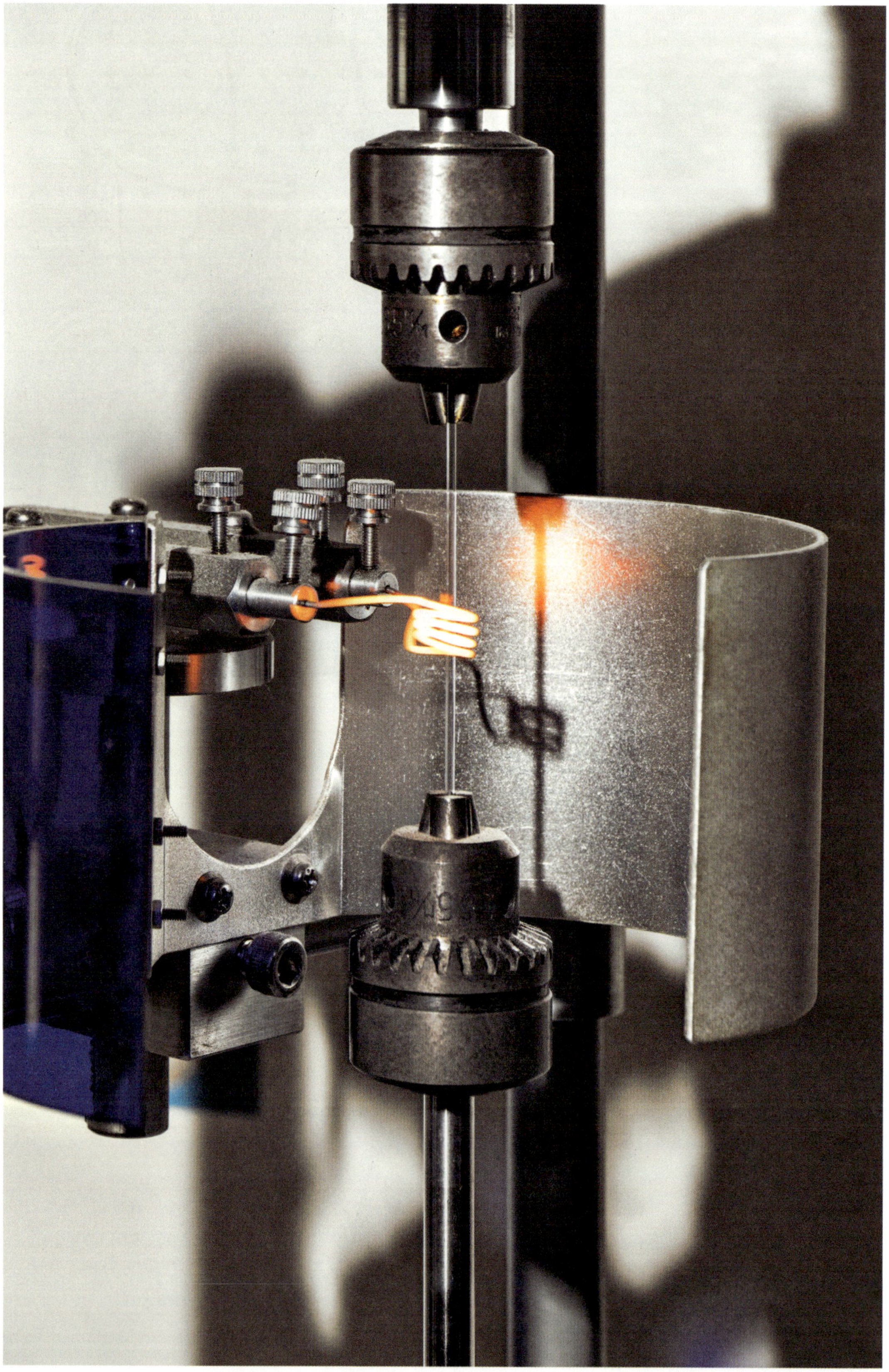

A vertical electrode puller. A resistive filament is heated to melt the glass while the weight is used to elongate the capillary. Upon breakage, a very thin tip is produced that allows impaling single neurons to record their activity.

That is a descriptive technique for sequencing mRNA, cell by cell, to genetically define the cell types. You can then see, for example, whether a cell possesses the gene that is required to manufacture glutamate. That is then a glutamate cell. Or you see the gene that is important for dopamine. That is then a dopamine cell. You typically look at thousands of genes and can thus define the different cell types. We then inject viruses into the brain of a genetically modified mouse, which in a highly targeted fashion uses genetic information in selected cells, which in turn program the cells to build light-sensitive proteins into their membranes. Using this method we can tell the neurons to express what we want: a calcium sensor, a dopamine sensor, or ion channels. For example, we recognize an active cell by the increase in calcium that is visualized with a fluorescent indicator that we can observe under the microscope in living tissue. To determine the dopamine concentration, as another example, we use a protein that binds dopamine and is coupled with a green fluorescent protein (GFP).

The more dopamine, the greener it gets?

Exactly. That enables us to measure the fluorescence in the brain of a freely moving animal using a thin optical fiber to establish whether the dopamine increases. But we can also activate or inhibit the cells with the light.

You can trigger the same reactions again using this?

Yes. Instead of giving an animal a drug we can directly stimulate the dopamine neurons with this technique. This leads to a change in behavior and ultimately to an addiction.

Apart from measuring, this technology can also be used to actively trigger things, which you can then observe under the microscope as fluorescence?

Yes. A mouse has an optical fiber on its head, which directs light of a certain wavelength into the mouse's brain and activates the channels. For example, when the mouse presses a button, a blue light is turned on and transmitted via the fiber into the reward center that stimulates the neurons more strongly than cocaine. With cocaine we observe that one in five mice become addicted; with this laser it is one in two mice.

So the mouse hasn't received any substance at all, and instead is just manipulated with the light?

That is correct. That is a model that allows us to examine the common features of all addictive substances with great precision and build a neuronal circuitry model of addiction. You can picture it like an anatomical map displaying the different synapses, along with the change in behavior that occurs as a result of these synapses being amplified. For example, it contains the hub in the nucleus accumbens, where amplification goes hand in hand with the locomotor sensitization. The initial changes in locomotor behavior are at first very simple and don't qualify as addictive behavior. With addicts you also have the craving and the development of strategies to obtain the drug again. It is more complex. On the other hand, even with highly addictive substances such as cocaine, only one in five people ultimately become addicted.

Because this cellular process only occurs in one of these people?

Exactly, the question is how this comes about. We don't have all the answers yet and are currently reaching the limits of our research. We need to understand this much better. But we know that the initial increase in dopamine in the nucleus accumbens starts everything. It is here that nerve connections from the prefrontal cortex and the hippocampus meet and control behavior.

Is this center composed of specialized dendrites?

The neurons here have dendrites with many spines, hence their name: medium spiny neurons. During the transition phase from controlled to compulsive consumption everything can still be stopped. However, as soon as the behavior becomes compulsive, we no longer find ourselves in the ventral but the dorsal part of the striatum. And then the region of the orbitofrontal cortex—which is located immediately behind the forehead above the eye sockets—plays an important role. That is the brain's administration center, which is altered due to the plasticity of the drugs and results in an addiction.

So the plasticity of the synapses changes?

Absolutely. Then you end up with a real addiction. We know that the activity in these circuits doesn't just correlate with the behavior, but that there are also causal connections. If we weaken this addiction circuit, an addicted mouse stops its compulsive behavior. Conversely, we can artificially activate the addiction circuit of a mouse that is not addicted and make it compulsive. We can shift the behavior in both directions. This enables us to talk about causality.

That means that with optogenetics you can turn the addiction on and off like a light switch?

You could loosely describe it like this, yes. But what we still haven't found out is why certain individuals become addicted and others do not. What happened before the first drug exposure? What proceeded differently in the brain? Or what happened during the initial drug consumption?

It is said that heroin leads to addiction after the first dose, while other drugs are easier to control.

Everyone becomes dependent on opiates. We distinguish between addiction and dependence. *Dependence* is defined by the fact that discontinuing the substance results in withdrawal symptoms. But not everyone becomes addicted. *Addiction* is defined as the compulsive use of a substance, regardless of the negative consequences.

But isn't the withdrawal a result of the suppression of dopamine by the drug, and so the body's own messenger substances are no longer available, resulting in the strong urge for the next kick?

No. The withdrawal probably doesn't have anything to do with dopamine. Naturally it is an additional motor for the consumption of more drugs, because one wants to prevent the withdrawal symptoms. We call this *negative reinforcement*.

In common parlance dopamine is called the happy hormone. *Is this where the positive feeling of intoxication comes from? Does the increased concentration of dopamine strengthen the sensations? Many chemical drugs such as MDMA are taken, for example, to overcome social contact problems or to increase sensitivity.*

Naturally we have a great many additional effects here. That is clear. Depending on whether somebody takes cocaine or cannabis, the experience is completely different. But this is not connected to the addictive potential, but with the other things these substances stimulate in the brain. Every pharmacological substance reaches the entire brain. There are no barriers. And wherever the receptors are located, there will be additional effects. At some point drug addicts don't have any happy feelings anymore, but still feel a strong need to take the drug again and again.

Because they are dependent?

No. That doesn't have anything to do with dependence. For example, there is no real dependence with cocaine. It is the addiction that rules here. Optogenetics allows us to look at these different circuits with great specificity and make these distinctions.

Nevertheless, you still can't answer the question as to what exactly happens in the drug experience?

Taking a drug is a human experience that we can't really observe with an animal because the animal can't tell us whether it has experienced a feeling of pleasure or not, or what else it has perceived or experienced.

The drug lowers the user's inhibitions and makes them more social. They can approach people better and overcome the inhibition to touch another person ...

This social component is not found to the same extent in all drugs. In the case of MDMA it is very pronounced, because MDMA doesn't just raise the dopamine, but also the serotonin, and that to a far greater extent. Every drug has its own receptor in the body, so to speak. For example, the most important receptor for cannabis is the CB1, which sits on almost every neuron in the brain, because there is an endogenous system, in other words the body's own system, in which cannabis is used as a fine tuner for the synaptic transmission. If a cell is strongly activated, then endocannabinoids are released, which diffuse onto the presynaptic neurons, which then activate the CB1 receptors. And this influences the fine-tuning of information processing.

These cannabinoids are very important for almost every biochemical and pharmacological process.

Yes. They play a major role in learning processes. That is why it is definitely not a good idea when young people smoke marijuana the whole day during the most intensive time of their education, as these subtle mechanisms no longer function. Then this can result in learning disorders.

So drug addiction is not a learned behavior; instead people have experimented with substances throughout their history only to finally find those that can stimulate the existing receptors. So the question of the chicken and the egg has been answered conclusively: it is the chicken—in other words, the receptor—that came first.

Yes, that's correct. And the idea is that at the end one has a circuit model that allows us to distinguish between positive and negative enforcement. Naturally, on the basis of the circuit model, you can also ask the question of how this can be exploited therapeutically.

For example, to reduce the severity of the withdrawal.

Exactly. To reduce the severity of the withdrawal, or even to reverse the compulsion. These are the questions that can now be posed very precisely.

Or even prevent addiction, for example, with a kind of vaccination against heroin?

I am not sure that a type of vaccination would be the right strategy. We strive to use strong techniques that we know can change the human brain's circuits. And deep brain stimulation is the most interesting for us here.

How is it induced?

Deep brain stimulation is a therapy that was primarily developed for Parkinson's disease. It involves the deployment of an electrode in a specific region of the brain. Our goal is to use light signals as opposed to an electric current for this.

Could this then be used to regenerate broken brain cells?

The cells are not broken. They are just a little more strongly connected. That is reversible. That is also an important finding of our research. The addictive behavior is not a result of the cells dying. The addictive substance does not result in cell death, instead the

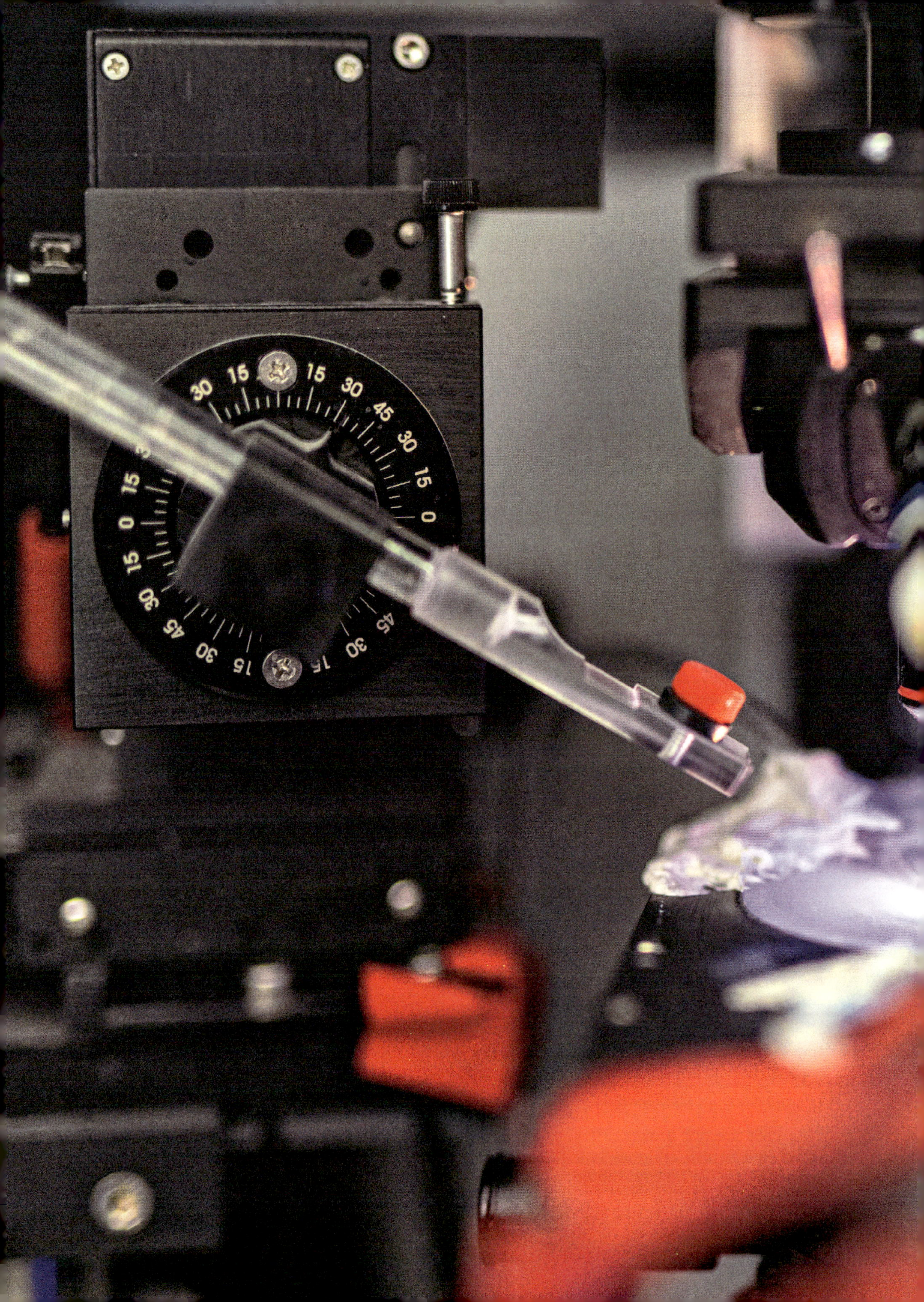

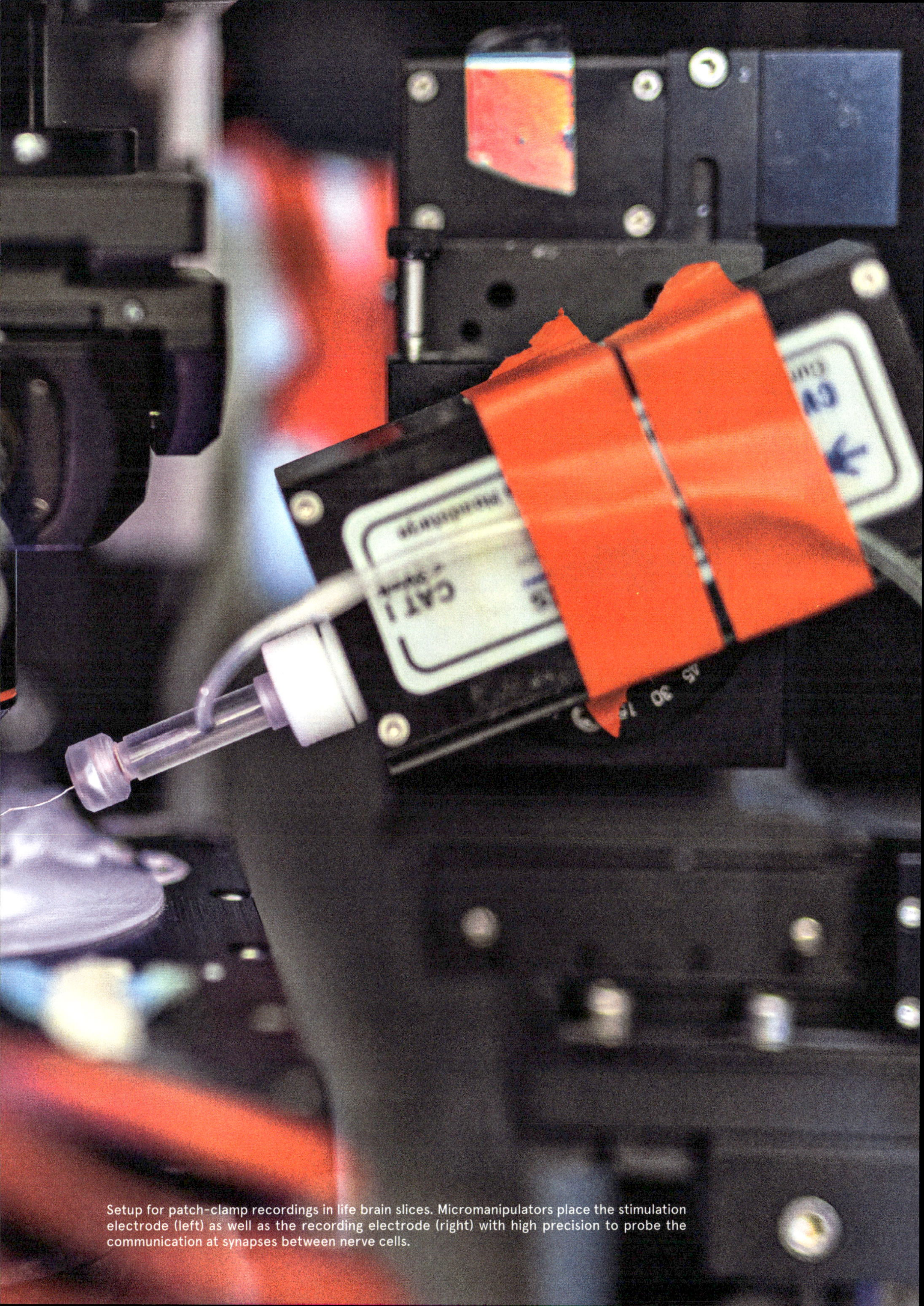

Setup for patch-clamp recordings in life brain slices. Micromanipulators place the stimulation electrode (left) as well as the recording electrode (right) with high precision to probe the communication at synapses between nerve cells.

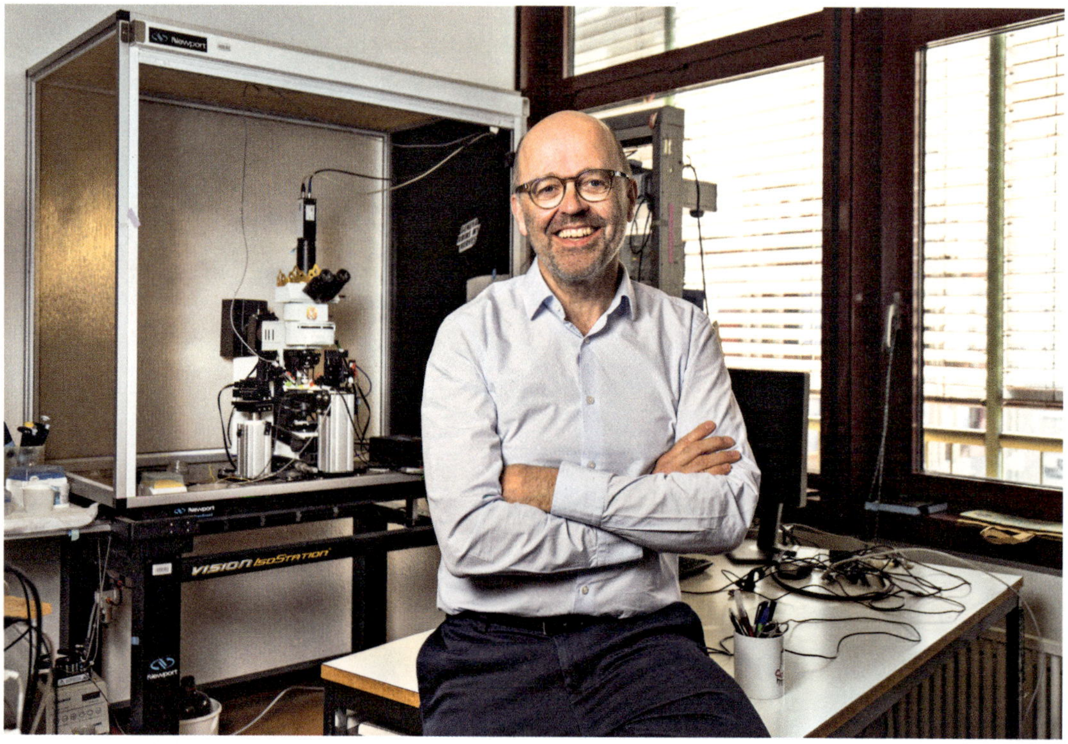

networking between the different cells is strengthened or weakened. This could eventually be addressed using such new deep brain stimulation protocols. We call this *optogenetically inspired deep brain stimulation*. Consequently, optogenetic findings could prove extremely valuable for many people.

There is a very broad spectrum of applications for this—for example, Alzheimer's.

Exactly. There are similar circuit models for other brain diseases that other colleagues are working on. In June we will be holding the fourth OptoDBS congress in Geneva, which will address the idea of using circuit models for new therapeutic methods. That is the line of attack. All interested disciplines will be gathered there. Naturally it is not just about addiction, but also about Parkinson's, Alzheimer's, and depression. It is about a diverse range of illnesses.

There have been a lot of experiments with the microdosing of LSD or ketamine for patients suffering from very severe depression who don't respond to normal psychotropic drugs.

Ketamine in particular is a very interesting drug. We currently have a study under revision where we asked what exactly happens with ketamine in these dopaminergic systems.

But that means depression must also have a cellular basis?

It does. But I am not a specialist in these questions. Colleagues working on this have observed certain changes in the circuits that can reduce the value of a reward. Although in principle something good has occurred, one no longer experiences it as positive. These are also circuits which can be normalized using deep brain stimulation.

But can ketamine biochemically change the biochemical composition in the synapses?

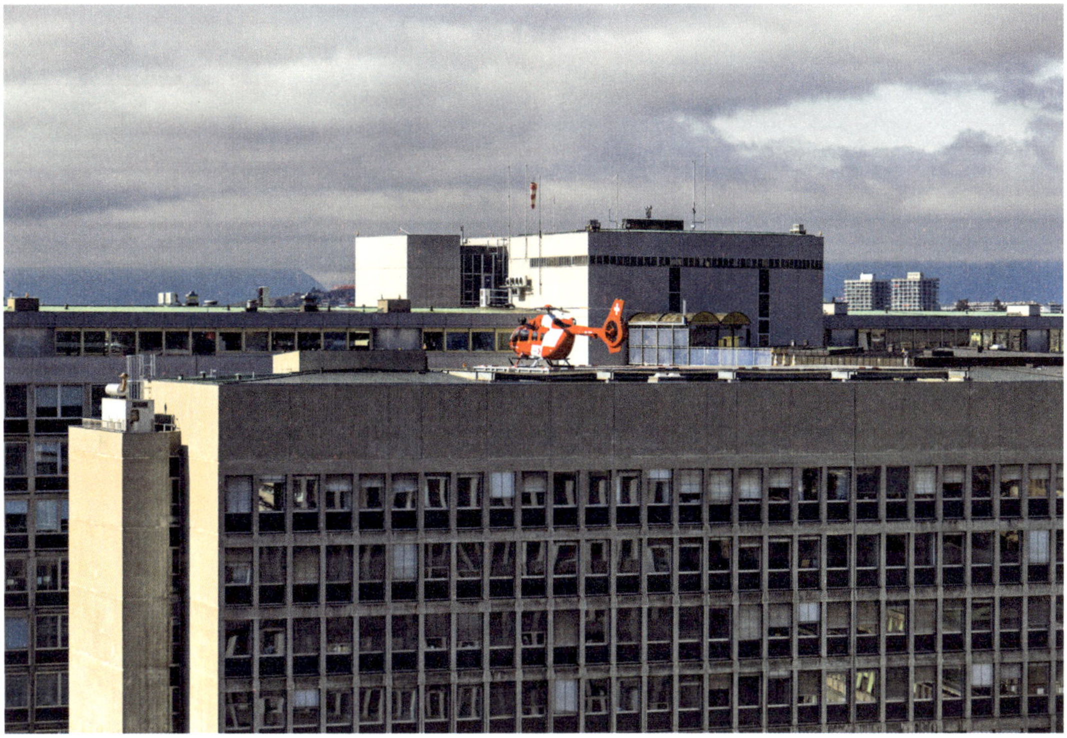

Ketamine probably also uses such plasticity mechanisms to combat the depression. This is a similar mechanism to that in other circuits and synapses.

So that the action potentials discharge more strongly again?

So that neurons can communicate better with one another again. For example, this also includes the hallucinogens. There are a lot of people who use psilocybin to wean themselves off smoking. That definitely has its justification. Through psilocybin one can open a critical period during which communication can be changed again, during which plasticity is possible again. This is a possible mechanism for combatting addiction.

One increases the potential, so that the neurons—

—can communicate normally with each other again. You have to picture it as follows: with addicted people the communication between certain centers in the brain, in particular the decision-making center in the orbitofrontal cortex and the downstream cells, is completely disrupted. What you are looking for are substances that make change possible again. In the case of real addicts the misdirected communication is so stable that is no longer that easy to change it.

Isn't it because the communication has become biochemically entrenched?

That is correct. The communication is biochemically entrenched. And the hallucinogens offer the possibility of dissolving this. The system involved here is serotonin. The serotonin receptors are the most important receptors for the hallucinogens. And in my opinion, there has been far too little research into precisely how theses serotonin receptors reopen this critical phase of plasticity. There is a lot to do there.

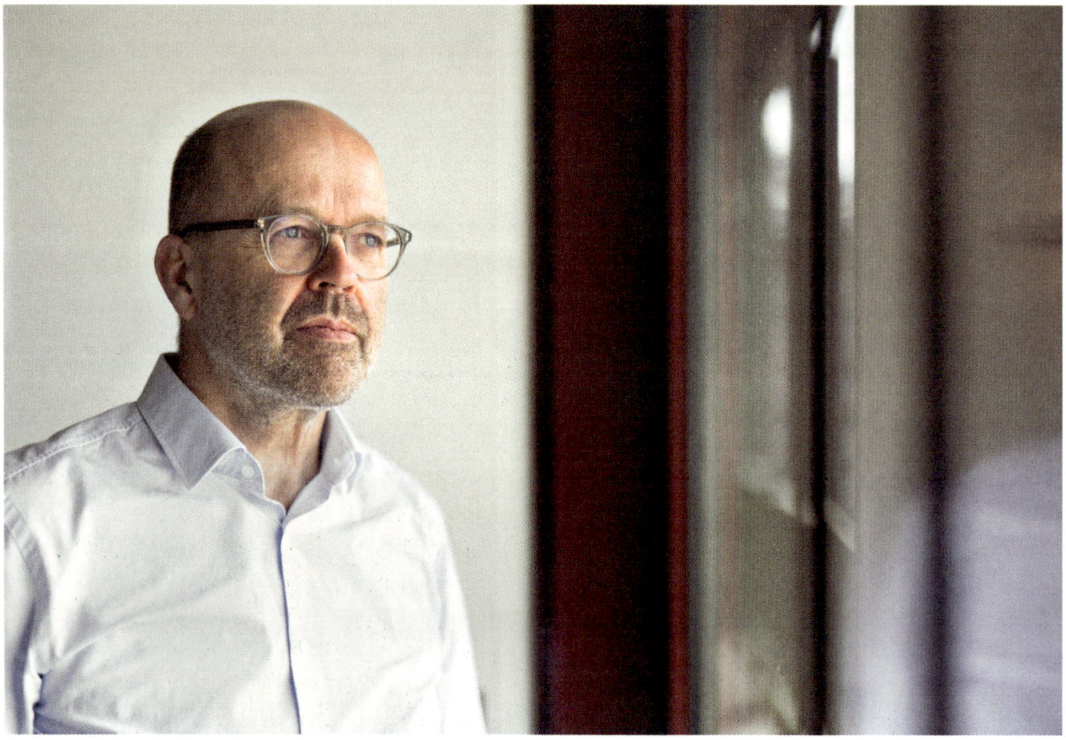

But in principle you now have a highly objective scientific detection method: What causes addiction? What doesn't cause addiction? Actually, you could now turn Swiss drug policy on its head.

We can say whether a new pharmacological substance has an addiction potential or not. Our research is possible because we make use of all these new tools and can observe and even manipulate these circuits with a far greater precision to establish causal relationships. As a scientist I can only provide certain pointers, but it is a political decision whether alcohol and nicotine are legal drugs and cocaine not.

Have you also examined alcohol and nicotine?

Yes, and one can see similar changes here. With nicotine the transitional phase to addiction is roughly as high as that for cocaine.

And alcohol?

In contrast to other addictive substances, alcohol kills cells. And that is naturally a big problem and a huge burden to society. But with alcohol it has been shown that the vast majority of people can consume alcohol their whole life without losing control, as we have regulative consumption in countries such as Germany or Switzerland. In Switzerland four to five million people regularly drink alcohol, but there are "only" 250 thousand alcoholics. Proportionately speaking that is a relatively small group.

Clearly this is simply accepted by society.

That is a political decision that I am not going to comment on. The 250 thousand people and their families suffer a great deal under the addiction and the costs are very high. If one could reduce alcohol addiction somewhat, then that would be a great benefit to the whole of society.

What is your personal motivation for your work on the theme of the drug problem?

My personal motivation is really a scientific one.

You never experimented with drugs as a student?

No, not at all. As a medical student I always wanted to conduct research, and then I went to a laboratory in San Francisco where we examined synaptic plasticity. This is where the idea emerged that dopamine could be responsible for this modulation. For me, addiction is a vehicle for studying cellular mechanisms that control behavior. It was not connected with any political ambition or personal involvement.

But as a result you have become addicted to addiction.

Yes, maybe. But not dependent!

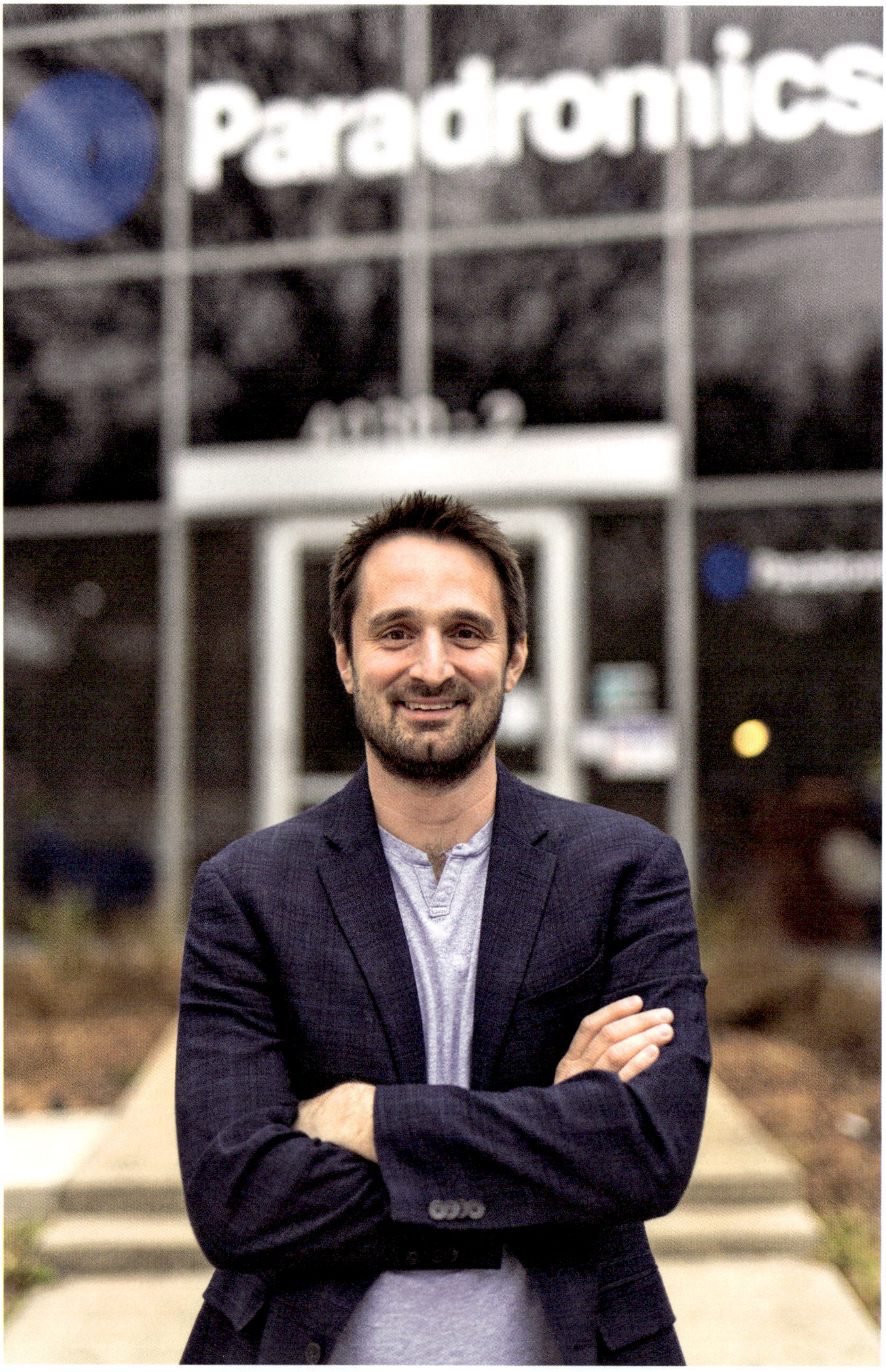
Paradromics

MATT ANGLE

"Therapies Not Cyborgs"

The founder of the start-up Paradromics is developing a human brain-computer interface whose data rate is sufficiently high to record and decode brain-signal data for an intended action such as speech or movement. Rather than creating some cyberpunk future, his dream is to use data from the brain to enable medical therapies. He hopes for a world where someone with a severe disability can seamlessly walk down the street or talk with you for an hour, without you even realizing that you shook hands with a robotic arm. Paradromics has demonstrated the feasibility of collecting massively parallel neural recordings in orders of magnitude beyond the current gold standard. By focusing on robust materials and form factors, Paradromics leads the field in developing a practical device for long-term daily service.

Tell me about how you started Paradromics.

Revolutionizing neurological health or neuroprosthetics wasn't our aim. The effort started out as a pure research endeavor that was looking at how to scale up in vivo chronic recording and how to get stable recordings using large arrays of microwires. As we were working, we realized that this kind of integrative, engineering-heavy project wasn't best served in academia. At the same time, we were looking at some of the really exciting brain-computer interface (BCI) work that was happening in different places and seeing the incredible results that you get with these hundred-electrode arrays called *Utah Arrays* and what people were able to do with them. We were developing this technology that could scale up to ten thousand channels, and we thought, "Imagine what would happen if we took this and brought it over into BCI." Those were the kind of ideas that were swirling around when we decided to start the company. It was a mixture of knowing that we wanted the technology to exist but realizing that academia was not the place to do it. And we knew that the technology is very valuable, so it made a start-up the kind of obvious conclusion. We were lucky at the time because it also corresponded to the peak of the Obama administration's BRAIN Initiative. So there was government money to be had from the start.

How far are you with your work? Do you already have the chip?

We are in the process of finishing out the implantable, what we call *cortical module*, which is a four-hundred-channel device. Every system has four cortical modules, so it's a total of 1,600 channels in the system. We plan to test out this cortical module in the spring of 2022, and the entire system will be done in the third quarter. Then we'll be taking that system into first in-human trials in mid-2023.

That's very exciting. How can I imagine the module? It has four hundred little needles?

Yes, that's right. You can think of it as almost like a hairbrush. It's a chip that's about eight millimeters by eight millimeters, and it sits on the surface of the brain. The needles penetrate about 1.5 millimeters into the brain.

The cloud of data that you will get is quite unstructured. How do you filter out the relevant information?

For communications or motor-prosthetics-related tasks, there's a training aspect involved. The patient attempts to do an action, and you record the neurosignatures of that action. They attempt to do the action again, and you record it again. Then you move on to the next action. After you've recorded the brain activity associated with these things that the patient is trying to accomplish, you can start to build up a model of intention.

You see an electric pattern, and when the patient makes a movement with his hand, you see a pattern in the information. So you know that this is the information for the hand?

Yes. The beginning of the training is trying to match what this person is attempting to do. Once you have done basic training, you connect their brain signals to, let's say, a cursor on a screen or to a text input, giving you a closed loop that involves the patient. The patient is learning how to use the device, and the device is learning how to interpret the information from the patient.

What I find very fascinating is that you view the brain as a data machine. We do not know enough about the brain, but we do know how complex the interplay between the chemical and the electrical information is. Is it enough to just take the electrical information?

You can think of the brain as a complex and hierarchically structured system. It's not the same architecture as a computer on your desktop, but in this they are alike. Computers have transistors at the lowest level in the hierarchy, and their physics and material science are well understood. You build networks of transistors to be able to do basic operations for logic. Then from logical operators, you can build simplistic commands, and you continue to build up until you have an operating system that's utilizing machine code and applications that run on top of the operating system. Sometimes you have other applications that are sitting on top of applications and talking to them. It's hierarchically organized. If you have a problem with Adobe Photoshop, no one would suggest that you take a focused ion beam and start cutting into the transistors and changing them at nanoscale level—that's not the right way to engage with a computer. You need to go into the menu and choose the necessary settings. Because you need to choose the right place in the hierarchy to tap in. And the brain is also a hierarchical system of that type. On one hand you have neurons that have a biophysical machinery that causes them to change their voltage relative to the outside solution. You have ion channels that mediate that voltage, you have receptors that are ion channels that change their conductance based on what's hitting them, and you have synaptic organizations. But if the problem is ultimately that you can't see and you want vision, the best way to tap in there is at a data level. Fundamentally you want to control the neurons in the visual cortex and give them data about the outside world. If someone's trying to communicate, it's not worth talking about the pharmacokinetics of the NMDA receptor. You want to just get the data out of the motor cortex, get the intention information, and allow them to get that out and type. The direction I personally think things are going is that even some neuropsychiatric conditions you'll be able to reframe from a standpoint of data.

I imagine that to treat a deaf person, you would take a healthy person who can hear. You would tell this person the words mother, father, baby, *and then measure the cortex to get the pattern for those words. You would repeat it many times so that the artificial intelligence can recognize the pattern of these three words. And then you would put the information that you measured into the brain of the deaf person, who would then be able to hear the words* mother, father, *and* baby.

Not exactly. There's not a good mapping from the brain of one person to another. For hearing, the auditory prosthetic takes advantage of the fact that the auditory system is

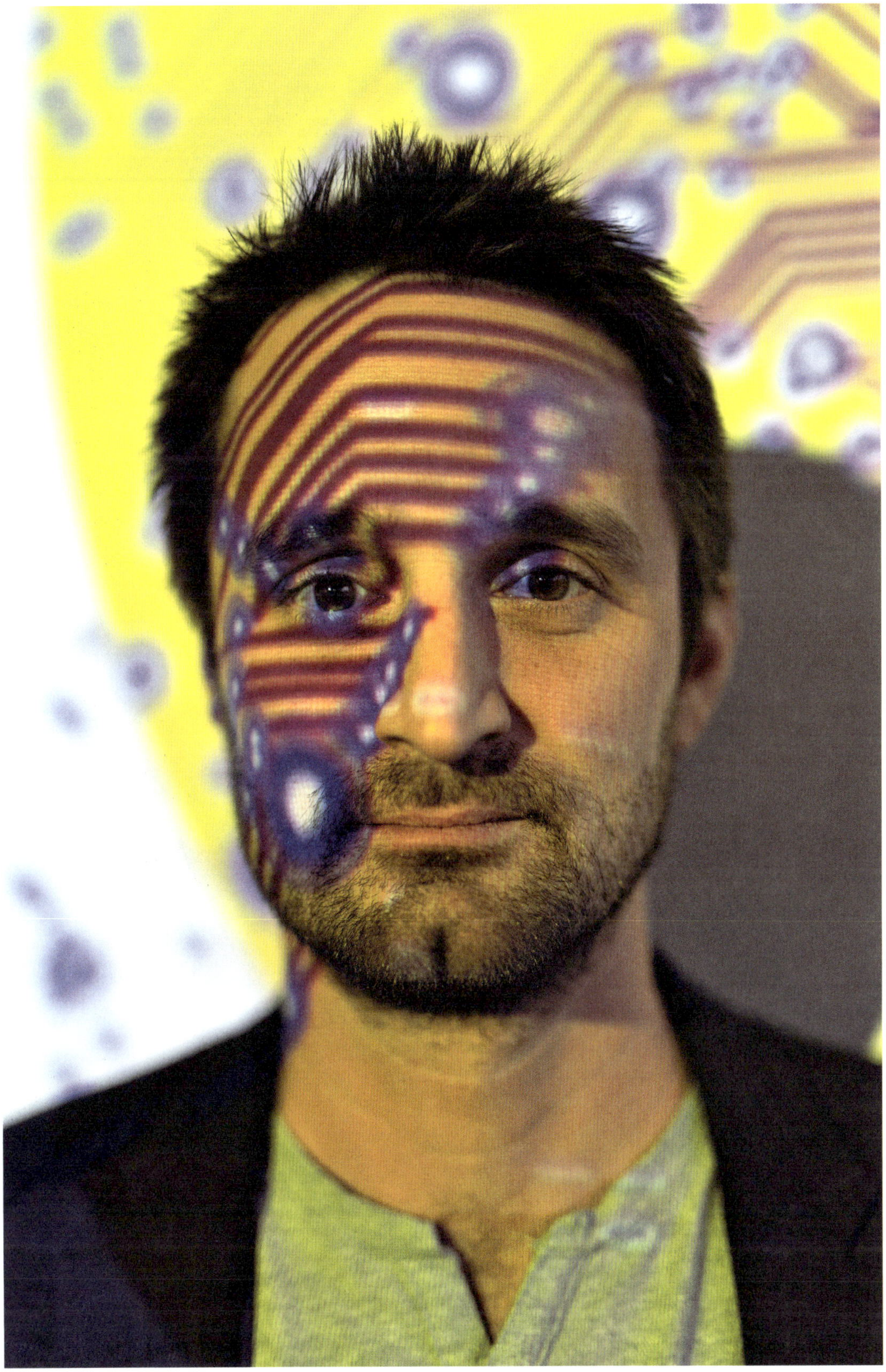

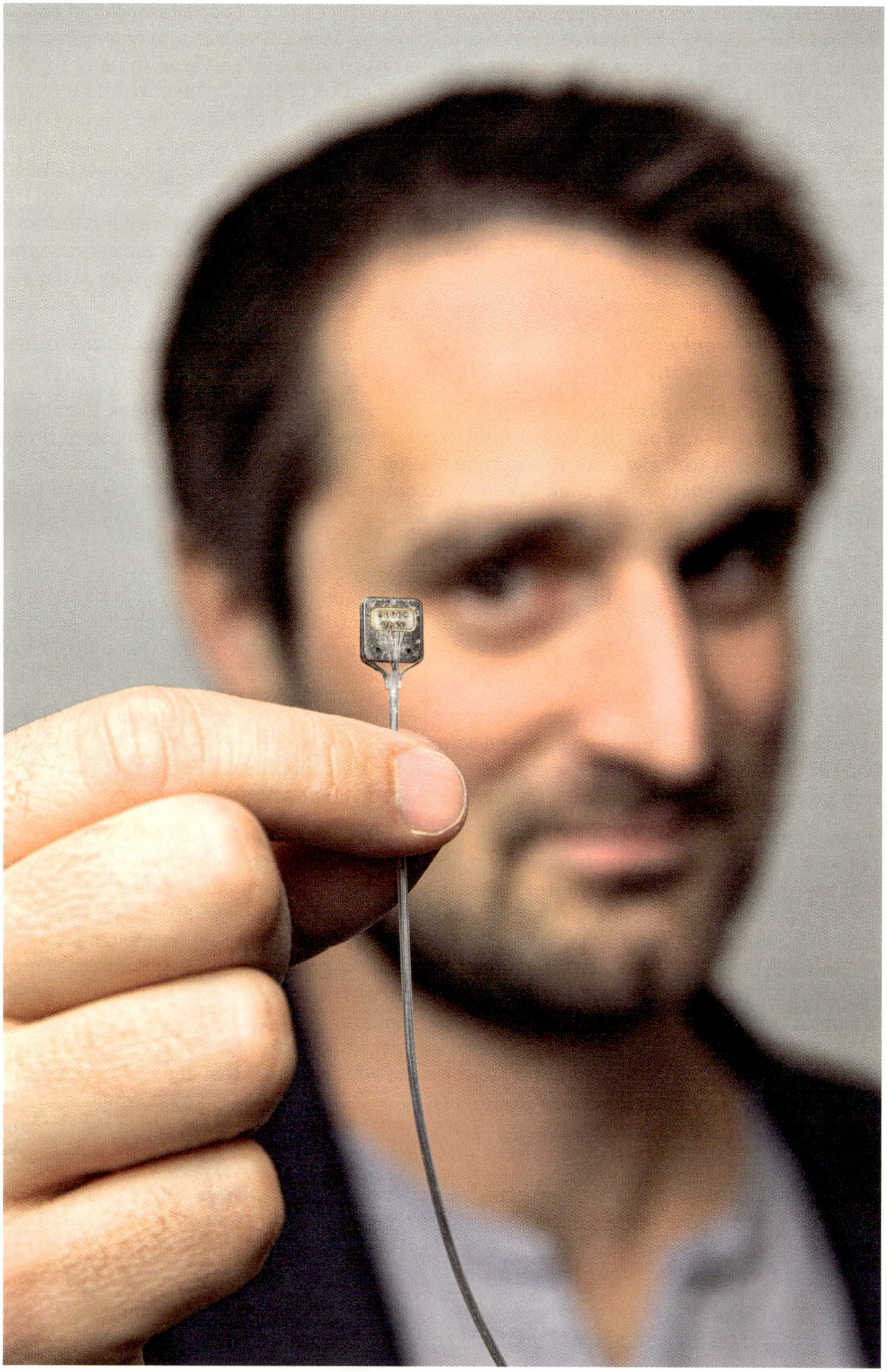

organized in what's called a *tonotopic* way, which is where the XY coordinates in the cochlea and the auditory cortex are organized in a map of tones. The way a cochlear implant works right now is this snake-like electrode is inserted into the inner ear and electrically stimulates the cochlea. Because we know the organization of the cochlea, we know where the high frequency sounds and the low frequency sounds are. So we can take a sound from a microphone, and we can break it up into the different frequency bands and present it to the cochlea in a way that allows people, even someone who was previously deaf, to even have a conversation on the telephone without doing lip reading. It's really scratchy and it's kind of imperfect. Another thread of auditory prosthetics builds what are called *Auditory Brain Stem Implants*. These are for patients who lose their hearing because the nerve to the ear has been severed. They put an implant directly into the brain, and then they stimulate the area of the brain that represents sound.

So you have to break up the sound, the same way my ear does it, and then you have to give the information directly to the brain. Is that correct?

Yes. What's interesting is that with the cochlea, we actually know how to break the sound up. We know where to put the different sounds, but when you go into the brain, sometimes it's more difficult to know exactly what the mapping is. The patient can communicate, so you can do a stimulation and then ask them to communicate what they hear back to you, and then you can start to build up a map that way.

The person will relearn hearing?

Yes. And it's not obvious that you're going to get the mapping exactly right. There's going to be a part of the computer that's trying to get the electrical signals to the auditory cortex to be as realistic as possible. But there's also going to be something that happens in the brain of the patient to learn that new representation.

Other experts told me that the information from the brain is not specific enough and the quality of information builds up in the muscle cells. The biochemical process in the muscle and the information from the brain together provide the complete quality that you need. How can you make sure that you get the right quality of information just from the brain?

It's true that the muscle is doing something. It has an advanced functionality. But it isn't true that you need to get the signals from the muscles to control a robotic arm. All of the signals that are necessary for you to control your arm originate from the brain. So it's a little bit like emailing a friend of yours and giving them some instructions to do something. It's true that your friend on the other side may have certain skills that enable them to implement what you've told them to do—you give them instructions to fix a car or something like that. You're relying on some prior information that's been stored with your friend, but the intention signal is coming from you. The instructions are coming from you. There's no reason why you couldn't have sent that information to another friend and asked them to repair the same car. So, in short, no. You could give a robotic arm to someone who's an amputee, and it's true that the Ottobock arm is going to have to recapitulate some of the processing that would've been done in the arm. But it's obvious that you don't need those muscles to be able to decode the neural signals. You could also build a synthetic system that would decode the signals as well.

There are different regions in the brain where all the different abilities are, so if I am walking through the forest with you and having a conversation, I don't need to think about my feet. They work automatically. Is it possible to get all this information only from the cortex? Or do you need to go to other areas as well?

The motor cortex controls the muscles for walking, but it's a different thing to say that the motor cortex can be actively controlling your movement. You may not be actively or

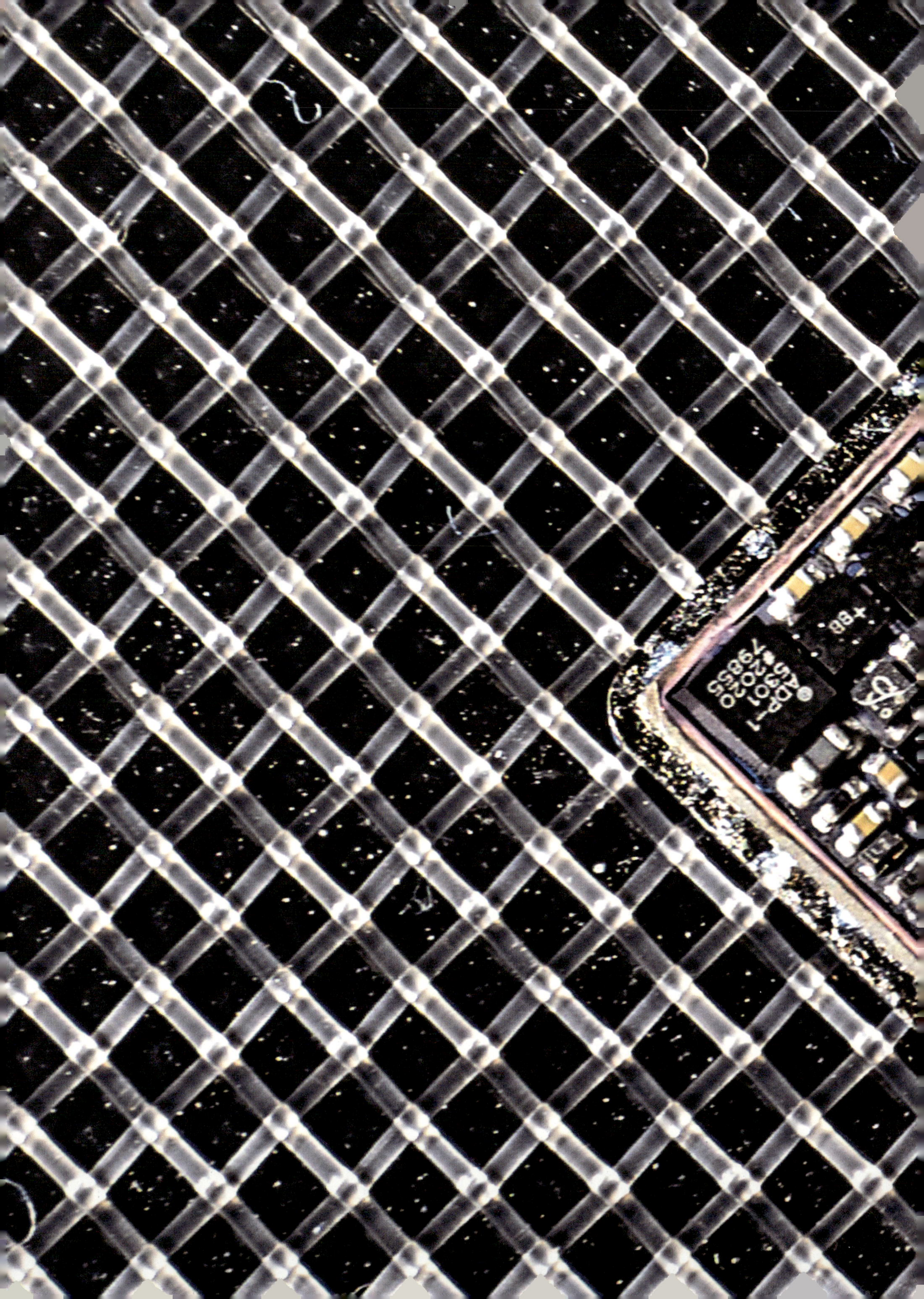

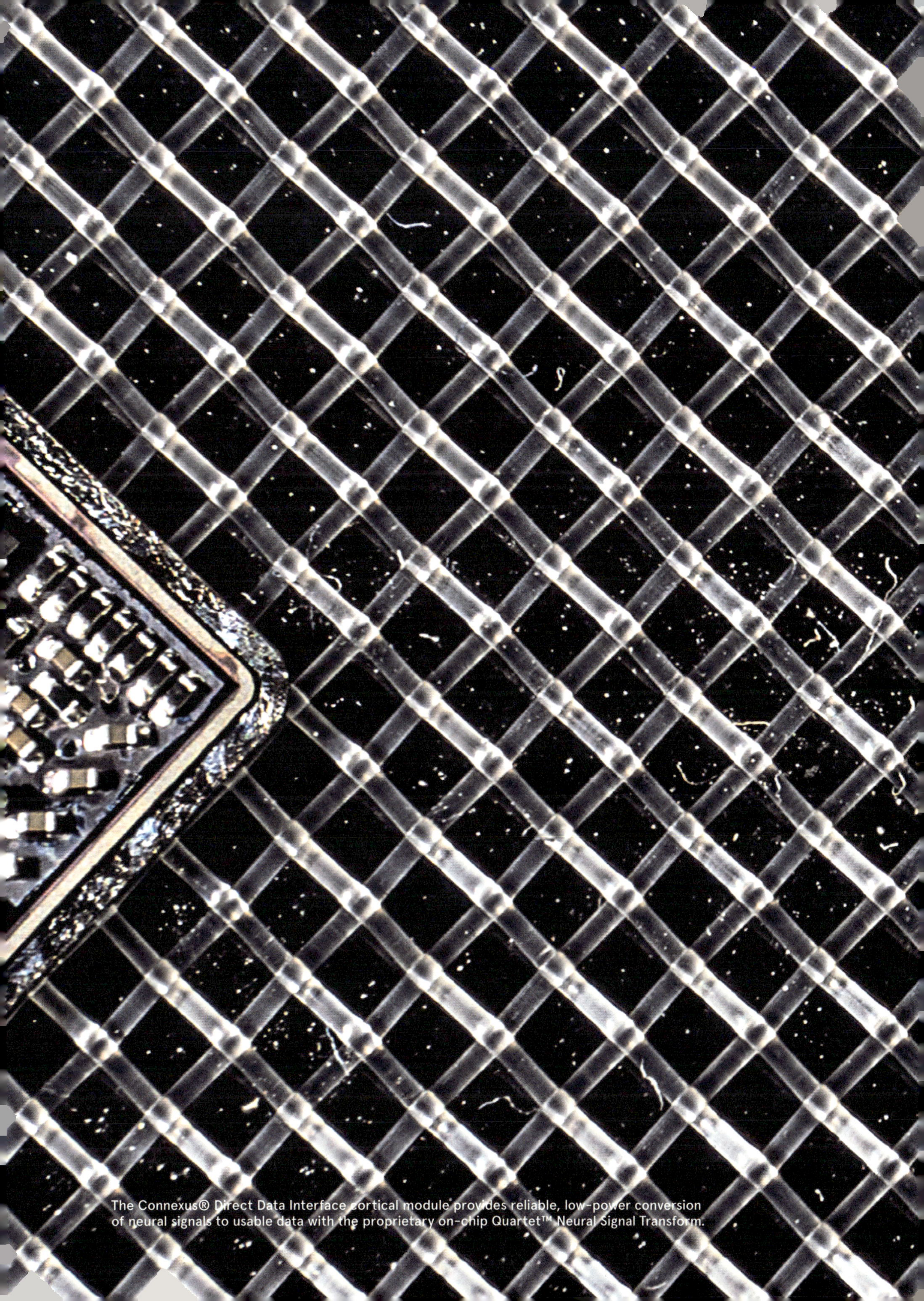
The Connexus® Direct Data Interface cortical module provides reliable, low-power conversion of neural signals to usable data with the proprietary on-chip Quartet™ Neural Signal Transform.

consciously attending to it. So I think one answer is that the motor system is enough to control robotic prosthesis. That's the system that sends out the signals, and that's where the signals pass through. It's kind of like the mailbox that goes to the body. And I think you raised a different point, which is an interesting one: for things that are automatic and very stereotyped sometimes some of that can be done by the prosthetic itself.

Of course, there are other companies building prostheses that are autonomous, like a robot leg. It walks your body, and you don't have to see.

Yes. And that gets to the point that I guess some of the people you talked to earlier were making: the spinal cord itself is a billion neurons. The spinal cord is a processor. This raises the question of how much you want to have the person directly controlling the thing in a very intentional way. And how much do you want to make it easier for them by providing them a little bit of leg autonomy or arm autonomy. That you guess what their intention is, and instead of taking the literal instructions, you do what you think it's supposed to do. Biology has come up with different solutions for this: the octopus's arm, for example, has a little brain on it. Our spinal cord has more neurons in it than a lot of animals that are walking around. So from a standpoint of how you tie in a device, like a Paradromics device with an Ottobock prosthetic, there's a lot of room for creativity there. BlackRock's Utah Array, for instance, is one hundred electrodes. We've seen amazing proof of concept demonstrations with a person controlling a robotic arm. But it's a little bit clumsy. The data rate is still low, so you can't gain access to all possible movements of the arm. With a device like that, you might want to put a little more intelligence into the arm. You might want to give high-level instruction and let the arm execute on a certain movement. As we start to move more toward the range of what Paradromics and Neuralink are creating—really high-data-rate interfaces—we can shift more of that back to the brain and give someone a more naturalistic control of prosthetics. Instead of saying, "Reach out and grab the cup"—and telling the hand to reach out and grab the cup, and then the hand makes the decision about exactly how it's going to grab it—they actually have control of the arm. And I think that you'll see a shift from the octopus version, where the arm has its own brain, to the human version, where the goal is to give someone lifelike control of an arm. I think that will unfold over the next five to ten years.

It's a big challenge to put hundreds of tiny electrodes in the right place and not damage anything. What kind of surgical procedure is involved? I understand that you are going to try this in patients next year.

Yes. Our first patient population will be patients like Stephen Hawking—people who can't speak and can't type—and we'll be giving them the ability to use a computer. We think that there's enough data that they'll be able to speak again, not through their own mouth and larynx, but by thinking of what they want to speak, and then a computer speaker will produce the sound. This kind of surgery is not something that you'd want to undergo to get better use of your smartphone. But if you can't move and this will give you the ability to communicate with your loved ones again, it makes a lot of sense. I think what will happen is that some years from now, when a lot of people with severe quadriplegia have implants and it's deemed safe, you're going to see someone who has had an amputation looking at the situation and saying, "Well, I can see that the complication rate for the surgery is low. And if I get this device, I would be able to have a lifelike prosthetic arm. Yes. I will do it." It will expand out from the patients of the highest need who therefore have the most willingness to try the surgery. As it gets safer and safer, more and more patients will consider using it.

When you test these extremely invasive devices in patients, you need the approval of the FDA, so you have to do a lot of experiments.

To prove that the device is safe, you test it in sheep or pigs. We're working in sheep now. The goal is that by the time the first person sees a device, you already know it's going to be safe. But what you're testing for the first time is that obviously a sheep can't speak or use a computer. So at that point you're really testing out if the patient can get the signals that you want in order to do the things that you want.

Elon Musk is investing a lot of money. He creates big narratives, and the financial markets follow him, investing billions of dollars. That's good for everybody, and we don't feel like competitors. He has prepared a field where we can also expand. Do you have contacts to Neuralink? Has he already tried to buy you?

It's a small field, so we know the Neuralink people pretty well. A lot of the original team has since gone on to do other things. We all know what the others are doing.

Is your company better than theirs?

There are two companies that can build high-data-rate brain-computer interfaces: Paradromics and Neuralink. With respect to the data, I would say that one of us is as good as the other. Currently our system has a higher data rate than the system that they showed in the monkey. We'll probably continue producing increasingly higher-data-rate devices. The third contender is very far behind Neuralink and Paradromics. The critical difference between us is device longevity. Neuralink's vision for BCI is one where it eventually replaces the cellphone, and accordingly Elon Musk thinks that patients will want to get their BCI replaced quite often. The material stack that they've used is one that's known not to last reliably for more than a year; they're using these polymer-based threads that have long-term reliability problems. For Neuralink, that's not the number one thing to optimize, because they don't think people will keep them for very long. We fundamentally disagree with that premise. We think these are medical devices and people will want them for a long time. So when we went about designing the device, we were saying what would we have to do to make sure that this device can be high data rate, but also last ten-plus years.

But how can I change the Neuralink device when it's implanted in my brain? Do I have to have another operation?

Yes. I think the idea is that you'd have them pulled out and then have a new one put in. I wouldn't sign up for that.

I can imagine that it's not so difficult to build a good device to handle all the electrical information, but isn't the software the decisive element?

No, I think that the problem is a little bit like getting oil. There are a lot of oil refineries that know how to refine oil, but the thing you make a ton of money from is if you can get the oil out of the ground. BCI is in that stage right now where we have the computational neuroscience—the machine-learning frameworks—to look at the signals when you get them. We know what we want to do with the signals when we get them. The problem is that the pipeline for the data is minuscule. You're recording from the Utah Array, which is a hundred electrodes. Neurons are slow. You know, the processor on your phone is a gigahertz processor, but a single neuron has a data rate far below that—it's probably in the five to a hundred bits per second, and it has a clock frequency that's effectively like a kilohertz clock frequency. The only way to get a lot of data out of the brain is with a lot of individual contexts. The problem right now is that we just aren't getting enough data out of the brain to run advanced applications. So, yes, I think it's much more about being able to build a really high-channel-count device and get really good data, and having that device be durable enough that people can rely on it as a medical device.

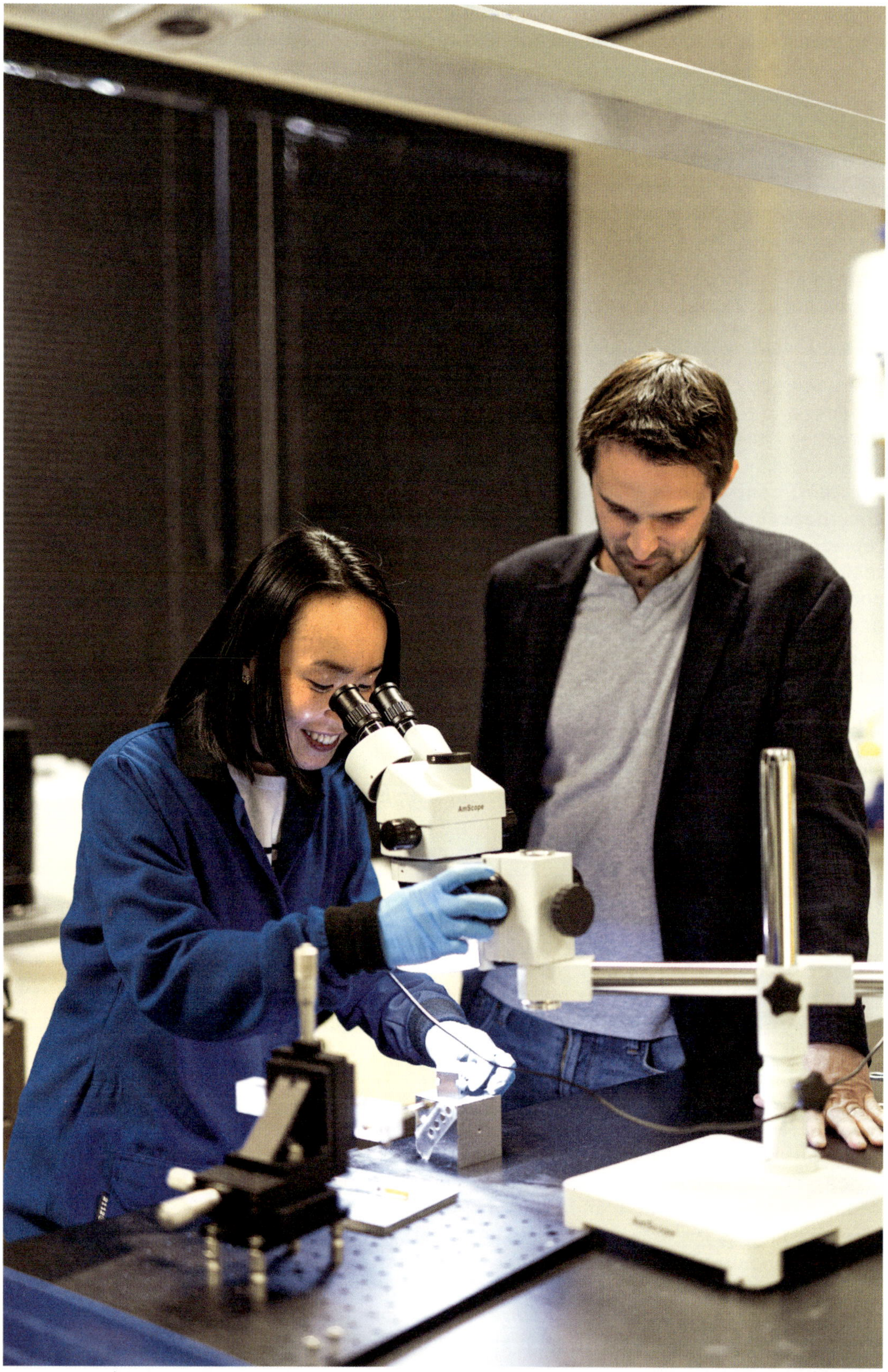

 Kimiko Nakajima, senior product development engineer, and Matt Angle

In the first scenarios for your technology, the aim is to get people to speak again and move again. What else can you imagine?

With a sufficiently high-data-rate interface, you can theoretically record and decode or stimulate anything your brain is doing. To use another analogy, I think of this a little bit like early internet. There are the applications that you could run in 1990, and then as download speeds got better and better, other applications appeared. We're really focused on communications, motor and sensory prosthetics, and mental health, which is the area that I think is probably going to be the biggest impact of high-data-rate computer interfaces. Because if you can start to turn a mental health problem into a technology problem, and we can start to leverage the tools that we have in terms of BCI paired with machine learning to diagnose and then even modulate mental illness. That would be massive.

You have announced that you would like to solve things like anxiety, depression, and schizophrenia. I think that you would get some resistance—at least in Europe—because people will be afraid that their brains and mental health can be manipulated with this device.

I don't know. I mean, you can already manipulate people's mental health with social media. I think people probably overestimate the dangers of a brain-computer interface and underestimate the ways in which their brain is already being tapped into that.

A lot of medicines depend on the interplay between the chemical and the electrical information in the brain. Can't you manipulate the biochemical processing in the synapses and between two nerves with medicines?

Drugs don't crawl into the right synapses and then act as traffic directors telling everything what to happen. They diffuse into every synapse and inhibit uptake of monoamine neurotransmitters, or they selectively block reuptake of serotonin—the drugs are not that clever. It's not like they're doing some sort of nanoscale-directed action. They're just diffusing and binding everywhere. Even the hypotheses that the drugs are designed around that are not right. The fact is that some patients respond well to these drugs, and therefore they're prescribed to many patients now. In Europe, I think that overprescription problems are probably not as big as in the US, but I think to present pharmacology as a more specific treatment than a brain-computer interface is dead wrong because it's less specific. There are fewer degrees of freedom.

Do you think that BCI is more precise and more effective than pharmaceuticals?

I think BCI will be more precise and more effective. To be honest, it doesn't exist yet. But I think what we can say is that some of the problems with drugs are that even if you design a drug that targets one receptor very well, the fact is that receptor is located in a lot of different areas of the brain. It doesn't have the same effect depending on the area that it goes to. And so, in and of itself, these sort of pharmacology-based treatments are imprecise. Then the layer on top of that is there are simply not enough degrees of freedom in a drug molecule to design them to really tightly bind one and only one receptor. The receptors all look kind of similar, and they tend to bind many different things and cause many different effects. The result is that you can find drugs that produce therapeutic effects, but they almost always—in the kind of therapeutic dosage window—cause some side effects. A brain-computer interface can be much more specific because these intercortical probes can target one or only a few neurons with a single wire. So you can really control the network with a high degree of precision.

I'm still thinking about what you told me about the neurons having such a low data rate. How can the brain execute such complex processes when it's so low?

Because you have eighty-five billion of them.

And they are connected to each other in a neural network.

Even though the individual neurons are slow. A computer is a serial processor, so it does very simple steps very quickly. For a computer to do a certain task might require a million steps, but for a brain to do the same thing may require three or four steps. And so, even though the individual units of the brain are much slower, the brain itself can often outperform a computer in certain kinds of tasks where the processing can be massively parallel.

So it's very complicated to exactly match the neuron you need?

Yes, that's true. The goal isn't to put the array in and match one for one electrode to neuron. The goal is to drop the electrodes into a field of neurons. It's almost like dropping microphones into a field of people. Because the other thing is that you're not actually that interested in motor tasks, for instance—the stuff that Ottobock might be interested in. Instead of what individual neurons are doing, you're interested in what the whole group is doing as a dynamic population. So to some extent there's enough redundancy in the crowd that you can drop them anywhere and still make a good inference about what the group is doing.

A lot of research has been done on the brain. Do you have an interdisciplinary exchange with brain researchers, and do you have experts on your team?

Yes. Most definitely. We're part of the neurotechnology community. There's a tight integration between the people who are doing these start-ups and the labs around the world that are developing new computational frameworks, for instance.

DARPA has invested a lot of money into your company. What is the military interested in?

Well, unfortunately the military produces a lot of prosthetic-need cases. For the last twenty years DARPA has been investing in, for instance, Dean Kamen's DEKA Arm, and a lot of the foundational work using the Utah Array at BCI was supported by DARPA.

I am wondering if quantum technology is involved in your product? I have spoken with quantum-sensor people who have measured the magnetic field when an electron passes through a muscle.

Not the sort of magnetometry that you're describing. But I think the effect of quantum computing would be very beneficial as we start scaling up to really high-data-rate interfaces. If you are running on conventional computing or a graphics processing unit, you have to make sure you keep the model light, so that it can run in real time. So you trade off performance and time and latency, but obviously if quantum computers became readily available and computation became super cheap and super easy, there'd be a lot of advantages to being able to put more computation at the image. I think people are going to be walking around with cyborg bodies before real practical quantum computing comes.

So with your product, the vision of the cyborg, which is already forty-five in science fiction, will take a big step toward that future.

I don't like the word *cyborg* because it has a certain implication to it. We don't want to create a cyberpunk future. We want to create a world where someone with a severe disability, someone who's paralyzed, someone who has an amputation, walks down the street and talks with you for an hour, and you don't even realize that you shook hands with a robotic arm. I think that's more the vision that we would see.

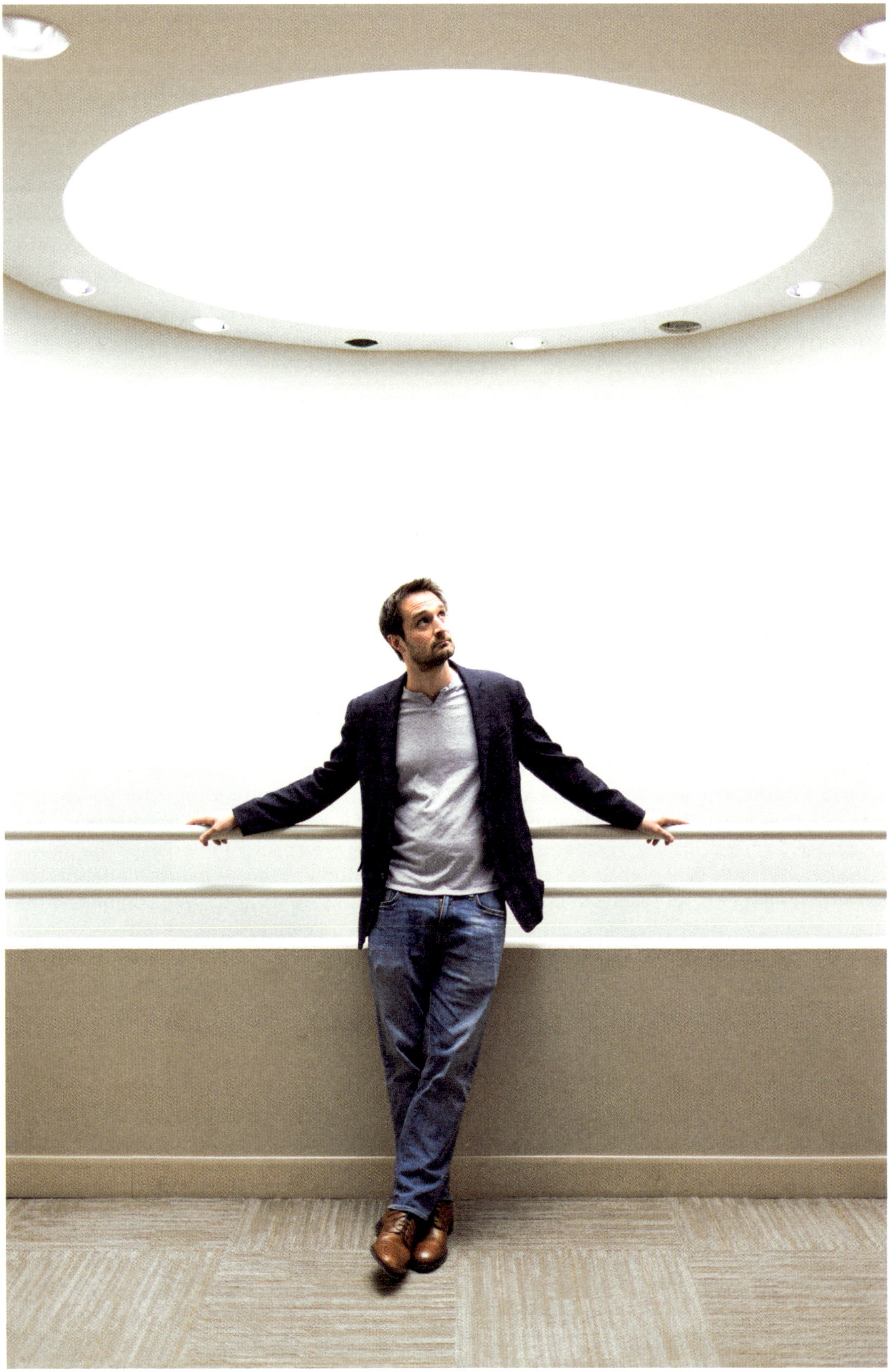

MARTIN BÖHM, HANS DIETL, ANDREAS GOPPELT,
BERNHARD GRAIMANN, AND GÜNGÖR KARA

Reflections on the Human-Machine Interface

Optimizing the human-machine interface is at the center of Ottobock's research. This is not only decisive for technological progress, but particularly for user acceptance of wearable robotics and prostheses. The overarching goal is the fusing of technical aids with the human body so that they are perceived as part of the body. The five experts from Ottobock discuss the current state of research.

Andreas Goppelt: With man-machine interfaces the individual interacts with the technology via a user interface. Christian Lüscher is conducting research into the effect of drugs and substances with optogenetic methods, in other words, with light. This could be very interesting for Ottobock. Up until now we have used electrical signals for the control algorithms, that is, electrodes that rest on the skin or are implanted. Implanted electrodes achieve greater spatial resolution and higher specificity. You can directly access specific sections of the nerve instead of reading off the whole nerve in a relatively coarse manner through the skin. We are already doing this very successfully today, for example, with the MyoPlus hand prosthesis controller, which measures and processes electromyographic signals, the electrical activity of the musculature. Hugh Herr also uses electrode conductors in his agonist-antagonist myoneural interface (AMI) to control an active ankle. These are highly concrete applications that have proven themselves in practice. The question is: which technologies will we be able to use in the future to further improve the control system? Quantum sensor technology could employ ESR signals, which are measurable in an external magnetic field by means of resonant microwave absorption or, as in the case of optogenetics, use light signals for control purposes. Studies from Hugh Herr's laboratory show that light control results in less fatigue symptoms as other axons are stimulated that don't require as much oxygen. Light is very fast and can be very precisely controlled. To do this, foreign genes are smuggled into the target cells, which lend the proteins light-sensitive and fluorescent properties, enabling cellular activity to be measured with light. The muscles are controlled using light. That is a very advanced form of control that is still in its infancy. But what fascinates me here, once again, is the increasing convergence of the life and engineering sciences. Ottobock is not active in this field of genetic engineering, but we will monitor this research very closely as the principle is very exciting, and theoretically it will enable us to significantly improve the quality of life. This can also be deployed in other areas. For example, Hans Dietl is currently engaged with optical cochlea implants, which are being researched by the neuroscientist Tobias Moser in Göttingen.

Hans Dietl: The classical cochlea implant is a hearing prosthesis for the deaf and a very widespread implanted neurostimulation application that generally has sixteen electrodes. A speech processor translates the speech into a signal pattern, enabling deaf people to communicate using rudimentary speech intelligibility, without having to even see their conversation partner. The group around Tobias Moser in Göttingen aims to modify the auditory nerves using genetic engineering so that they are light sensitive. As a result, instead of the

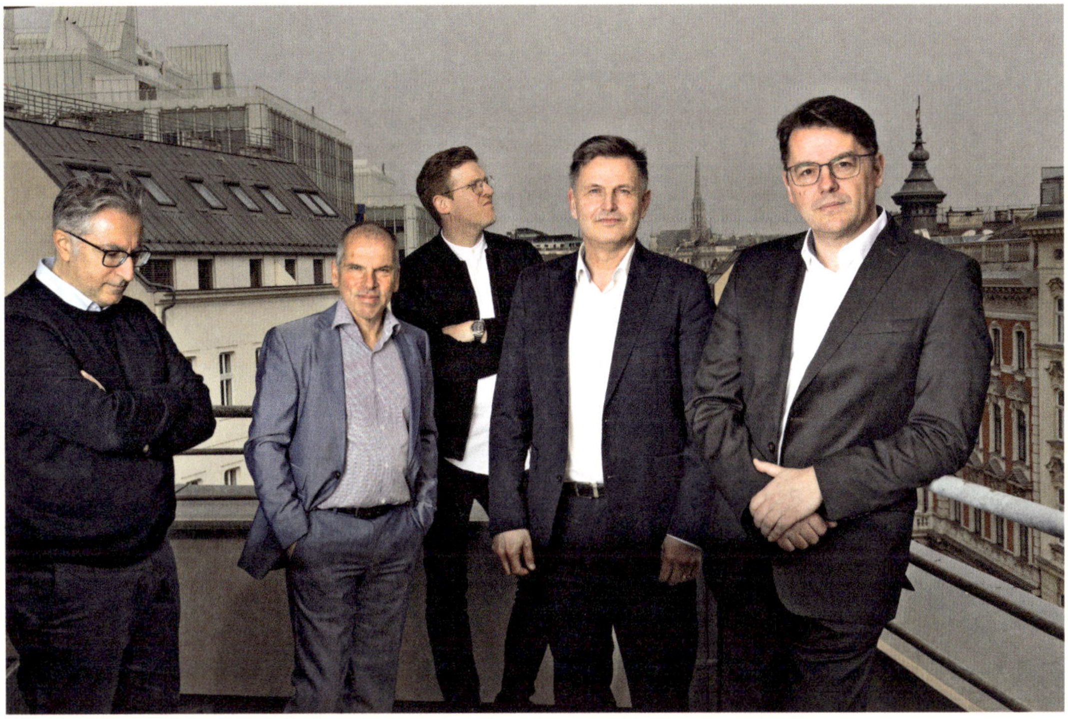

Güngör Kara, Hans Dietl, Martin Böhm, Andreas Goppelt, and Bernhard Graimann (from left to right)

electrical electrodes, they will introduce a light conductor with sixty-four exit points and modulate the speech as a light pattern. The hope is that with sixty-four channels the speech quality will be substantially improved in comparison with conventional cochlea implants. At Ottobock stroke patients with difficulties lifting their feet would be an interesting application case for optogenetic procedures, as a means to improve their walking. The circulatory disturbance of the brain tissue caused by the stroke means that the brain is not able to send enough impulses—or sends the wrong impulses—to the foot-lifting musculature to activate them. Instead of activating the foot-lifting musculature with functional electrostimulation using electrodes on the skin or an implant, the muscles could be optically stimulated via the surface and more precisely addressed. Ideally an implant would not be necessary.

AG: Stimulation with electrical signals is often associated with pain and scarring. The electrical resistance increases as a result and the quality of the signal deteriorates.

HD: Electricity through the skin is not optimal for the user, but light through the skin does not cause pain. That is a possible vision.

Bernhard Graimann: As in the past, long-term stability is still a very big problem with invasive, high-resolution nerve interfaces that go directly into the nerve.

AG: Controlling using light signals could preserve the long-term stability.

BG: As in the case of the cochlea implant, you could remain on the surface of the nerve. A light source can be much better focused than an electrical field, and stimulation is thus more selective and precise. An optical electrode has a far higher resolution than an electrical one. Furthermore, I wouldn't apply it to a muscle but to a nerve.

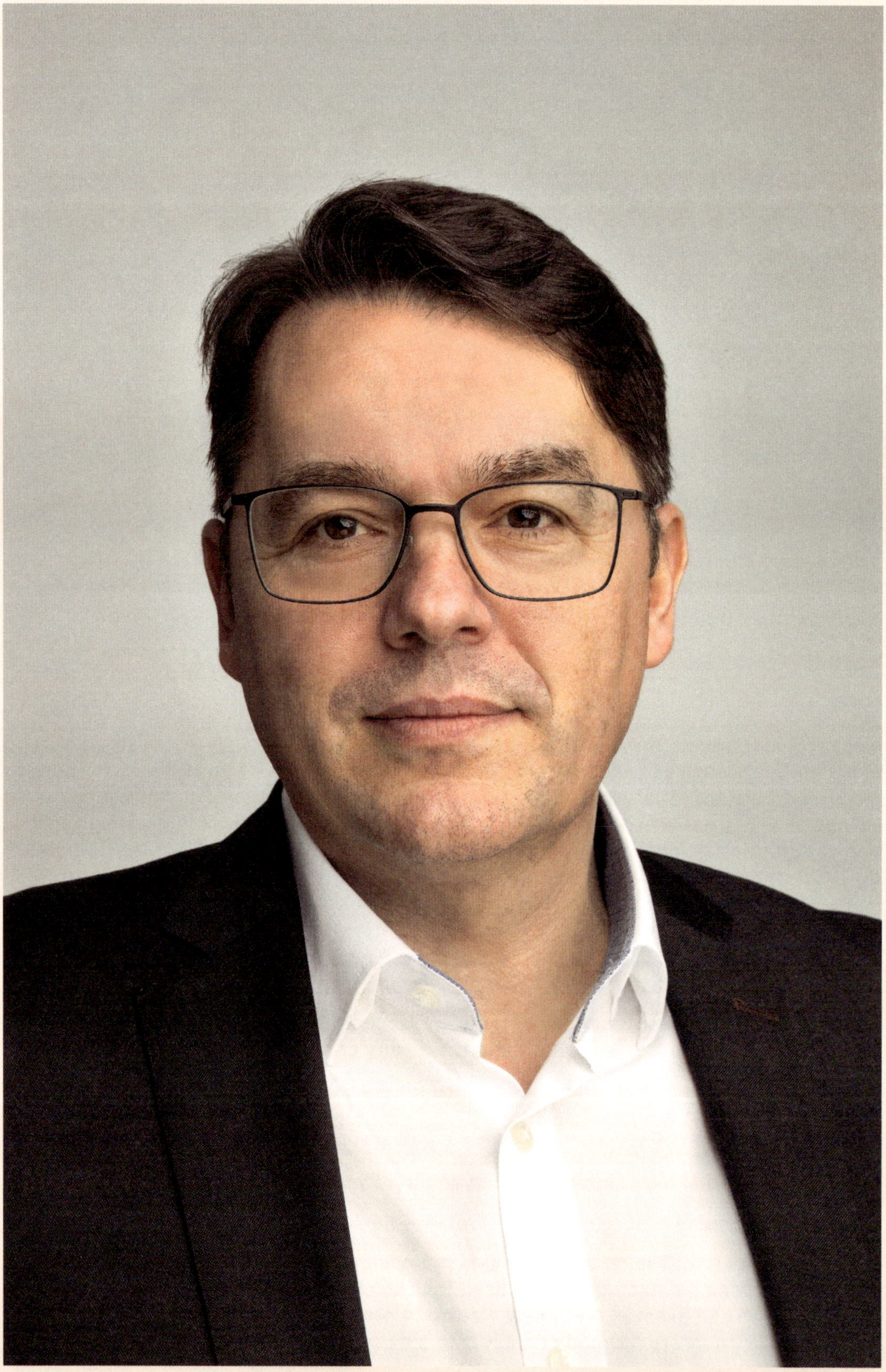

 Dr. Bernhard Graimann, Vice President Global Research & Development, Ottobock

AG: Hugh Herr also applies it to a nerve.

Martin Böhm: I am not a physician and a complete layman in this area, but I read a lot and like learning. With the nerves we can reach the "cables" but not the "processor" in the brain. However, if I want to address all the senses, then I would actually have to go to the top, the brain?

AG: This is what Elon Musk is doing. His vision is to enable people to see again—and his approach begins with the brain, where the signal-to-noise ratio is unfortunately highly undifferentiated. In contrast, the nerve signal is more differentiated at the periphery. This is better for our purposes.

BG: Ultimately, it is the application, the needs, and the user acceptance that decide what is the best approach. User acceptance plays an especially important role here. Just to give a brief overview: there are invasive and noninvasive systems. We can use this to measure motor information or stimulate somatosensory information. This can take place at the central or the peripheral nervous system. Brain-computer interfaces measure or stimulate noninvasively on the head, invasively on the cerebral cortex, or directly and intracortically in the brain. At the peripheral system we can pick up myoeletrical signals noninvasively on the surface of the skin, or invasively on or in the muscles. In terms of performance, invasive systems that pick up motor signals directly at the peripheral nerve or stimulate somatosensory sensations are especially interesting. The big challenge here is the required long-term stability of reliable connections at the interface to the nerve. Optogenetic approaches promise higher selectivity together with improved long-term stability.

HD: With an electrical electrode you can measure and stimulate, but optically you can only stimulate. I can't measure anything because the nerve does not light up. The information does not come back to me. It is a one-way street.

AG: However, in principle, molecular biology is also capable of making the nerve light up.

BG: For some applications stimulation is very important, for example, deep brain stimulation in the case of Parkinson's. It is always a question of what problem we want to solve. In the case of the hand prosthesis, the prosthesis user wants to be able to control the hand. In current prostheses we only have a restricted number of six degrees of freedom. That is already quite respectable. However, with the natural hand there are twenty-seven. This also raises the question as to which interfaces can be usefully employed here. And we also have to consider other issues such as costs and user acceptance. We have to design the product to be economical and viable. Only then will the product also be available for clinical applications and help patients. And here a noninvasive system is definitely preferable to an invasive system. You can't control things as precisely and with such variability, but the "good enough" principle could be the killer argument for the expensive invasive system. We always have to assess whether and how we can continue to develop noninvasive systems. Quantum sensor technology, which remains on the surface of the skin and can be more selective than other technologies, could come into play again here. Hugh Herr inserts small magnetic capsules into the muscle—that is a minimally invasive system that measures muscle movement and can thus control prostheses.

AG: We already have a lot of experience with all these technologies, for example, with stroke patients who suffer from foot drop and frequently stumble or drag their foot behind them. Today we are working very successfully with the surface stimulation method FES (Functional Electro-Stimulation) using cotton sleeves. Electrical signals are passed through the skin transcutaneously. Naturally this can also cause pain, and that is why you have to consider the advantages and disadvantages or rather the costs and benefits. And with costs I don't just mean money, but also possible pain, sweating, and so on. And what distinguishes our neuro-implant ActiGait is that it stimulates in the fibula nerve. In addition to lifting the foot, users can also control it. This is a very nice example of the advantages of the selective stimulation of nerve sections, as this enables you to do things that you can't do with surface stimulation. Ottobock has actually proven this on hundreds of patients through to market readiness. But at the end of the day the cost-benefit ratio is decisive, which is why we haven't pursued this technology any further, although we have learned a lot about human-machine interfaces.

HD: In clinical studies conducted with Oskar Aszmann we have also had experience with implanted myosensors for the control of prostheses. These are small sensors that are implanted onto the muscle and are supplied by an external coil. I believe that the future of the man-machine interface will be with implants.

AG: That is definitely correct in terms of the technology. Nevertheless, the overall superiority of such invasive systems compared to surface stimulation must be so great that the associated acceptance problems can also be overcome.

BG: And I would like to emphasize once again: cost-benefit aspects are decisive here.

AG: Cost is not just measured in dollars, but especially in acceptance. After all, implants are surgical interventions that can also go wrong.

BG: The peripheral system functions sufficiently well for orthotics and prosthetics. But we could simplify the interface through a selective nerve transfer or bionic reconstruction—in other words, through a neurosurgical measure. It could be noninvasive—that is, just on the surface—or minimally invasive.

AG: Up until now we have discussed the human-machine interface in a highly focused fashion. How do I establish an optimal nerve connection or signal control? Do we go through the muscle or the nerves? Invasive or noninvasive? Brain or periphery? But we should consider the issue of human-machine interfaces more broadly: a wheelchair that can be used without any discomfort or an individually tailored socket—these are all themes that Ottobock is addressing in order to improve the user experience.

MB: We talk a lot about new technologies and research. To be honest, this is too engineering-driven for me. "Patient first"—in other words, the psychological acceptance of the physical coupling—is not given sufficient attention in my opinion. The users must also connect with their prosthesis psychologically; they must also accept a foreign device as part of their own body.

HD: That is embodiment: a question of physical manifestation, how this system is received by the human body as a whole. What do the users feel? Can they perceive the foreign body as part of their own cognitive system?

Güngör Kara: This is already a part of conventional patient care. It is called *rehabilitation*. Unfortunately it is frequently gets a raw deal since the classical patient care business tends to be more interested in pushing the patients through as quickly as possible, as rehabilitation is not reimbursed to the extent it actually should be.

MB: People are all striving for physical reintegration and functional and mental enhancements. That is why technologies such as augmented and virtual reality are being increasingly deployed in the rehabilitation field.

AG: If you take a look at the start-up CTRL-Labs, which was recently sold to Facebook for an astronomical sum and also collaborates with Dario Farina, it's a real beacon. Mind control for consumer electronics for all people, who, for example, control their laptop with gestures or a keyboard. Technologically a relatively simple product—but clearly a huge market.

BG: The system from CTRL-Labs functions very well if you still have a healthy hand, in other words, if the complete sensomotoric hand system including proprioception is still intact. If that is no longer the case, because the hand has been amputated, controlling the remaining musculature is considerably harder.

AG: However, we can further extrapolate the solution to problems that arise in our core area of orthotics and prosthetics (O&P) in the direction of the consumer and make better use of all our experiences. Ottobock is a gold nugget because it is permanently working on such themes and if they were translated could have far greater leverage. CTRL-Labs has a target market of eight billion consumers. In principle what we are working on and how is also suitable for this. Here it is not about repairing the human body but augmentation.

BG: That is a way off though.

AG: Nevertheless, we can still think about the people who don't primarily have a deficit. Ultimately this is transhumanism, or the fusion of humans with technology. In the future the technologies that we currently carry on our bodies as wearables will be inside of us. The human will finally develop into a cyborg. That is a broad field. But it also offers unimagined market opportunities for Ottobock.

Wearable Robotics / Bionic Reconstruction

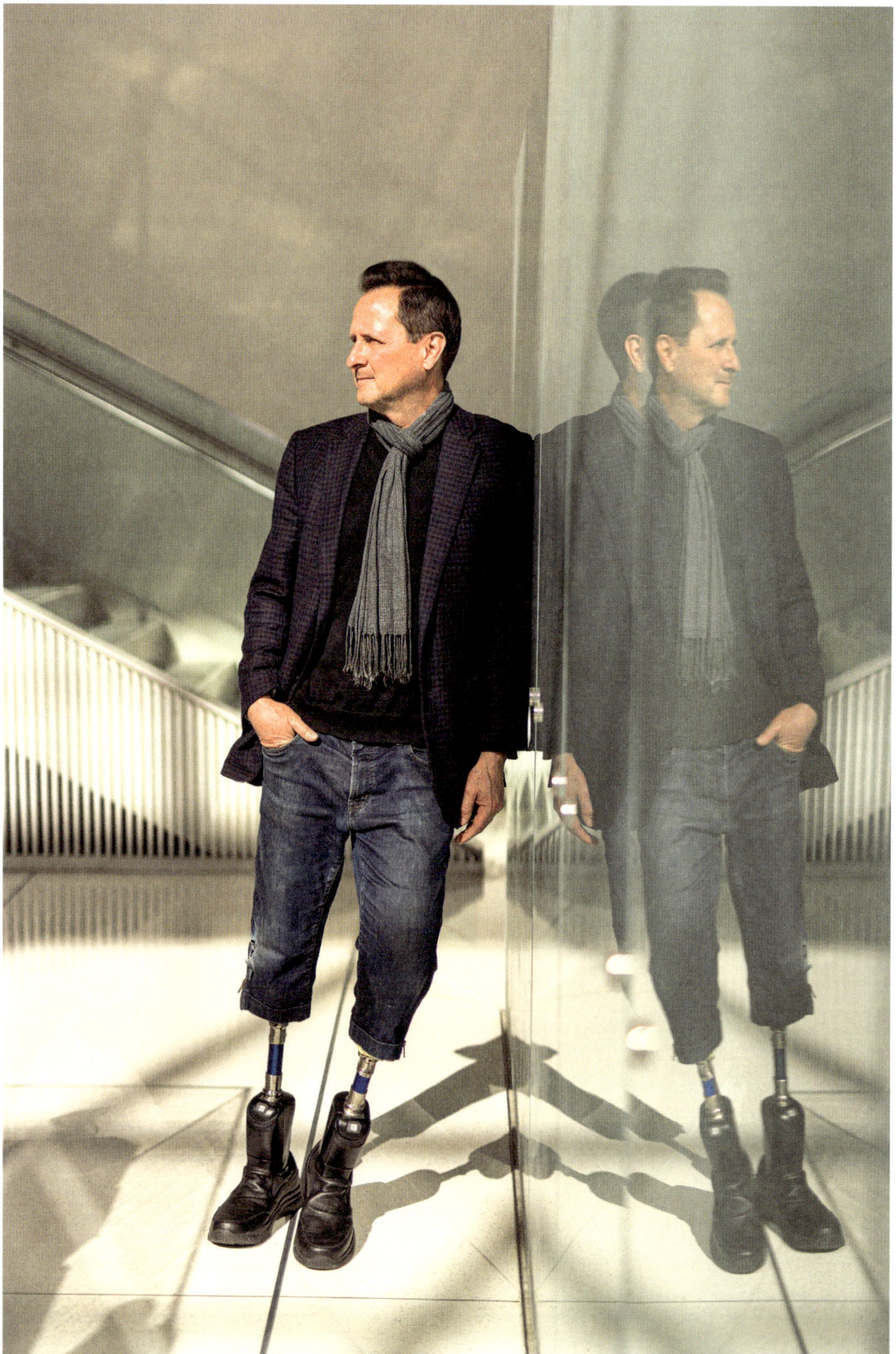

HUGH HERR

"We Fundamentally Change Human Identity"

The famous pioneer of the Bionic Age creates a real cyborg by closing the loop between synthetic computation and the nervous system. A double amputee himself, his concept of neurological embodiment is profound, and he has developed the most comprehensive proprioception feedback framework in the world. As a professor at the MIT Media Lab and coleader of the K. Lisa Yang Center for Bionics at MIT, he would like to overcome the narrow view of beauty and intelligence that exists today. His hope is that the bionic revolution will enable a magical transformation of our bodies and greatly expand human diversity.

The last time we spoke three years ago you told us about your breakthrough in sensation feedback in prostheses, and I hear that you already have around two dozen patients. How is your work progressing?

In 2014 I invented a technique called *agonist-antagonist myoneural interface*, or AMI for short. The AMI comprises two muscles that are linked mechanically in series such that when one muscle contracts, it mechanically stretches the other. For every prosthetic joint or degree of freedom, we build at least one muscle pair, or AMI. When a person thinks and moves their phantom appendage, those muscles are fired and move dynamically as if there were a real physical joint, and due to proprioception, the person feels the full range of motion of their phantom limb. When that person then looks down and sees the mechatronic prosthesis, those movements correspond to their natural movement sensations. Through a collaborative effort between MIT and Brigham and Women's Hospital, AMI amputations have now been implemented at all anatomical levels: below knee, above knee, below elbow, and above elbow. We continue to gather enormous efficacy data, that is, relating to the performance of the device or technique.

What do the patients report on the feeling of this new technology?

It's very interesting. When patients are neurally linked up to the prosthesis in a bidirectional way, they can direct the motors to move by thought and at the same time feel sensory information from the prosthesis relating to its position, speed, and applied force. The bidirectional link makes patients say that they have their leg back, their body back, their life back. I refer to this as a neurological embodiment, as opposed to a technological embodiment. In the literature, the classic example of technological embodiment is a person with seeing impairment who is technologically embodied with a cane. However, if you ask that person to explain their body, they won't include the cane. With neurological embodiment, the embodied person actually includes the brain-interfaced device when asked about their body. That is just very extraordinary because it indicates that we're at the point in human history where we're transitioning from a mere tool use to something far deeper and profound where there's an actual integration of a designed construct with human physiology. Humanity is now at the cusp of a new technological age that will enable the design of the human body itself. Through such physiological sculpting, humanity will design itself and fundamentally change human identity.

So you actually created a real cyborg?

That's right. My definition of cyborg is that bidirectionality piece: when you not only can think and affect synthetic computation, but synthetic computation actually inputs information into the nervous system, closing the loop, if you will.

That's a huge leap since our last interview.

Yes, we now have the most comprehensive proprioception feedback framework in the world. We're extending it to cutaneous and touch as well. Through my new MIT center, I now have funding to do a large clinical trial of a bionic reconstruction of the leg that involves cutaneous feedback, proprioception, and full brain control. Arguably, it'll be the most advanced bionic reconstruction ever conceived and implemented.

We took a picture of you in the first book that showed you moving your hand with your eyes closed, and then you said that a patient should also feel the position of the prosthesis.

Just as if the prosthesis were made of flesh and bone. Neurological embodiment is a very deep idea.

Tell us about your center.

It was launched here at MIT on September 20, and it's called the K. Lisa Yang Center for Bionics. Lisa Yang is a well-known philanthropist at Harvard and MIT. She and I have a close relationship, and we built that relationship over some time. She invested twenty-four million dollars to launch the center. We're going to spend ten million in the next four years, and then fourteen million will be endowed. So I now have a fair bit of money to do first in-human clinical trials for a number of technologies impacting amputation and paralysis.

How many participants do you expect? Are there hundreds of patients?

No, typically in bionics, patient numbers are on the order of five to ten due to the expense. It's distinct from large pharmacological studies.

Have you built individual prototypes? Or is it already in a more manufactured phase—because there are so many people who need these prostheses. How can you build them all?

I design bionic concepts with a knowledge of translational challenges, including regulatory and cost issues. My designs have translation built into them. For example, we were able to perform the first AMI amputations in humans without calling the Food and Drug Administration, because you can perform an AMI with only soft tissue manipulations that do not necessarily require synthetic implants in the body. This enabled us to move very quickly toward proof of concept in humans. We are transitioning from these surgical techniques and marrying them to minimally invasive implantables. We're only implanting passive beads—like tiny little pearls in the body without any wires—that require just a very small incision and insertion tool. Practically, the beads can be inserted in an outpatient basis.

Do you cooperate with a particular hospital or operation center to do all of this?

The model is the Center for Bionics at MIT, and we developed the concepts and advanced them preclinically using animal models. We typically team with a hospital under the Harvard Medical School umbrella. We've done tremendous work here in Boston with Brigham and Women's Hospital and Dr. Matthew J. Carty. We plan to expand that to additional hospitals in the area, and then from that foothold, we expand to Walter Reed National Military Medical Center in Bethesda, Maryland. We're teaching the AMI amputation to folks in the western United States, Europe, and beyond. We've done over thirty patients, and so far, all patients have been largely reimbursed by standard insurance. If that's maintained, all that's required to translate this globally is teaching. Dr. Carty taught a

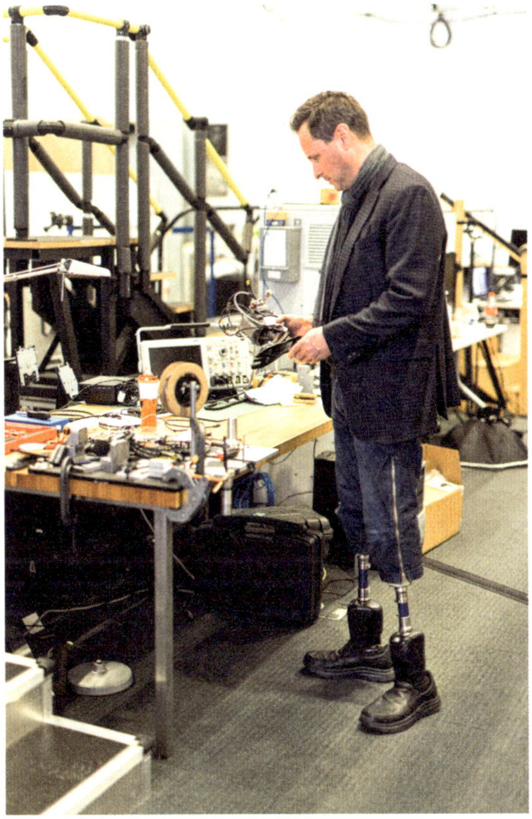

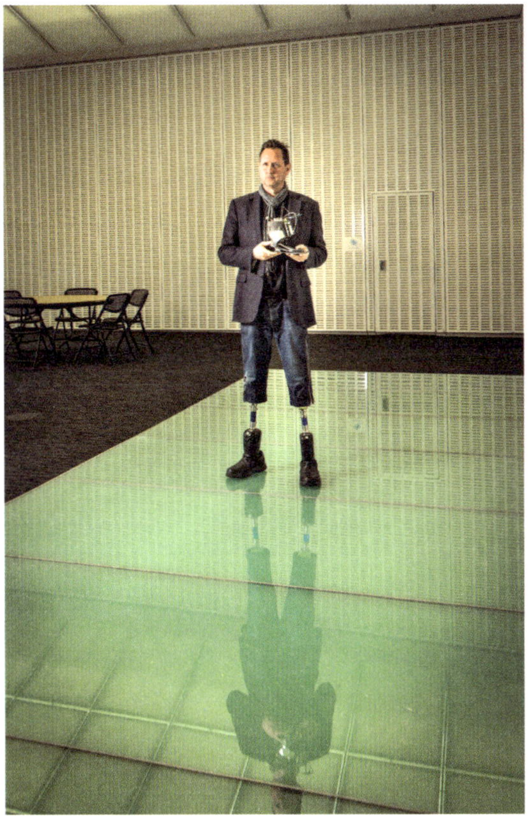

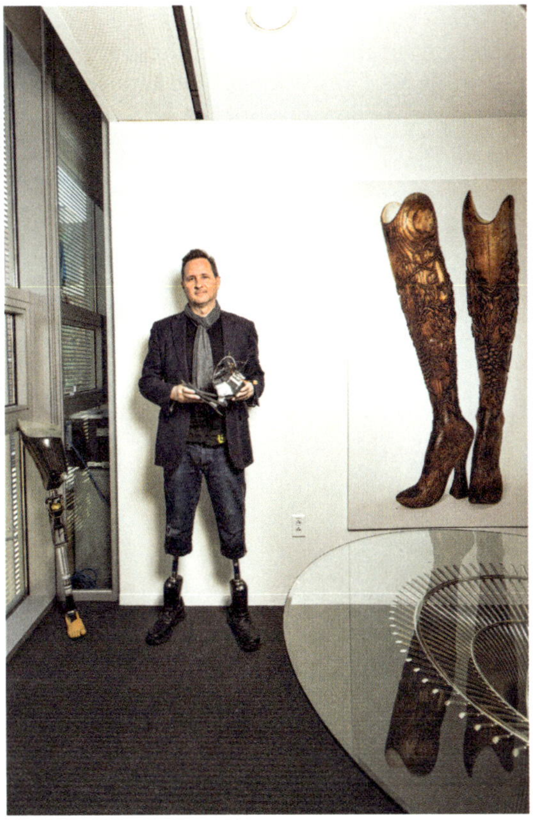

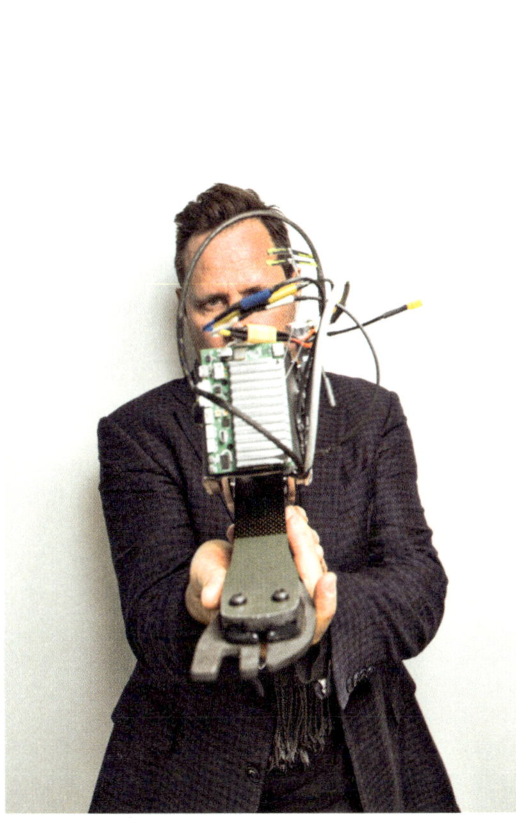

fellow surgeon over a Zoom call, so we're confident that this form of amputation is highly translatable.

Anyone can do it?

Sure, with training. We want to go beyond surface electrodes, which is the standard neural interface for prostheses. Electrodes can be uncomfortable, and the signal quality is poor. There's only so much control one can do with a surface electromyographic (EMG) sensor. It basically tells you whether the muscle is on or off. What we want to know is the length, speed, and force of the muscle because we want to control the position, speed, and torque of a prosthetic joint. So we've invented a technique called *magnetomicrometry*, in which we implant small three-millimeter magnetic spheres into a tissue—in this case, a muscle. And then with electronics on the skin, we have an algorithm that runs the mathematics to compute where those magnets are in three-dimensional space from the fields that they create. Then we compute the distance between the beads, so as the muscle shortens and lengthens, we have a submillimeter accuracy and high precision measurement of what the muscle is doing in terms of length and speed. If you combine magnetomicrometry with the AMI amputation, I believe we'll see a step function in clinical care.

Is that because you can precisely communicate muscle function to the prosthesis?

Yes, it doesn't require any active technology within the body. It's minimally invasive, since there are only passive elements implanted within the body. This means that I can implant an eight-year-old, and I never have to go in and open his body up again to fix anything, even if he lives to 108. We want to measure what the muscles are doing and then do a master-slave control from the muscle dynamics to the prosthetic joint motors. We build AMI constructs that comprise native nerves and muscles to the degree of freedom that we're controlling. So, for example, if we want to control the ankle, we use the *gastrocnemius* and *tibialis anterior* that natively move that joint, and in so doing eliminate the need for training. It's completely seamless. So when a person with an AMI amputation thinks and they move their phantom ankle, they feel that movement, and the bionic limb is directly commanded in those precise movements, effectively coupling movement with perception.

What is the role of artificial intelligence? Is it an algorithm that just uses statistical machine learning systems?

If you biologically recreate agonist-antagonist joints within the amputated residuum from the native nerves and muscles, and you have a direct measure of those muscle dynamics, you simply do not need AI to restore natural biomechanics. You're tapping into the biological intelligence, the human brain. If you give it the right percepts or afferents, the brain knows exactly how to control the prosthesis.

How long does the brain need to learn how to control it? Does it take weeks?

It's immediate. Because even when your leg is amputated, your brain still remembers how to control your ankle or your knee. It's like riding a bicycle.

Wow.

That's where we're heading at the Yang Center for Bionics. We want to translate this rapidly into humans. My dream is to have a person with an arm amputation playing Mozart on the piano at normal speeds. My dream is for a person with a leg amputation to be able to dance ballet. Such artistic performances to me represent the extremes of human expression both artistically and in terms of human physicality. My ultimate dream is to express such limits of human passion through electromechanics. Bionics need not be rigid, cold, and devoid of humanity, but rather should embody what we cherish and hold dear.

Last time we had a discussion about what it means for a human being to be converged with technology: we live longer, we use technology not only to repair the missing part of the body, but to enhance human powers and abilities. Everything is possible—we just have to imagine it, right?

The limit of human augmentation extends to the outer boundaries of humanity's collective imagination, constrained only by the laws of nature, a limitless volume for which we cannot foresee or comprehend in its entirety.

Does this mean that the vision of transhumanism is getting closer?

Indeed. That brings us back to neurological embodiment. We've now conducted a study where we image the brain. We have a person in a functional Magnetic Resonance Imaging (fMRI) machine with completely normal, intact limbs, and we have them move their limbs in certain ways as we're imaging their brain at high magnetic field strength. We then put persons who are matched in terms of age, height, and whatnot, who had an amputation, both standard amputation and AMI, and we image the brain with the same phantom-limb movements. It turns out that for a person with an AMI amputation, the proprioceptive activity in their central brain is indistinguishable from a person with physical limbs and completely distinct from a person with a standard amputation. It's almost as if the central brain doesn't even realize the limb has been amputated in the AMI case.

So that means the quality of the information is not only in the brain, it's in the muscle.

It's being sensed in the extremity and provided to the central brain. One day we'll be able to amputate a limb and have such perfect signaling into and out of the central nervous system, that that central nervous system won't even realize the limb is amputated. And then, the titanium, the silicone, the synthetic computation will be equivalent to flesh and bone in terms of brain representation. We're already seeing glimpses of that phenomenon with our current work. We want to deeply understand embodiment. How does one produce it? How does one recognize it? And what does it mean for humanity? We don't want prostheses that are driven blindly by motors and AI. We can capture elaborate human dynamics with machine learning, but to the human who's sitting on this technology it'll feel like they're in the back seat of a car. Whereas if we directly link the brain of the human bidirectionally to the mechatronics, the human will feel that their hands are not only on the steering wheel but that their body is embodied with the car. In my view, it's very critical that we move in that direction in the field. Now, will AI be important? Yes, sure. We can augment once we have beautifully, elegantly, and seamlessly linked the nervous system to synthetic computation. And once we have these glorious bionic reconstructions and can match normal physiological function to bionic function with embodied electromechanics, then sure, we can use AI to augment a person beyond innate physiological levels.

But this is not the kind of connection you developed.

In my view the first step—pun intended—is to link the brain to bionic limbs. We need to first harness this computational center, and the value of doing so is embodiment. Our goal in the field is to rebuild humans and have them feel that, although they suffered from a disease or a traumatic injury, they can return to their normal pre-disease or pre-injury physicality if they so choose.

You learned a lot about how the brain works with the extremities. The information from the brain is only complete when the embodiment is united with the muscle and the sensory information.

Your hand would cease being your hand if you couldn't feel it, right? If you couldn't feel it, it would be more tool-like. The fact that you can feel it and move it makes it you. My thesis

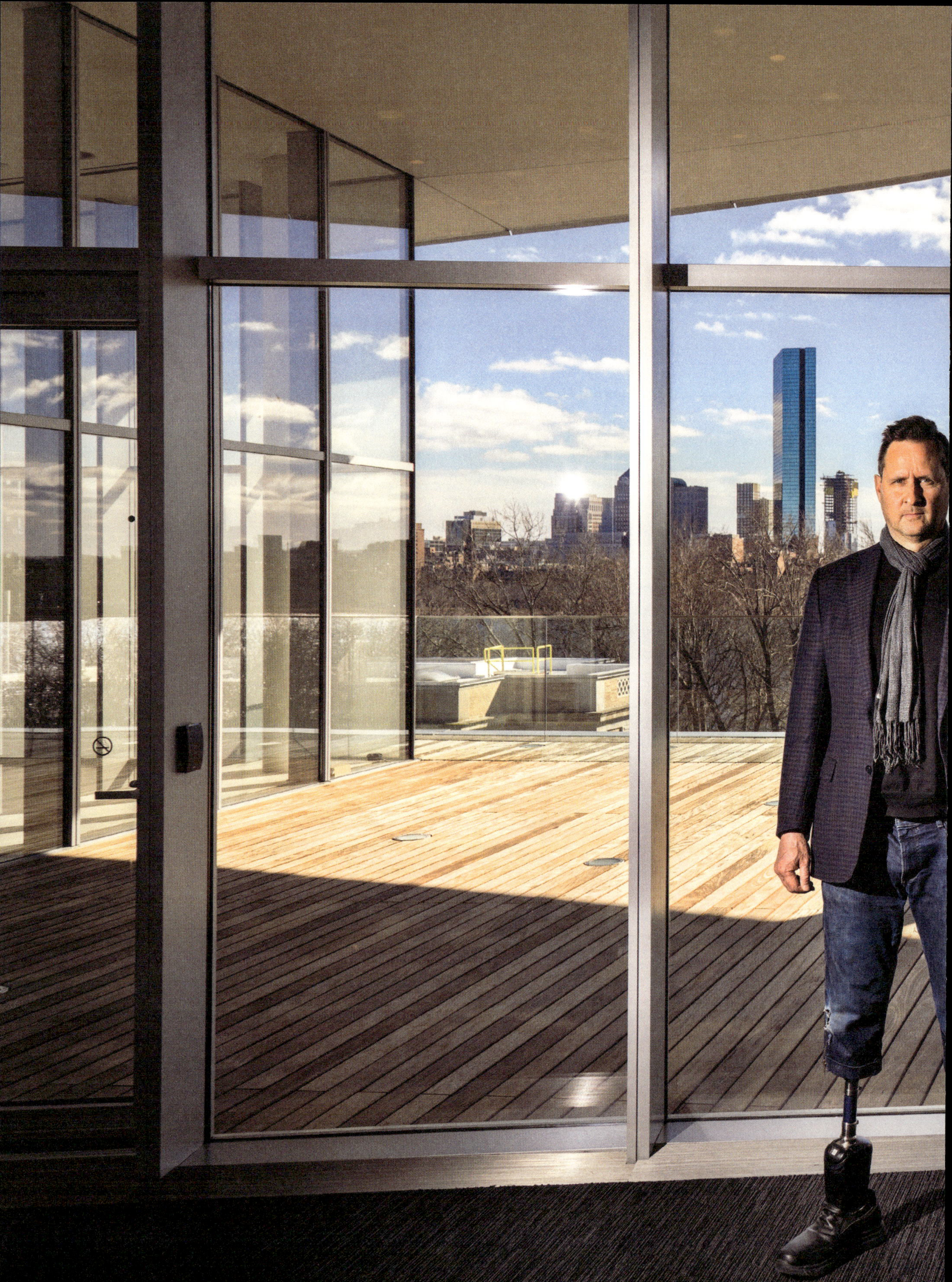

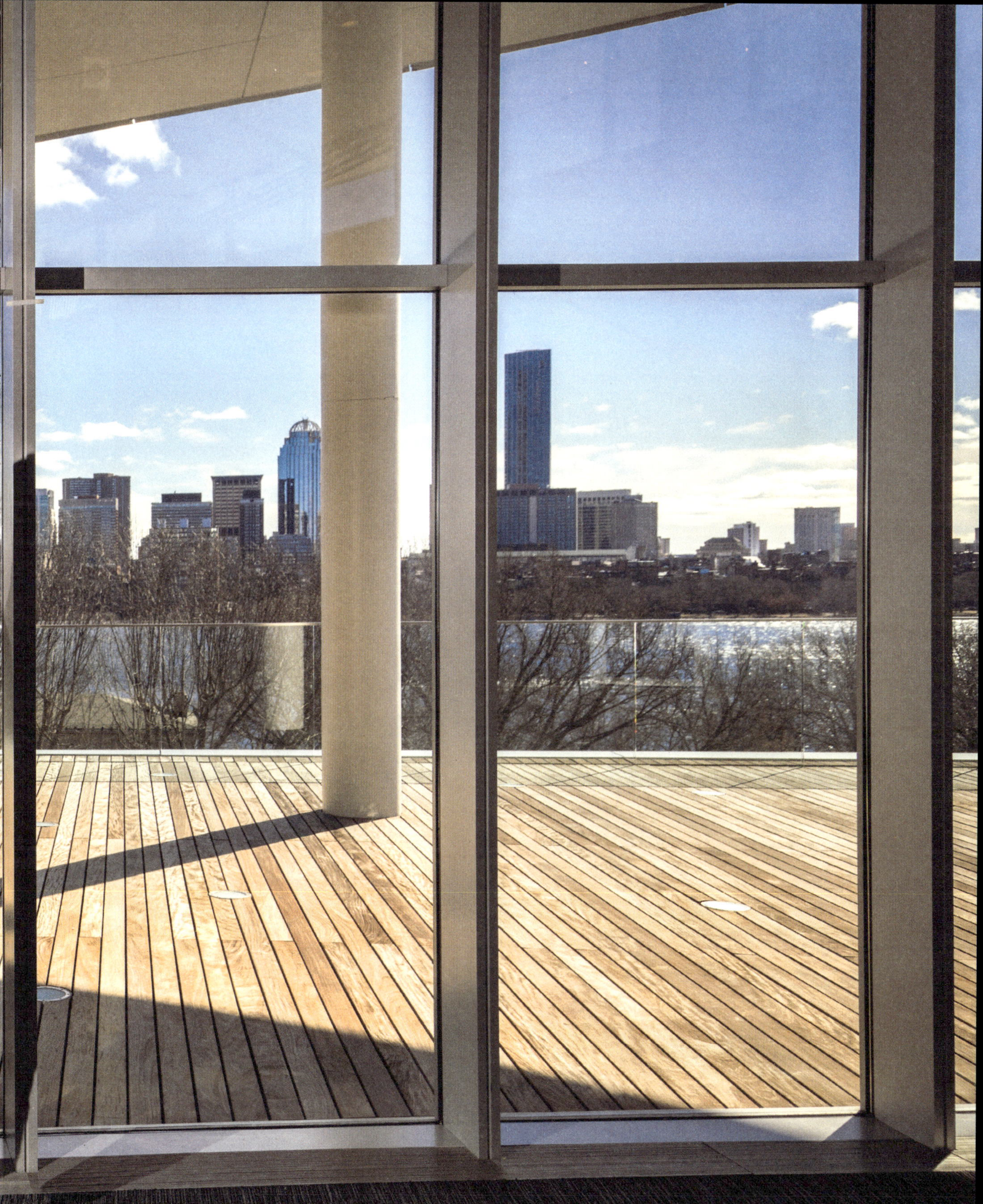

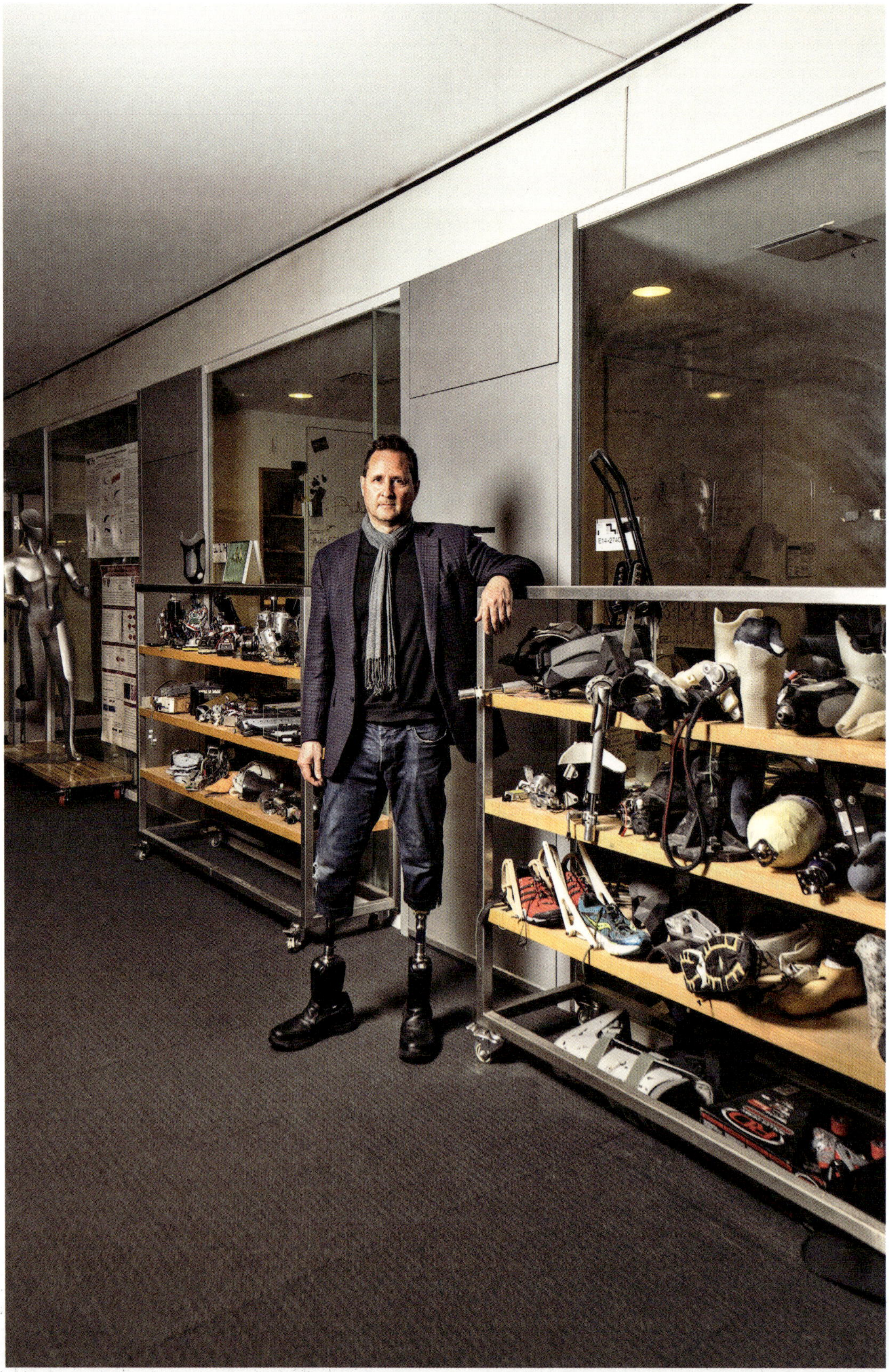

is that even nonanthropomorphic technologies can be embodied within the brain, like wings or a third arm. In a hundred years, human morphology and dynamics will be unrecognizable from what exists today. If we can sculpt our bodies and have designed constructs neurologically embodied within our nervous system, the level of human expression that we'll see will be utterly remarkable and surprising.

There is also a military side of turning humans into warrior-machines. Is there any progress in that part of the industry?

In 2014 my lab here at MIT was the first to build an autonomous leg exoskeleton that augments human walking. We now have an exoskeleton where a person with completely "normal," intact limbs can put it on and walk and run with reduced levels of metabolic energy. Bionics used for augmentation will go across many business sectors, military included. Industrial warehouse workers will use exoskeletons. Athletes will use exoskeletons.

So you can also develop bionic devices for everybody just to enhance their abilities? It would be a much larger market than amputees.

Sporting bionic apparel will be a massive business opportunity. Imagine augmentation technologies that enable the athlete to not injure. Imagine running your best marathon time when you're sixty, when you have been running marathons since you were eighteen. Augmentation technology will allow your innate athleticism to not degrade with biological age. So yes, it's a massive business opportunity that we're just seeing emerge now.

What do you think about Neuralink, Elon Musk's new project? He wants to take the information as an interface in the brain and expects that the quality of the information will be good enough to use in all the other projects. He's a genius. His ambition and activity will bring a lot of capital and attention to the field, which is good for all of us. But do you think his project will work? Have you talked to him about that?

No. We need interfaces to the nervous system at all levels: muscles and nerves, spinal cord, and central brain. For someone like myself with a healthy brain, spinal cord, and nerves out to my amputated peripheral system, it's probably fairly unwise to put chips in my brain when all you have to do is connect to my muscles and nerves. But, of course, we need bidirectional interfaces to the central brain for locked-in patients, for high spinal cord injury patients, and whatnot. The winning approach to interfacing to the central brain is not yet clear. Elon Musk and Neuralink are taking a fairly traditional approach to brain interface. Their approach is to place fine wires into brain matter, actually piercing the tissues, near the brain's surface. Such an approach is highly invasive, lacks specificity to individual cell types, and has poor spatial targeting—only superficial areas of the brain can be addressed. A very different approach is to enter the brain through the brain's vascularization. Imagine nanoelectronics entering very deep parts of the brain through the vascularization pathways, and then using external power sources, one could stimulate via the nanoelectronics electrically, optically, or ultrasonically in order to target individual cell types throughout the brain volume.

Could it be used for rebuilding the neurons in Parkinson patients?

Yes, for all kinds of applications. Also, for the treatment of inflammation. I have a colleague who's developing nanodevices, linking them to immune cells to deliver pharmacological interventions directly to those areas of the brain to treat Alzheimer's and many other disorders. So, again, Neuralink is taking a traditional approach. There are other approaches that may be far superior. It is rather unlikely we will be inserting wires into the brain three decades from now.

If he invests billions and billions, there will be an outcome.

Sure.

But our brain is not a computer interface that can be expanded by an extra terabyte.

The brain is remarkably complex. It's believed to have tens of thousands of different cell types. If you look at the electronics of a smartphone, there are only a modest number of distinct electronic chips. The brain may have twenty or thirty thousand distinct cell types that do different things, that talk to their neighboring cells in different ways. Conditions of the brain impact differing cell types; for example, schizophrenia affects basket cells. So we need a brain interface with high specificity. Ultimately, brain interfacing technology not only needs to target an individual cell type, but ideally, a particular cell within that cell cohort that isn't signaling properly, or an individual circuit that isn't operating effectively. Thus, techniques like optogenetics where you can target cell types will be critical to the future of brain interface. Again, Neuralink is taking a very traditional approach to an organ that's deeply, profoundly complex.

It's not only about electronic signals; there are biochemical signals, enzymes, and a very complex system of electronic and biochemical processes. So it's not enough just to take the electronic signal from the brain tissue, is it?

If you put a chip in the brain of a person, for example, with no arms or legs, and have them control the robotic hands, sure, it'll work—once you have solved the risk of putting the implant in, the risk of causing a seizure, and the issue of scarring. So if you throw enough money at it, sure. But if you want to solve the brain, if you want to have a revolution in brain science and a revolution in the treatment of brain conditions, again, you need a far more sophisticated approach. It has to have that particular cellular specificity. The interface has to be seamless again, exploiting the vascularization of the brain where you're not piercing the brain with synthetic wires. So it's unclear. I do not believe the winning solution has been discussed in the media yet. But I do believe it's being worked on.

Have you seen it already?

All these elements are being conceived and developed.

Will mRNA technology play a role in pharmaceutical strategies to help them destroy the wrong cells?

Genetics and CRISPR will play a paramount role in the menu of technologies that will lead to the elimination of disability and disease. You might naively think that the MIT Center for Bionics is a bunch of robot-assisted AI experts. It's geneticists, neuroscientists, tissue engineers, surgeons with a sprinkling of roboticists and machine learning specialists. We're dry, but we're also very wet.

It's not enough to think in terms of robotics; the body is much more complex.

Because we're not robots. We're soft and mushy, chemicals, electricity, and mechanics.

What are the psychological consequences of constructing ourselves in our brains depending on how we see ourselves? Isn't it a problem for our self-identity the more technology we have in our bodies? Or is this changing? Is the brain flexible enough to accept that you have, say, thirty percent technology in your body, but you're still a human being? Is it only for very stable people with extremely fit minds? Or does it work for everyone?

It's for everyone. Today we have a very narrow view of a beautiful man and a beautiful woman, and we're just beginning to go beyond this gender profiling.

Your vision is that I can go to bed as a woman and wake up as a man. Identity politics from American universities have started big discussions in Germany, which is actually cementing the roles. If you are a biological female, can you change your role with technology? Your vision is that you can be anything.

Yes. In the realm of gender, the phrase *body alchemy* has emerged. I love that phrasing because it'll extend to all forms beyond gender of human expression. You know, the definition of *alchemy* is "a magical transformation." To use technology to create alchemy of our bodies is wonderful.

This was our headline last time. I love it, and I told everybody about it. We have a lot of discussions about gender fluidity in Berlin, and I always say, "Wait, technology will free you. It isn't necessary to cut your dick off. You can be anything."

We have a narrow view of beauty and intelligence today. My hope is that the bionic revolution will greatly expand human diversity.

Are we ready for that? Is the human brain ready for that?

I believe the mergence of electromechanics with human physiology will be a massive lever for human creativity. We're so limited in today's world. We can think of things but we cannot always make them happen. In tomorrow's world, we'll have these incredible levers for our own imagination. If you want to run around the rings of Saturn, beautiful. If you want to have a third arm, fantastic. I hope that in fifty years when one walks down the street of any major city in the world, one will see this extensive human diversity that is the consequence of human imagination manifested into physical form. A world in which there is no longer a social construct of human difference from a norm—where difference *is* the norm. A world in which each human being is unique, cherished, beautiful and has absolute authority over their own expression. This nonsense of what a beautiful man or woman is, of what intelligence is—all these narrow views we have today will be shattered. That's my hope. Humanity could go in the opposite direction where there is a collapse of human diversity. In my view it would be a disaster for human diversity to allow parents to design their future babies based on some narrow view of beauty and intelligence. But I think society will get it right, and we'll finally break out of these rigid molds that we now find ourselves in.

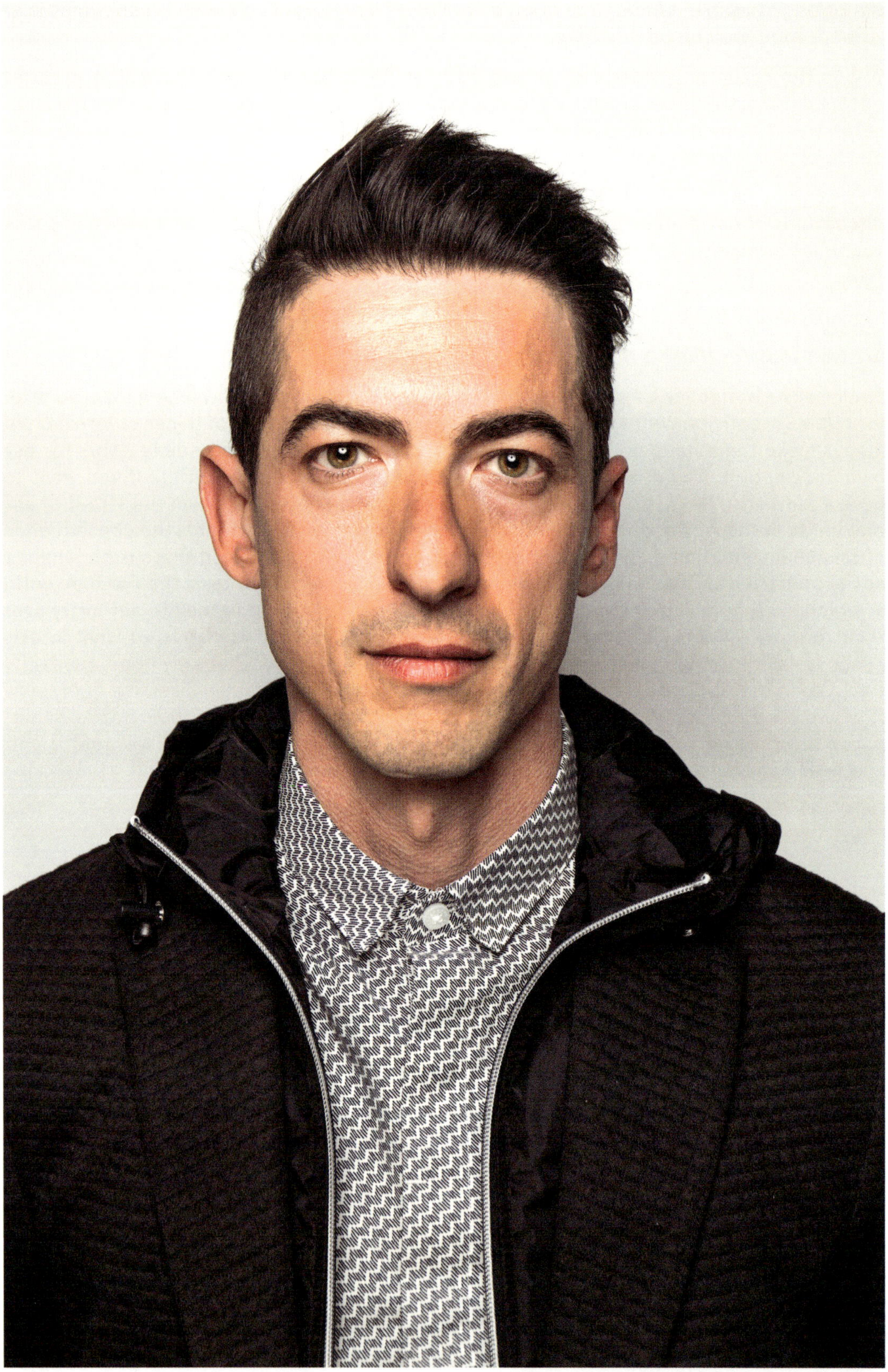

TOMMASO LENZI

"There's No Bigger Problem for Engineers to Solve Than Empowering People"

The professor of neurorehabilitation engineering and director of the Bionic Engineering Lab at the University of Utah has loved robots since he was a kid. The young, passionate engineer aims to fuse artificial intelligence and mechanical intelligence to make robots more human and humans more robotic. His prostheses and exoskeletons are not based on the blending of nerves and technology, and his robots are not interested in the human brain—they merely use the neuron data for their own autonomous work. AI can predict what the user wants to do and controls every movement. With their custom electromechanical engines, actuators, and batteries, the devices are more powerful than most people's biological legs. They are also extremely lightweight. Most of all, these devices empower people to move and regain their independence.

Cyborgs have been around for forty years in science fiction, and they are now becoming a reality. Technology is increasingly becoming part of our bodies and brains—my iPhone is already part of my body—and this will continue to intensify. What does this mean philosophically and culturally? Tell me about your vision of robotics, which you combine with artificial intelligence.

Things are evolving. Ten or twenty years ago there were "normal" people and "disabled" people, and the concept of mobility and function was very limited. Now we're seeing that the ability to move is part of a spectrum. Most people are somewhere in the middle, but there are definitely those who can do more and those who can do less. Technology can help people move along this spectrum by giving them abilities they don't normally have. Our ability to move within this spectrum and become stronger or faster has changed.

The technical history of mankind is full of these examples. If a tool is so efficient that anybody can use it, they will, and it will change everything. Just think of the invention of metal knives ten thousand years ago. Stone Age man already used cutting stones and thus developed from a herbivore to an omnivore. This marked the next stage of evolution and finally separated man from the animal kingdom.

Technology is becoming one with the biological body. Normally you get stronger and faster as you grow up, but at some point you start to slow down and get weaker. Technology can change that, and it's going to have an impact on what people do, knowing that they have the ability to move and stay on top of the ability spectrum. The same applies to injuries: you're a perfectly able person, then something bad happens that moves you down in the spectrum, and you're not able to do what you could before. Technology can bump you up again, giving everyone the ability to become faster, strong, more mobile. For example, it can help stroke victims walk again. I want to give people the ability to achieve their goals, which might just be to be independent and strong enough to go about their lives—or to run marathons. My goal is to empower them to do whatever they want.

Empowerment *is a central word. It's also the title of our book:* Futuring Human Empowerment. *The issue at hand is no longer only to replace lost limbs and repair the body, but also to empower people with and without disabilities to strengthen their abilities. What makes your bionic leg so robotic? What kind of artificial intelligence does it use?*

We develop prostheses and exoskeletons. A prosthesis replaces a missing body part, while an exoskeleton uses technology to make a person's limbs stronger and faster. These technologies are very similar in terms of robotics AI. Your questions about how much of it is mechanic and how much is the brain is good because many types of behavior that we perceive as being intelligent are actually what we call *morphological* or *embodied* intelligence, meaning that the shape and mechanics of the body have a huge impact on the way it functions. If you think about walking legs, there isn't much control going on.

The brain is not very involved in walking. Is it a routine activity?

It's not just a routine. Your legs are affected by certain forces. Artificial intelligence and mechanical intelligence are not separate, but should work together. We're trying to fuse them, which has a huge impact on what we're able to achieve with our devices.

Your bionic leg doesn't need the brain, and it works on its own so that I think it's transporting me. I'm not involved because everything is automatic and artificial intelligence does all the work. I don't have to steer it with my thoughts.

The way we think about the artificial intelligence of the leg is that we look at what you are doing and try to understand what you want to do. We try to predict what you are going to do, even before you do it, so that it feels like you are in control. This is important because as soon as there's a disconnect between your intention of movement and what the technology is doing, there's frustration and failure. These technologies are meant to become part of your body, and you have to feel one hundred percent in control of the device, as if you were controlling your own limbs.

I have the feeling that I'm still in control, but it's an illusion?

It's an illusion that we've worked hard to fine-tune—and we're still working on it. It's a research project based on what we call *shared control idea*. What the leg does is completely autonomous, and you have no idea that it is happening. Part of what the leg does depends on what you do. If you're walking and there's an obstacle, you want to step over the obstacle so you can keep walking. Artificial intelligence enables us to react whenever there's an obstacle.

That's a good example. Two people, one of whom has an artificial limb, are walking through the forest and talking. As they talk, the person with the artificial limb has to watch the ground in order to avoid stumbling. The goal is that the person doesn't have to interrupt his conversation or even care about the ground because everything is automatic.

That's what we want, but users also need to feel in control of the device. They don't want to think about it, but they must feel in control. So if you're developing something like the example you gave about stepping over an obstacle, it is very important that you're given the option to fail in your artificial intelligence system. If you want, you can kick the obstacle and trip because that's exactly how your body works. I don't want to constantly focus my attention on something. I want to be able to control it, maybe even drop it—otherwise you don't feel in control. The same applies to the leg: users don't want to always have to think about what the leg is doing. It's a very fine line: your leg isn't a completely autonomous

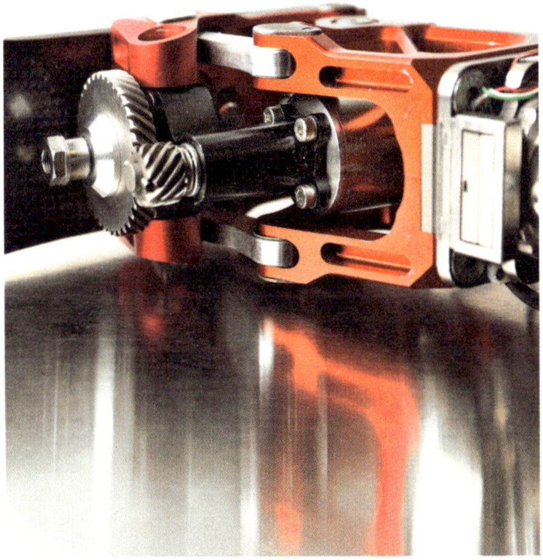

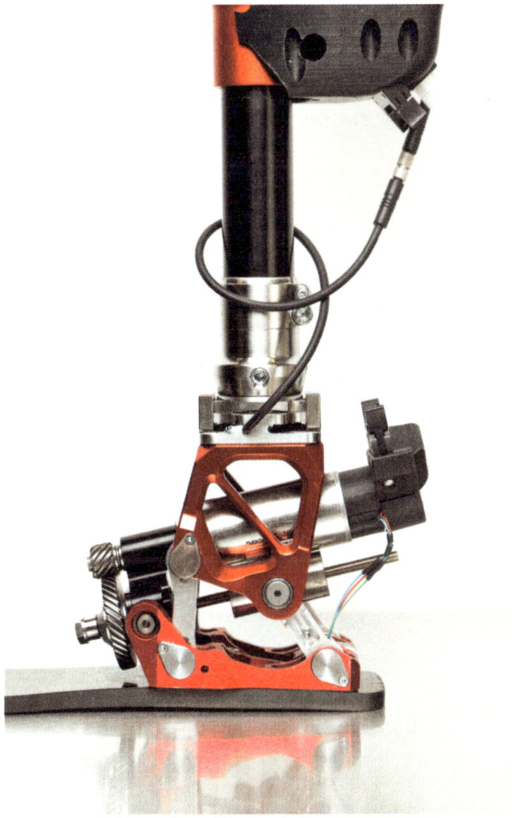

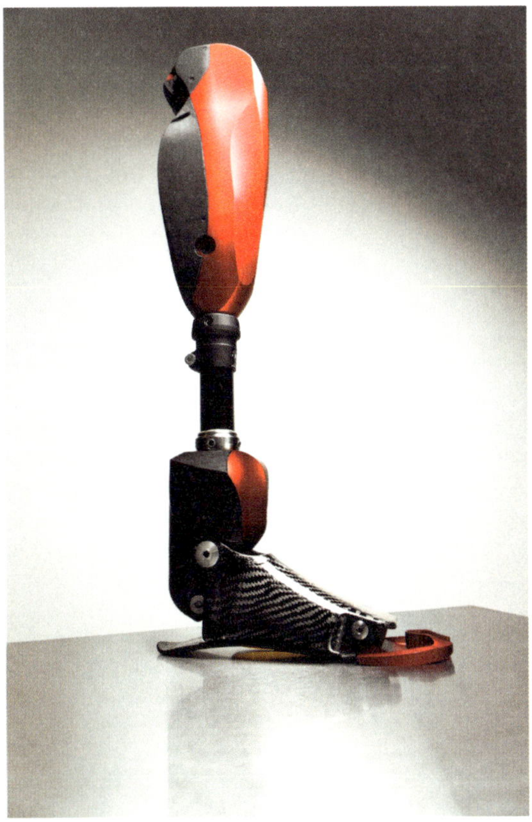

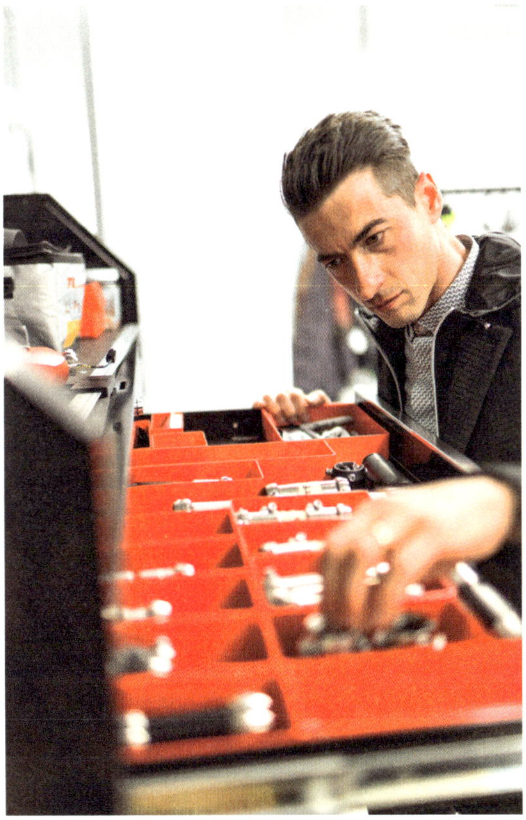

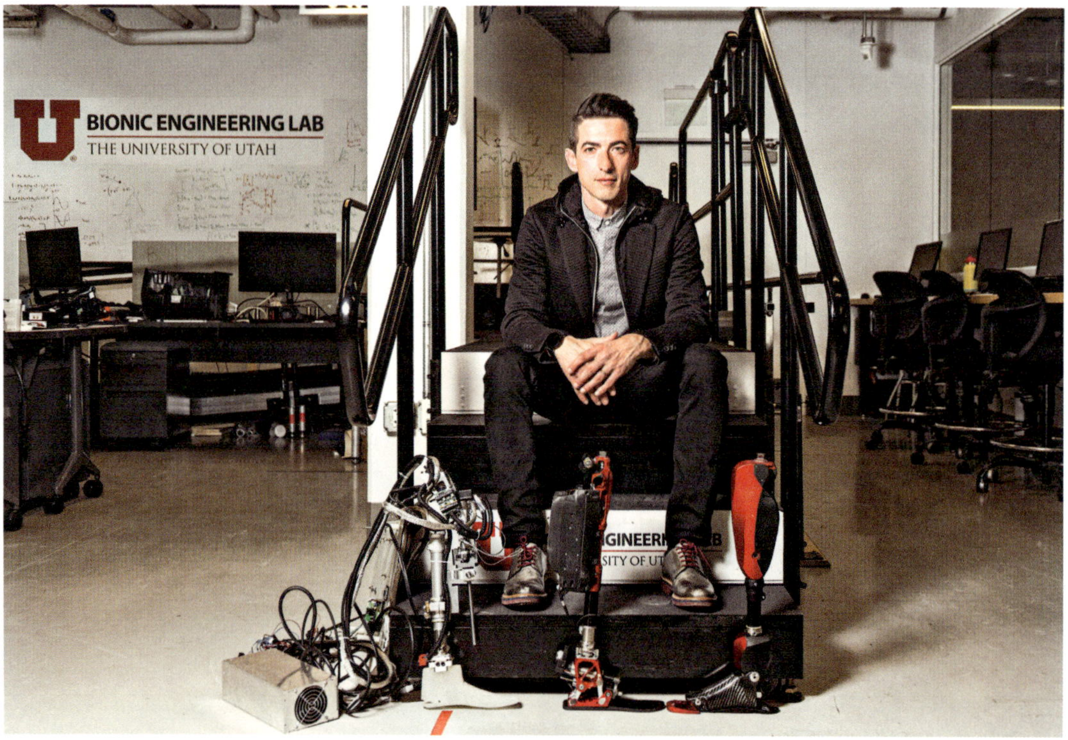

system, but you want it to be. If you make it completely autonomous, people will consider it part of their body.

Won't the brain be annoyed about what the leg is doing autonomously?

It would if it didn't feel like it's in control. That's true for all the controls that we developed in every kind of movement. We want to make sure that the leg is not always doing the same thing. Industrial robots, for example, just pick up something and move it in an incredibly accurate way. That's not how biological limbs work. The control of our devices happens in very different ways. When you walk with our robotic leg, every single step is different. This kind of variability is built into the system because that's how biological legs work.

But is that artificial intelligence? Isn't it more like machine learning, where you gather a lot of data and compare the situations?

No, it's not. I would say that it's more artificial intelligence. We make extensive use of machine learning wherever we use neural signals. But otherwise, it's more like an artificial intelligence system where there are programmed behaviors. For example, as the leg moves, it is continuously trying to optimize the smoothness of the movement. That's a principle it's using, but how to do it depends on intuition, which is continually changing. So if you're climbing stairs, the principle that the AI is following is to maximize the smoothness of the walking movement, because that's what biological legs do. But how do you change it for a specific situation that's managed by artificial intelligence?

Artificial intelligence involves patterns, so how can you identify the information and the patterns that lead to this or that movement? How do you translate that into an algorithm?

We start with an extensive set of ambulation data, with data from all kinds of people, with or without amputations, with or without disabilities. So we have a huge amount of data that tells us what the gait looks like for walking, stairs, or any kind of activity you can think of. We try to reduce these activities to basic principles and create mathematical equations that represent how we want the leg to behave.

How do you get the information? Is there a camera that measures how high the steps are?

In one of the studies that we did maybe five or six years ago, we used cameras, but we realized that you don't really need to look at the stairs in order to climb them. We have many sensors in the leg that tell us about our contact with the ground, about acceleration, and about the movement of the leg. These sensors tell us pretty much everything we need to know about where you are based on what you're trying to do.

How deep are the sensors? Are they on the surface or in the tissue?

For now, we are focusing on noninvasive technologies. We feel like that's the fastest way to translate these technologies to the market. We have two kinds of sensors for reading neural signals. One is the simple electromyography (EMG) sensor that has been around for many years. Recently we began studying sonomyography (SMG), which is basically ultrasound technology. It is like having a scan of the leg in depth, so we know exactly how muscles contract inside the leg, for example, or where the femur is. We use these readings with mechanical sensors, fusing all this information together to understand the right movement for the leg.

The ultrasound echo reflects muscle structure and muscle contraction, but not the nerves?

We combine them, and the electromyography signal detects electrical signals coming from the muscles and nerves, while ultrasound reveals the structure of the muscle itself. It's almost like you have a physical measure of how much the muscles are contracting.

So by measuring the signals with electrodes, you know which signals are required to move a leg? Do you need data from thousands of people?

We don't need that many people. Apart from weight, age, and gender, the differences between people are quite minor. Height and weight affect the energy, torque, and power that goes through the joint, but otherwise we're quite similar in the way we walk.

The big problem with a passive prothesis is the high energy expenditure. A person amputated above the knee requires almost seven times as much metabolic energy, making a two-hour walk the equivalent of a fourteen-hour workload. It's very tiring. Many people avoid using their prostheses because it's so exhausting.

The biggest difference between the devices that we are developing and what's on the market is that ours are all powered. We have custom electromechanical actuators and batteries that are quite strong. They are actually more powerful than most people's biological legs. The prosthesis becomes a stronger leg.

A normal leg can't keep up?

No. When people stand up from squatting, which is a pretty taxing activity, they put more weight on the prosthesis side, just because they don't have to generate that energy. The robotic leg does all the work.

It's more comfortable?

It makes it easier. That's the principle that explains how people move. They always optimize their energy consumption and try to optimize movement in a way that minimizes the effort

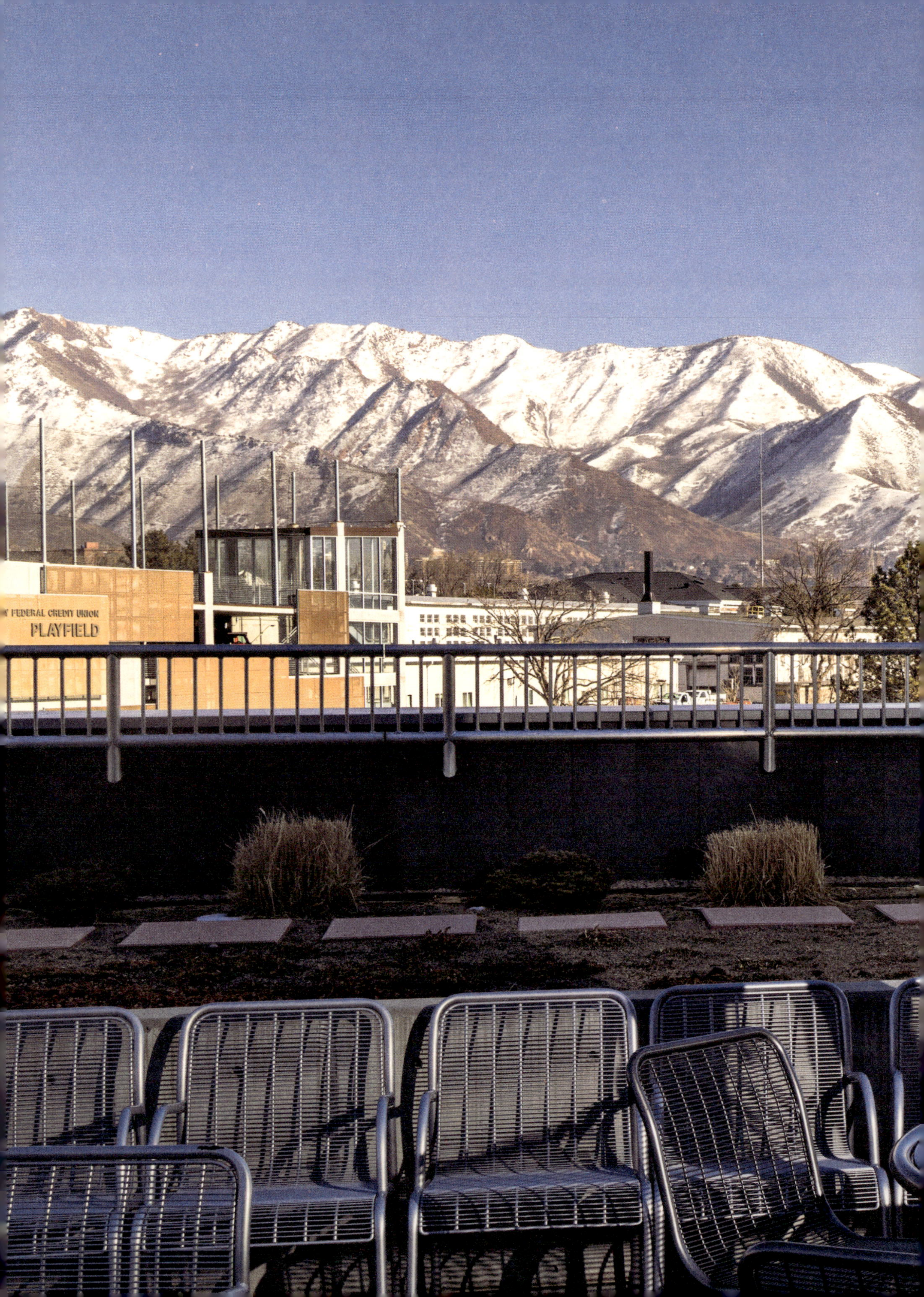
FEDERAL CREDIT UNION
PLAYFIELD

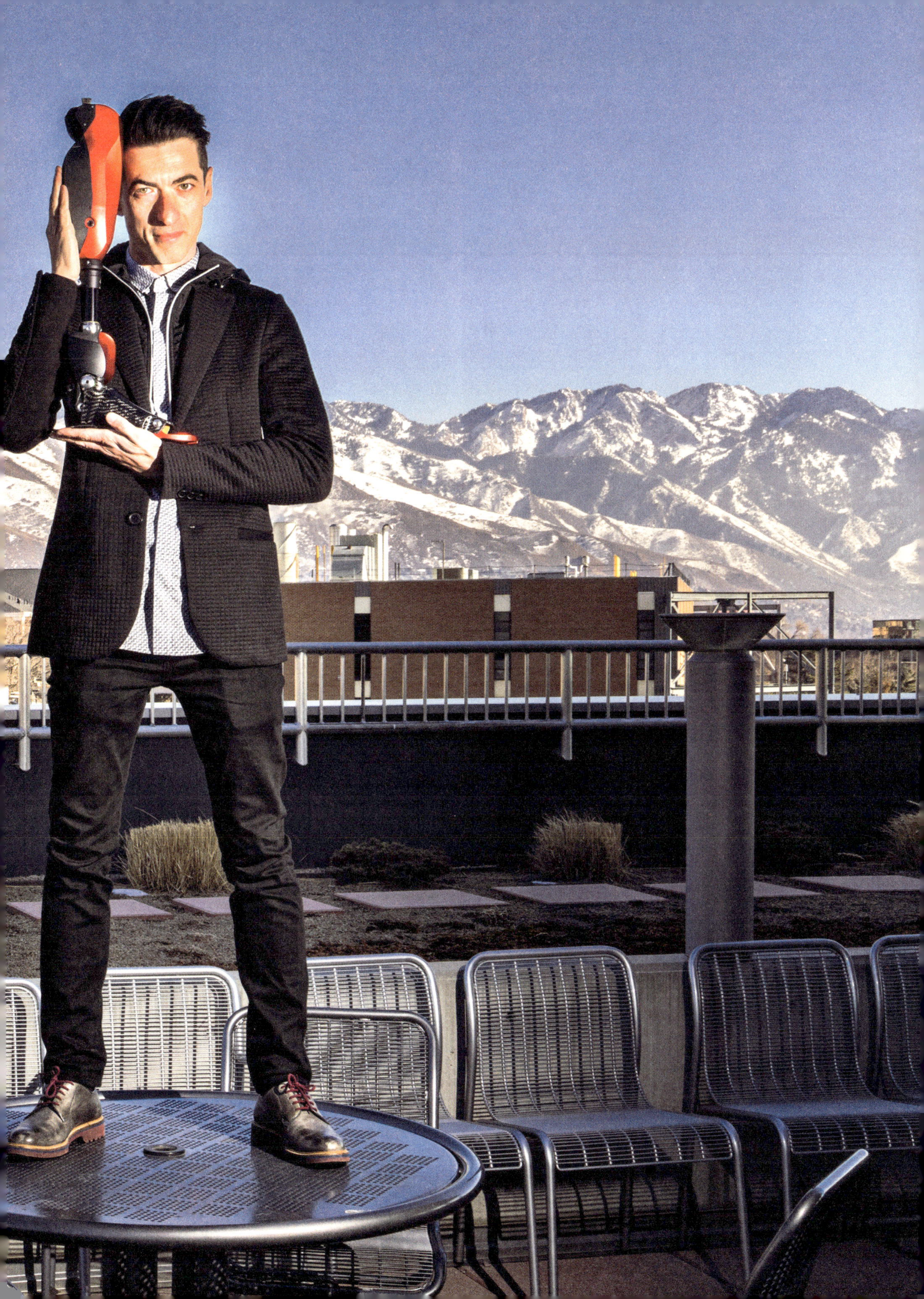

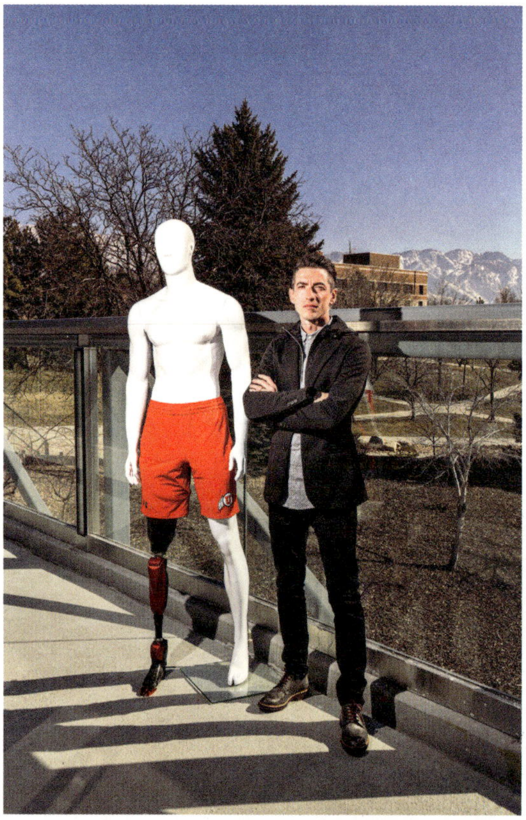

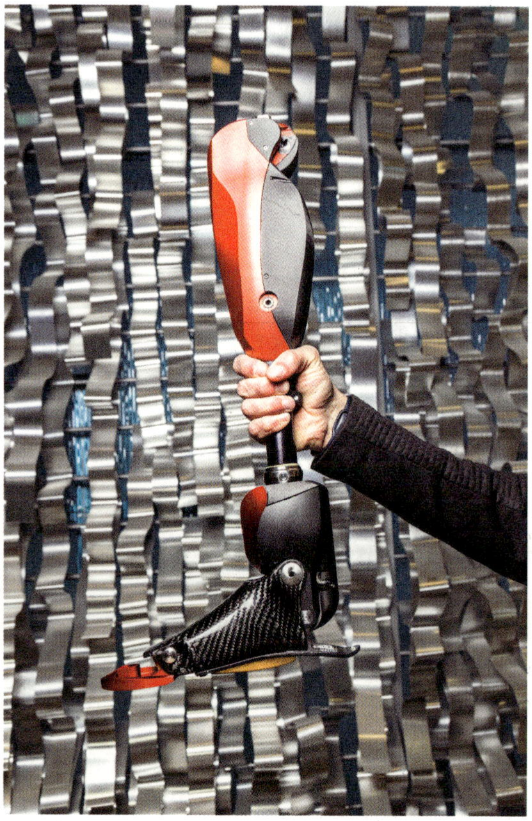

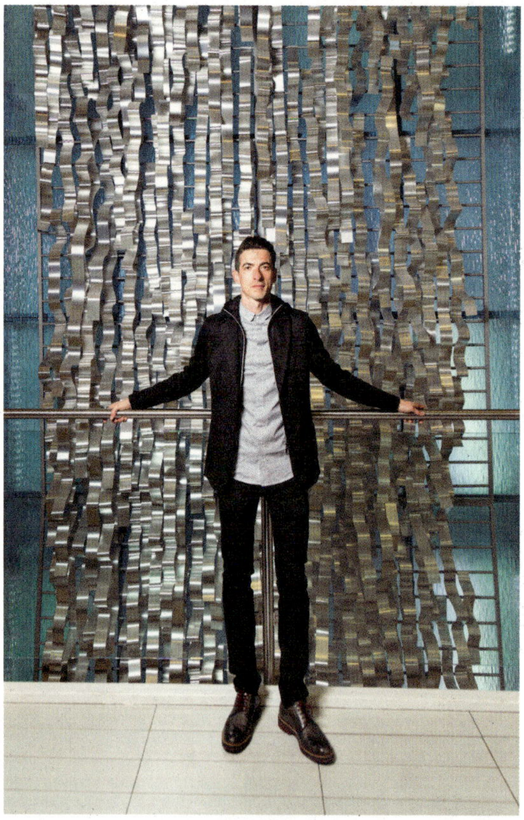

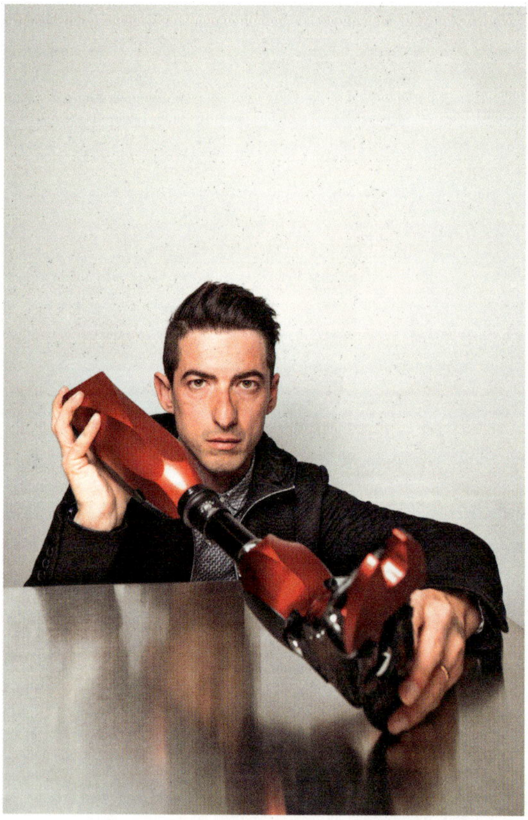

that they are putting into the movement. Their robotic leg overpowers the biological leg, so they rely more on their robotic side than on their biological side.

We know that people using the power foot are taking more steps. The gait pattern is not necessarily different, but they are able to walk much longer. Do any patients have two bionic legs?

Yes. Last summer we did the first study with a bilateral above-knee amputee climbing up and down stairs on bionic legs. What we found is that we don't need to actually synchronize the legs. The legs don't need to communicate with each other because the bionic legs are synched to the user's residual limbs.

I saw a video of a man who was walking up stairs with a robotic leg. The robotic leg always took two steps, not one, and the normal leg had to keep up.

Stairs are an issue for robotic legs. We had one person yesterday testing the robotic leg, climbing up and down stairs, one step at a time and then two steps at a time. People are very emotionally connected to the experience that they have with these devices; I've had subjects bring in their mothers or grandmothers so that they can show them what they can do. It's understandable if you haven't been able to do something for so many years and now you can. It really changes your perspective. That's why we have one-hundred-percent retention of all our subject participants because they love the feeling that they are involved and that their help will make a difference.

Does that motivate you?

I have been working in robotics for about fifteen years, since I started school. What I love about engineering is that it solves problems and creates possibilities. There's no bigger problem for engineers to solve than empowering people. The field has really evolved in recent years. I wouldn't have believed it if you had told me fifteen years ago what we would be doing today—and I'm an optimist. The pace of the technological evolution has been quite amazing.

It's quite unusual for someone in the field of robotics to focus on the limbs and not on industrial robots or machines. How did you get into this field?

It was a long process. I've loved robots since I was a kid, and it was obvious that I would go on to study and build robots. My PhD involved robotics, but with an industrial angle. Then I realized that I wanted to work with bionics and build cyborgs. So I moved to the US and worked with patients at a hospital for a few years. Talking to patients and learning about the impact these technologies have on their lives motivated me to change from the theoretical or industrial side of robotics to more applications. It's not a dream anymore; it's very real. There's a lot of work still, but the devices already exist. Ten years ago, we were imagining how it would be if we had a robotic leg that could climb stairs and walk. Now we have a bionic leg. We're still at the stage where it works well in the lab for a few hours but can't really be used by everybody—yet.

Is Boston Robotics watching what you're doing? What impact does your work have on robotic science?

The most important thing is for robotics to have an impact. It is also important that these fields converge. Only one third of the people in my lab are roboticists, and the rest of them do biomechanics, human factors, and controls. So my lab reflects the convergence of the clinical side of things and biomechanics, human science and human movement science, and robotics and mechatronics.

Doesn't bionic *mean the blending of nerves and technology?*

Bionics is about building things that behave like biological systems. The connection to nerves is definitely one aspect, but mechanics is also very important. The goal of replicating the incredible ability of people to move does not mean that we have to copy biology. Your brain may have ten thousand kinds of cells, but you don't necessarily need them all. The whole point of engineering is to find alternative solutions to solving the same problem and doing the same thing. One thing that we've spent a huge amount of time on and is still one of the biggest challenges is to make devices that are robust but lightweight and functional. You don't want to carry a fifteen-kilogram bionic leg around. What makes our device unique is that it is lightweight as a passive prosthesis.

How much does it weigh?

The knee is about 1.6 kilos. It's the same as Ottobock's C-Leg, which is the best-selling knee ever. The whole leg weighs about 3.2 kilos, including batteries and control.

That's not much for a leg.

It's about the same as a passive leg. It's a huge thing for us because there are only two options in robotics right now: passive and lightweight or powered and heavy. That makes it unique, and its lightness affects how it moves.

The fact that you don't need sensorial sensations makes this project much simpler. You don't need the neurosurgical stuff to actually connect the nerve to the prosthesis.

That's right. When you're interfacing with a very sensitive nerve, you're trying to record thousands of tiny signals. People don't want to have surgery every six months. That's why we're going in a different direction philosophically. If I want to replace what your brain does, I don't have to use biological cells in a nervous system. I could probably do it in a different way. And you can extend this to the device that we're using. We are trying to replicate what you can do with bones, ligaments, muscles, and blood, and we're using batteries, gears, springs, and other mechanical systems. So I don't think that we have to copy biology, but we certainly need to understand it. We need to interface with the biological body, but we don't have to replicate biology. Because our building blocks are different. There's something about our physical bodies that you would never get rid of. Only things that move have a brain. Things that don't move don't have a brain because they don't need one. The fact that you have a brain is because you're planning physical actions and physical interactions with the world. So I don't think the mind and the brain will ever become disconnected from the body. It's not something that people will want. I just give people what they want. More than anything people want independence, and the ability to move is so important for independence and being able to do whatever you want to do in your life. If it's taken away from you, it's a huge loss. People don't want to lose their independence. That's what it boils down to. But the physical world is connected with the mind and body, and I don't think that people will want to change that. I'm not saying that it's impossible, but I don't think people would want that. People are not going to base their decision on nerves and or sensors, but on how well it works and how empowering the technology is. And it has to be noninvasive.

In the field of robotics, people are working on building robots that resemble humans. I interviewed the outstanding Japanese professor Hiroshi Ishiguro in Osaka. His android Erica has already hosted a TV show, and we did an interview with her too. But the cultural difference is huge. The Japanese want robots with children's faces, big eyes, and a big smile. Germans and Western Europeans think that they should be a gray box with one arm.

I think many roboticists here in the US will disagree with that. Many of them are in favor of humanoid robotics, because everything in our cars, houses, and working environments is designed for things that have two feet and two hands. Bipedal locomotion is extremely hard to achieve in robotics. So it is much harder to develop a robot that walks with two legs than one that just uses wheels. But a robot with two wheels cannot do exactly what you do.

The robotic industry is trying to make robots that are more human, but what about doing it the other way around: making human bodies more mechanical or more robotic?

I want people to start perceiving the robot as part of their body, so, yes, maybe in that sense I'm trying to make humans more robotic. But I'm also trying to bridge the gap between this idea of the android robot, an external thing that doesn't really do what I want. There's this fear about the robot revolution, with robots rising up against us, and in this sense you have a robot that is like a servant who amplifies what you're doing. There is a big difference. There is definitely a lot of interest, and the technologies that we are developing could fundamentally work for humanoid robots. We're building legs that are very strong and powerful, extremely efficient and lightweight. I've talked with Samsung and other people working in humanoid robots, and there is an appealing synergy there, but I need time to think. I have decided to stay out of it for a bit and see where it goes.

How do you build the best limbs for a humanoid robot? Do you build the best legs and then test them on the human bodies?

That's right.

That's why Boston Robotics will come to you because you have the best legs.

Even Tesla will come for our knees. You forget it's a prosthesis. It is technically stronger yet lighter than a biological leg. So it already exceeds what a biological system can do. That's very appealing because you could have humanoids that potentially exceed what the biological body can do. A humanoid robot is not just like a human; it's stronger than a human. I do think that human robotics can do very good things, although I also see a dystopian world where it does really bad things, especially for military applications.

A robot with three hands is much more effective.

That is why I've stayed out of military applications so far and only worked on medical applications. I want to fully understand the technologies that we're developing and fully understand what they can and cannot do, and how they can be controlled and used, before deciding if it makes sense for them to become part of military projects.

Can you tell me about what you're doing then for the Defense Advanced Research Projects Agency (DARPA)?

We are now working directly with DARPA, and it has been fantastic. They want us to push these technologies. They just gave us a large grant for pilot-testing robotic devices at home with three subjects, which enables us to monitor them for a couple of months and see how the robotic legs change their daily lives.

Is this for veterans?

We work with a lot of veterans, who have a high rate of limb loss, but it's really for everyone. We have a wide spectrum of subjects. Our youngest subject to test the power legs was twenty-one, and the oldest was seventy-four or seventy-five. It's interesting to see how different expectations are and what these power legs can do. A lot of people think that these devices are great for just the youngest, strongest people who want to do a lot. But that's not necessarily true. There are a lot of people who are in their sixties and seventies who really struggle with simple things like getting up from a chair and walking around their houses, and I think that robotic legs can be a great solution for them as well. It can make a big difference because their residual limb is actually much weaker than is the case with a young amputee. The leg of a seventy-year-old with diabetes and other comorbidities likely has other problems such as osteoarthritis, which is very common because you tend to overuse your biological leg.

What is the impact of your work on people involved in artificial intelligence? Do you use standard artificial intelligence, or are you making a real contribution to this field since you are working with human beings and the signals of the muscles and the brain?

We don't really look at theoretical contributions. We're doing what we can to solve the problem at hand, and we are always finding new problems. We focus on specific problems and try to see how they can be used for our goals. We change and adapt them as we go along, and that's where the contribution is: to really fuse artificial intelligence with machine intelligence and have a real-world impact.

So instead of replacing the brain, artificial intelligence is just a coordinating tool to organize and synchronize the mechanical processes?

Yes, definitely. We don't want our legs or devices to start making decisions on their own. Sometimes people ask me if the robotic leg uses artificial intelligence. My answer is that our leg *is* artificial intelligence.

Is your work with Ottobock a kind of cooperation, or do you just talk a lot? Are there any projects on the horizon?

It's a little bit of both. It started with the whole group providing us with a lot of support in terms of composites, components, and parts, but also know-how about understanding the clinical problem. We've been keeping in touch to see how our technology has changed and matured. Hopefully we'll have a chance to collaborate more closely as our technologies become viable from a clinical standpoint. That's the plan. Hans Georg Näder, the owner of Ottobock, has been involved in this process lately, really pushing for us to work together.

How did you meet him?

I met him in an elevator in Chicago, at the hospital I told you about earlier. I literally gave him my elevator pitch. It was like forty-five seconds. After two days, I got a call from his assistant who asked me if I would like to come to Berlin and Vienna. So I met him in Berlin. We had a really good conversation. We spent a few hours thinking together and everything. He gave me his personal phone number and told me that if I ever needed anything I should just text him. Five years later, when we were developing our devices, I did text him. I told him that we had a wonderful device and asked for his help in building our new lab. He put everything in motion. So hopefully we'll have a first official collaboration and maybe our robotic legs will become reality and help many people.

His support of technology, which is different than Ottobock's approach, is really amazing. I like the way he tries to bring that together. A company can't cover all the research things you are doing. It's an investment in the next generation.

But it takes someone like him to buy into the vision. Nobody else believed it when I said I was going to build a device that is as lightweight as a passive prosthesis and stronger than a biological leg. But Näder believed me. When I showed him the prototypes, he was very enthusiastic about making it happen—not just in the lab, but to make it available to people who need it.

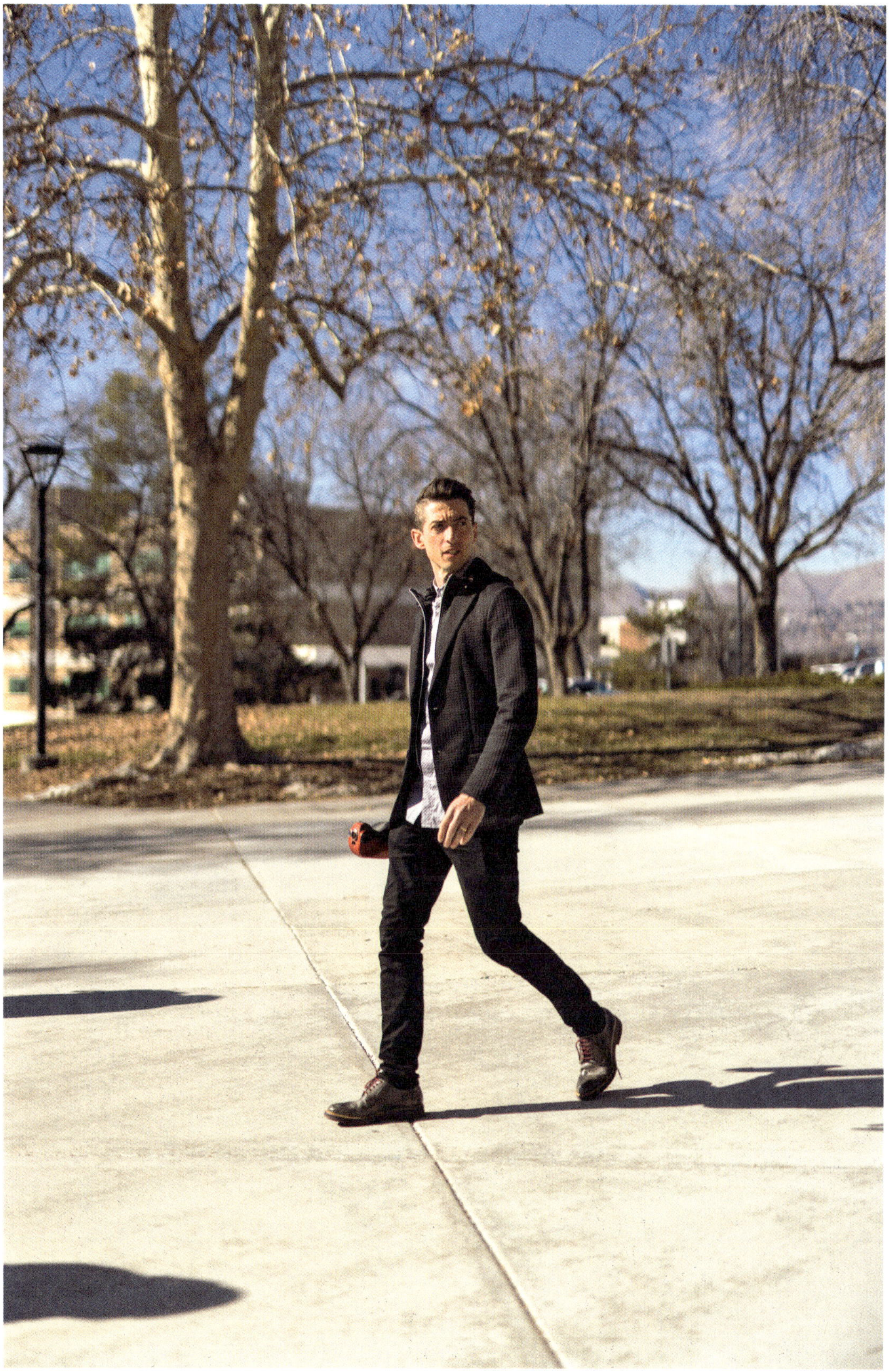

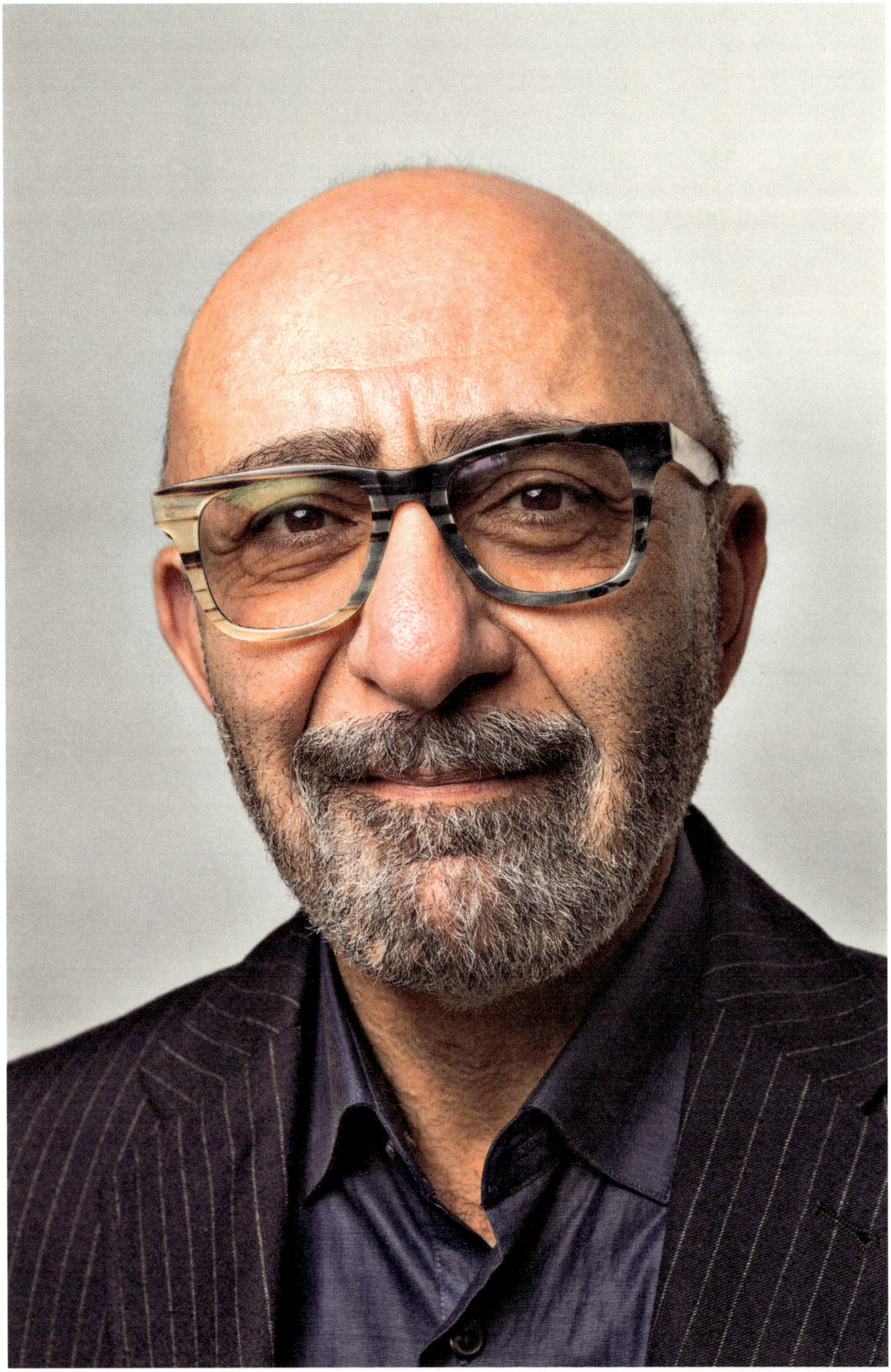

HOMAYOON KAZEROONI

Dancing with Robots

He is not a manager, but a leader. The charismatic professor of mechanical engineering at the University of California, Berkeley, and founder of several robotics companies inspires his young students and staff through his vision of our future life. His Berkeley Robotics and Human Engineering Laboratory, also known as KAZ LAB, is the birthplace of a great number of exoskeleton technologies. The wearable exoskeletons that his latest company, suitX, produced in a building resembling a warehouse in an industrial park in Berkeley are aimed at the young and sporty generation that is closer to future technologies. In addition to being super practical, these exoskeletons are modern looking, intriguing, stylish, trendy, and oriented toward diverse lifestyles. Ottobock acquired suitX in November 2021, becoming a world-leading provider of occupational and medical exoskeletons. Kaz will remain with the company as the company's chief scientist. He is impressed by the quality of the German engineering at Ottobock and pleased about the powerful resources Ottobock can offer to suitX. He is confident that this acquisition will lead to a much larger potential for suitX's engineers and scientists to get their technologies out for public use.

Can you tell me a little about the future of devices with embedded intelligence?

Today's generation wants embedded intelligence. The technology has been changing faster than ever—every four years is one generation—so it won't take long before we see all kinds of embedded intelligence in all technologies including wearable exoskeletons. The embedded intelligence in an exoskeleton essentially modifies the device behavior based on the user and environmental actions. For example, in a trunk-supporting exoskeleton such as backX, the embedded intelligence changes the device power not only as a function of the weight being lifted, but also as a function of the dynamics of the user such as the user's speed. In the exoskeleton world, sport equipment suppliers will begin offering wearable exoskeleton modules such as smart knee braces that boost their wearers' power and prevent injuries for various sport activities such as hiking and running. An example of the embedded intelligence in these knee braces is a feature that can differentiate between "walking downstairs" or "squatting on a level ground," although both maneuvers require knee flexion. In general, the embedded intelligence in exoskeletons can vary from simple features such as power modification in the presence of load variations to complex features such as stabilization and balancing of an exoskeleton and its wearer. In some cases, the embedded intelligence incorporates a great deal of changes through a long period of time such as gait alteration as the wearer—or patient in this case—improves in locomotion rehabilitation. In summary, we should see and must focus on a great deal of embedded intelligence and less emphasis on the hardware development.

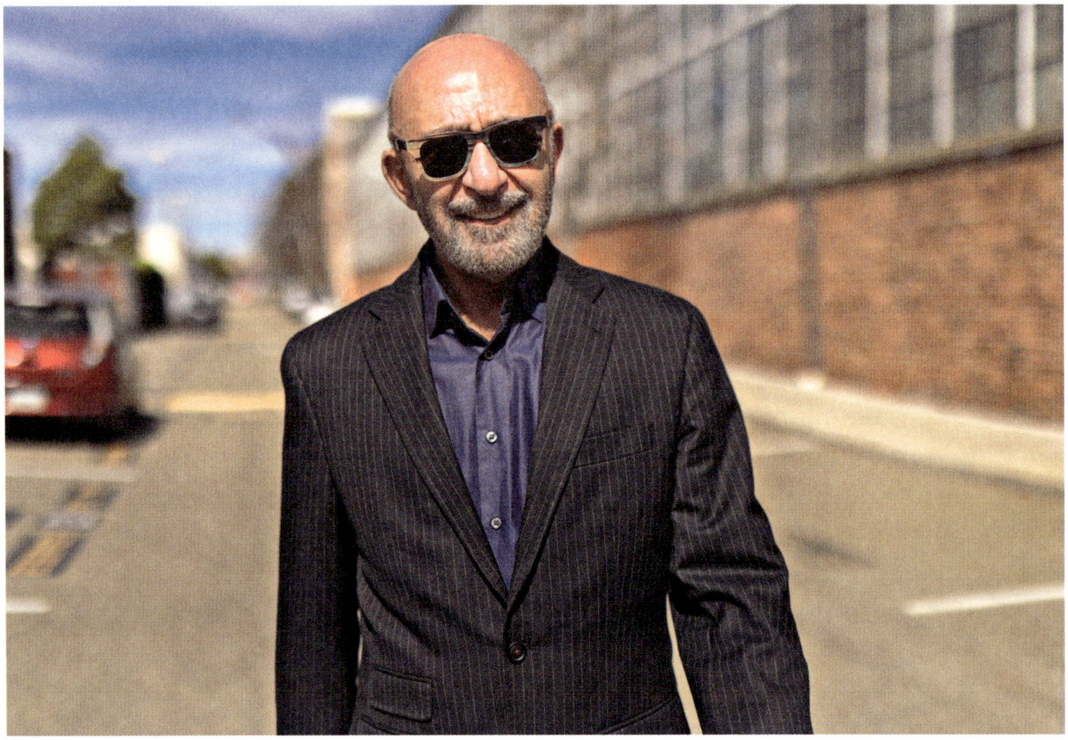

What about military exoskeletons?

In the 1960s the United States Department of Defense funded various projects about augmenting soldiers. The notion of using exoskeletons to physically empower people in a substantial way—as it was portrayed by DoD—created an image that has been repeatedly reflected in science fiction books and movies. The military exoskeleton state of technology is in its infancy. I believe the exoskeleton technology will initially be deployed for various logistical military applications. For warfighting uses, I predict new technologies are needed to create partial—as opposed to full-body—exoskeletons to support the war fighters in some specific military war fighting maneuvers. We have designed and built exoskeletons that reduce the forces, due to the heavy weight of the backpacks, on soldiers' backs and shoulders. A successful example is HULC (Human Universal Load Carrier), which enables a soldier to carry backpacks that weigh over fifty kilograms without feeling any substantial amount of force.

Have you seen the movie Alita: Battle Angel? *It's like a lineup of all possible kinds of cyborg technology.*

No, I haven't seen it. Almost all science fiction writers and filmmakers use exoskeletons to show physical superiority. These movies often involve physical battles where exoskeletons are used to create power for their wearers. During fights with their adversaries in these movies, the exoskeleton wearers climb walls and barriers, break down doors, jump long distances, move heavy obstacles, and carry out all kinds of supernatural maneuvers. This is an image that people have had for a long time. We need exoskeletons to provide physical power to people who need a little help in their daily tasks. People with mobility disorders need the physical support of exoskeletons to be upright and walk. Workers in various work settings such as manufacturing facilities and construction sites need exoskeletons support

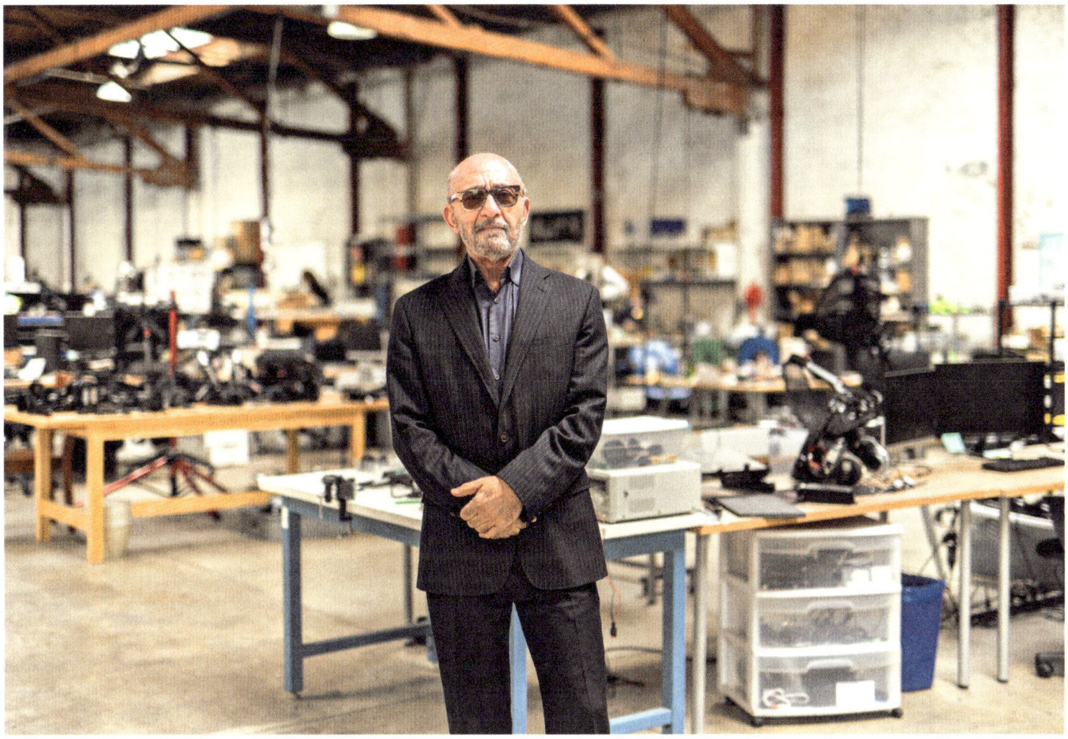

to reduce the injury risks. Warfighters need lower-limb exoskeletons to reduce their backpack weight on their spines.

Do you know Philip K. Dick? He was a fantastic writer who lived in a suburb in California, somewhere around here. He never left his home, did absolutely no research, and never went to the library. He just took LSD and wrote fantastic books like The Man in the High Castle. *The movies* Blade Runner *and* Minority Report *were based on books by him.*

I saw *Minority Report*.

That movie inspired engineers at Apple. Steven Spielberg rented an entire hotel for a week and invited sixty scientists from all disciplines to describe what the world would look like in 2045. The movie is only twenty years old, and we already have this future with predictive policing, widgets, device swiping, and augmented reality. All these fantasies have become reality.

Our approach in technology development for exoskeletons always starts with people's needs. For example, workers who move heavy objects for ten hours a day need low-cost, accessible exoskeletons that substantially reduce the risk of back injuries. The aging population that has some difficulties walking also need the support of exoskeletons. Creating a better of quality of life—that is where exoskeletons are needed. I founded suitX and Ekso Bionics to just do that. You know, when I founded Ekso Bionics and suitX, I dedicated myself and surrendered many other opportunities for creating a better quality of life for people who have hard jobs or hard time doing their daily tasks. I have a good feeling about Ottobock taking over suitX, and I'm happy about the work we are doing with exoskeletons. Our goal is to combine the exoskeletons from both Ottobock and suitX to get the best exoskeletons to market.

Which one is better? Yours looks much fancier than the Ottobock exoskeleton.

Ottobock exoskeletons have sold well but didn't meet all requirements of some industrial clients. Ours had some other deficits. So now we're combining them after the acquisition, and customers like these devices with combined features. Of course, we're going through a transition right now, and very soon we will codevelop various exoskeletons with superior characteristics.

But your brands continue to exist?

We are going to use our brand—backX, shoulderX, and legX. *PA* actually stands for "passive exoskeleton." But Ottobock is going to keep that brand for other products.

Did you ever see the Exopulse Mollii Suit that Ottobock acquired from the Swedes in September? It's for people who have neurological disorders.

Yes; it electrically stimulates the muscles.

It's fantastic. It has fifty-eight electrodes that stimulate different groups of muscles, and the electrical impulses remove the spasm.

I hope I can get an opportunity to work with them. SuitX now has a new CEO from Ottobock, Sam Reimer, and a chairman, Sönke Rössing, who are quite knowledgeable in exoskeletons marketing. Our future activities are mapped by them. I'm the chief scientist in charge of the product development.

You are a very busy man. Do you still teach and doing research at the university?

Yes, I'm a professor. Every now and then I would like to take the technology we develop to market. I work efficiently, and I make sure I do not lose work quality. I do not waste time either.

The American system is very good. It's different in continental Europe.

Well, European professors are untainted by industrial work; in general, they don't get involved in technology transfer. In the US we have more freedom to get involved in technology transfer and entrepreneurial activities. Of course, most European professors receive funding for their research, but in the US we need to hunt for it. We must raise our own research money, either from funding agencies or from industry. For me, technology transfer is a natural progression of my university research. Since I've been doing technology transfer, I've become a better teacher. Dealing with real problems, people, and business aspects of technology transfer has made me a better teacher.

When did you start?

I have heard a few times that people call me the father of exoskeleton. I was doing a lot of robotics research in early 1980s. I realized soon that the robots and the state of robotics were not ready to be adopted broadly in manual work settings such as factories, construction sites, and logistics. Robotics is effective mostly when the job involves repeated tasks, which means you can program the robots to work. But for unstructured jobs, like loading boxes on a truck, one needs human intelligence, decision-making, and adaptivity to maneuver things and do the job right. For these unstructured jobs one currently can't use a robot effectively. Robots are strong, and humans are rather intelligent; if you combine the strength of machines and intelligence of humans, you have an effective combination that empowers the person, resulting in a better quality of life and work.

So you don't believe in artificial intelligence?

I believe in artificial intelligence, but I am not expert in it. We simply cannot wait until AI reaches to a level of maturity that can be employed in various work settings. The state

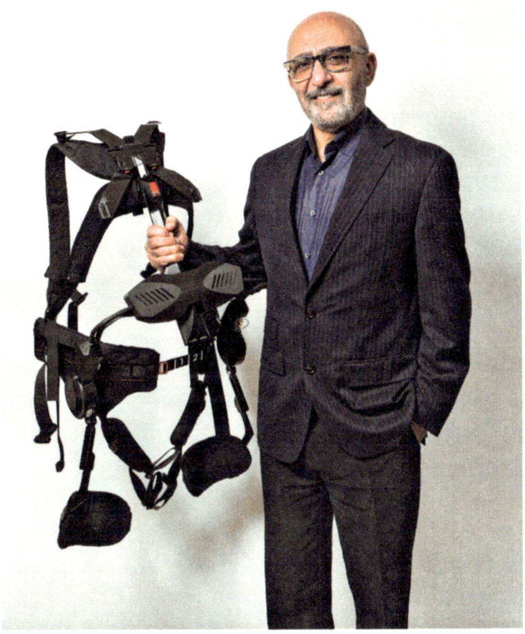

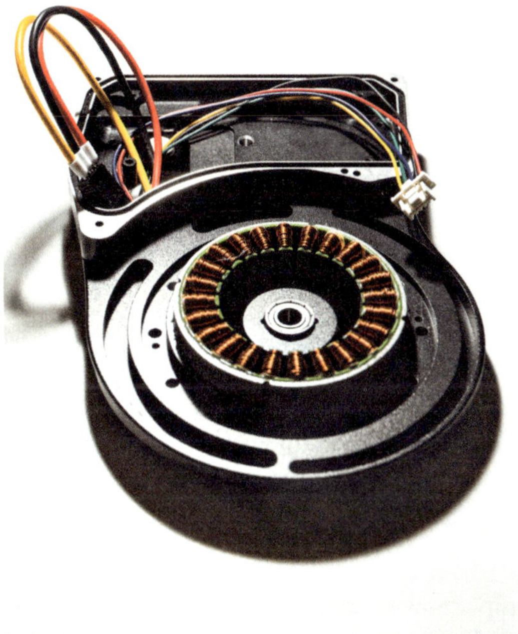

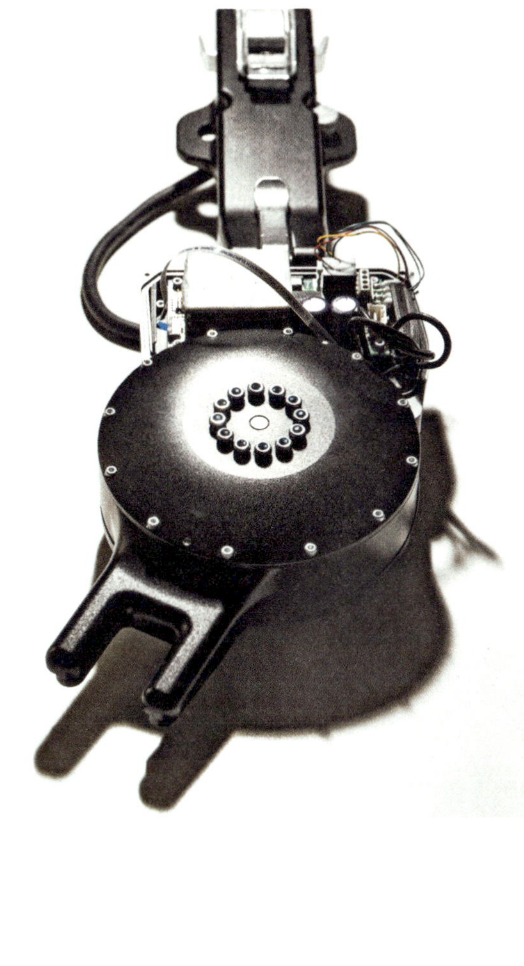

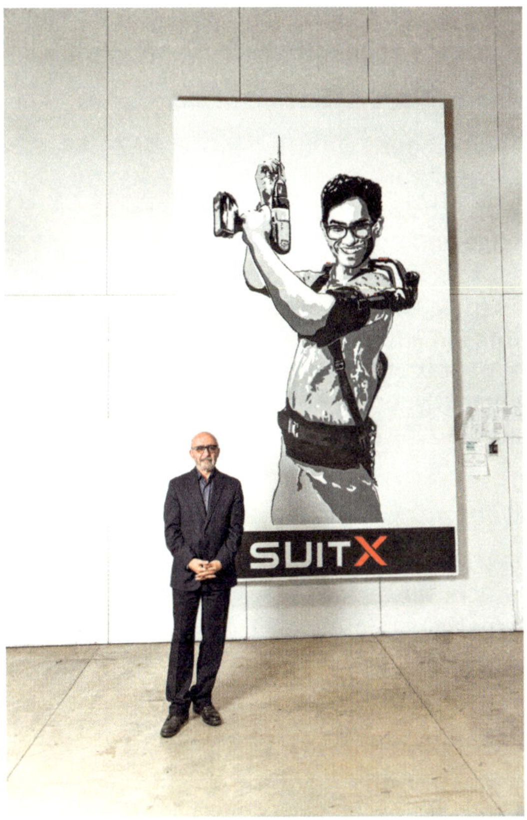
SUITX

of AI has progressed immensely, but currently it is not mature enough to be employed in many existing work settings. In many unstructured situations such as search and rescue operations humans will prevail. So I'm not against artificial intelligence; I simply think that in many unstructured applications the combination of the human intelligence and machine intelligence is preferable. I was disappointed in robots because they can mostly be effective for repeated jobs. I come from a family of workers, so I have a soft spot in my heart for workers. It was natural for me to get oriented to the technology that protects the workers instead of replacing the workers. Currently the personal protective equipment for workers comprises safety glasses, hard hats, steel-toe shoes, earplugs, and gloves. It is just natural to consider exoskeletons as personal protective equipment. Hopefully the state of technology in robotics and AI will reach a point that we can employ robots to do all these difficult and demanding jobs. So my inclination to use exoskeletons to protect workers does not show any skepticism in AI sciences; it simply shows my rush to use the current technology to protect workers. I've been very lucky to be among many smart engineers and students who think similarly and prefer to combine the strength of machines and intelligence of humans in many unstructured industrial jobs.

There are a lot of students here?

The interns here in suitX are all students from Berkeley. I founded two companies in the Bay Area. In 2004 I founded Ekso Bionics, and in 2011 I founded suitX, which was originally called US Bionics. Many students work here as interns and that helps their academic venture.

How many jobs have you created?

We have about thirty-five here employees in suitX and probably about sixty at Ekso. More importantly, people in other countries get inspired to do the same thing and initiate companies that create jobs. Have you heard of German Bionics? It's a German company that is basically copying us. Our name was US Bionics; they called themselves German Bionics. It's actually a good thing because people are inspired to continue things that I had passion for.

And KAZ LAB. I like that name.

I work closely with students at KAZ LAB. When students come here, they know little about exoskeletons, but they have a great deal of potential. It's very easy to dismiss the students with little experience for doing research. If the students have little experience and knowledge and come to my lab, I'll try my best to get them inspired to learn and create. After four or six months, they learn how to be creative and use their engineering knowledge. Further, they learn quickly that passion plays a great role in their jobs. I deliver all my lectures in a formal way based on a solid foundation in engineering. As I am getting older, my expectations from students are increasing; I must lower my expectations a bit. Students, currently, are under a great deal of pressure as the technology expands considerably. So, I am learning to lower my expectations a bit to allow them collect broader knowledge instead of more focused subjects.

What was your first prototype? Did you start with exoskeletons?

I was involved in a simple strength augmentation of moving my arm in one direction. The second prototype was moving my arms in a horizontal direction. Then after these two experiments on human-machine systems, we created two arms that were used to move boxes, because we thought we could apply that. Later we designed a device to load and unload an aircraft.

Did you use small electromechanical engines?

Yes, we used computer controlled electric motors to power the exoskeletons.

You didn't get them from the airplane industry?

Well, we bought some of the standard components, but we had to make the rest to make sure that they would work with humans. We made two arms to pick up heavy objects. After working on augmentation of the upper extremity exoskeletons, I got attracted to the lower extremity exoskeletons. The upper extremity exoskeletons are for object manipulation and the lower extremity exoskeletons for walking.

Aren't arms and hands more complex?

Yes, but the lower extremity is a lot more interesting because there are more applications. Walking and running are complicated, and if you don't perfect them, the person will fall or will be impeded. We made several prototypes and several large exoskeletons. Then in 2010, I got attracted to smaller modular devices. That's why I founded suitX. Instead of the whole system, we offer the smaller components: backX, shoulderX, and legX. They have great potential because the prices are more reasonable. I want to get the price of a smart brace for hikers to one hundred dollars—no more than a pair of shoes—by 2025. The devices will be sold at sport stores. We started with very simple devices for the proof of concept, but the more complicated ones were too expensive. So we returned to the shoulder and back modules. It's like the parts of a suit: there are pants and a jacket—that's what suitX is. That means we are trying to make a suit for you that has some element of sensors in their augmentation. So now we're doing modulars and then we met Ottobock, and they're doing similar thing. So it was very natural for them to acquire us.

Do you hope to raise the standards with the resources from Ottobock?

I'm so glad you bring this up here. We are very thankful to be supported by such a big international corporation. The new budget is good, but I was hoping for a lot more. It's always been money that has limited me and my team, not the intellect or passion.

Yes, it's all about passion. You're such a charismatic person.

Thank you. I am also able to carry people with me. My students are my colleagues now. They're all here. We can achieve the goals that you showed me within five or six years. After that, someone else will take over. When my time is over, then all will recognize there are passionate people at universities that want to bring technology from the university to other applications. This is important. Universities cannot be separate from society. That's why I don't like pure academics. It would be easy for me to go to my office, close the door, and write papers—well, it's intellectually challenging but doable. But there are no discussions.

Yes, it's lonely being a writer. I'm so happy when I am traveling and doing interviews.

I agree with you. I want to be involved with people and solve engineering problems. The University of California hired me to teach to create engineers and do research. If I don't transfer that to industry, then my work would have been incomplete. It would just stay in the library. I want it to be useful, which is why I founded these companies. I rented this place with very little money. It was an old, dusty warehouse building; with a great deal of effort we were able to gather talents and work hard and eventually make exoskeletons that are commercialized internationally. Putting a company together from scratch is not easy. The baby has been born, and the raising began when we were acquired by Ottobock. We're just ten minutes from Berkeley. We need to take advantage of this resource and create more technology. I need to bring more experts here. One of my jobs as a leader is to create the leaders of tomorrow. We need to give them opportunities to be involved and lead. I have to help them establish a level of confidence so that they can lead. We have to show them the required knowledge, how to collect the future knowledge. There is a saying, "I don't want to give them a fish. I want to teach them how to fish."

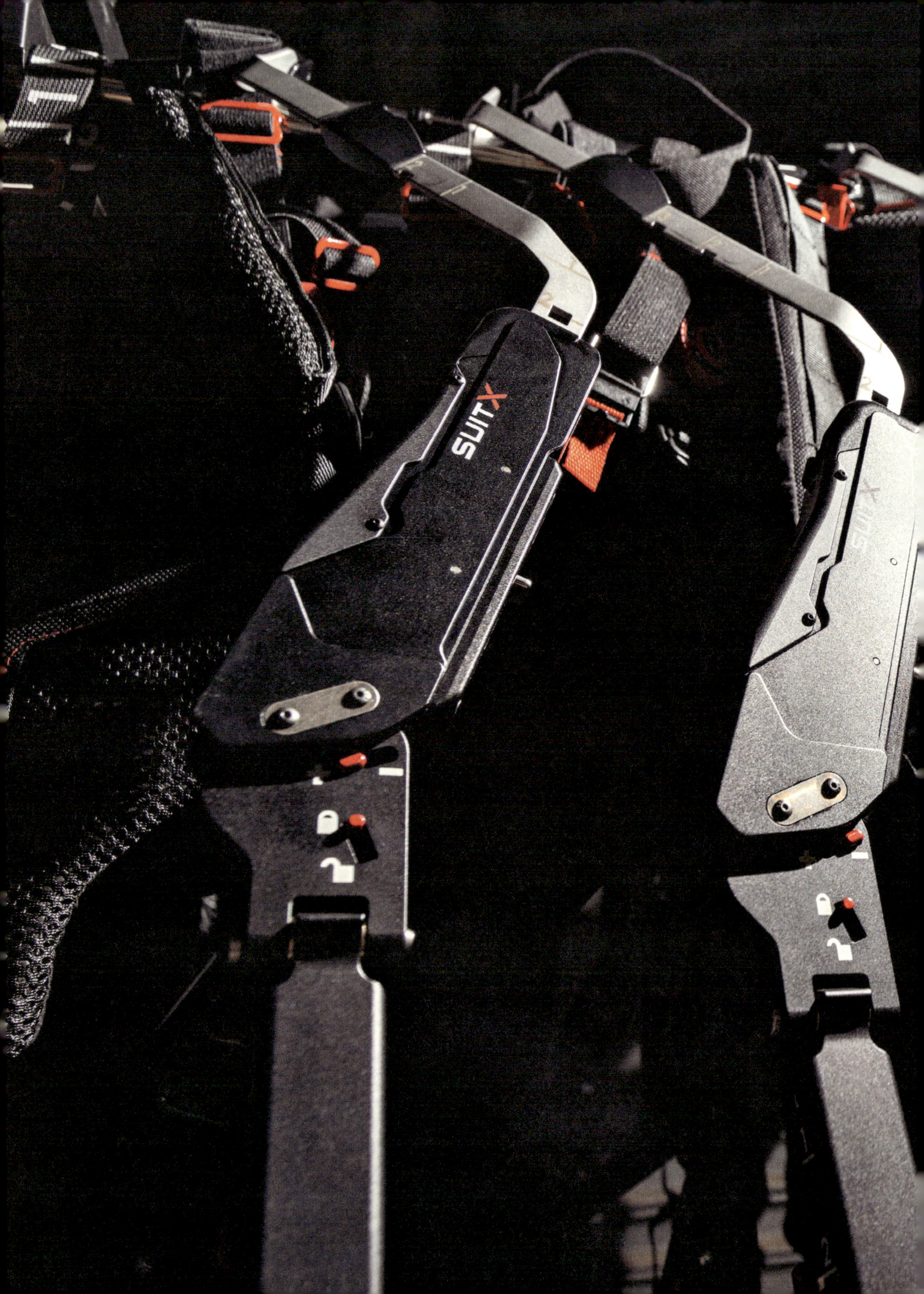
SUITX

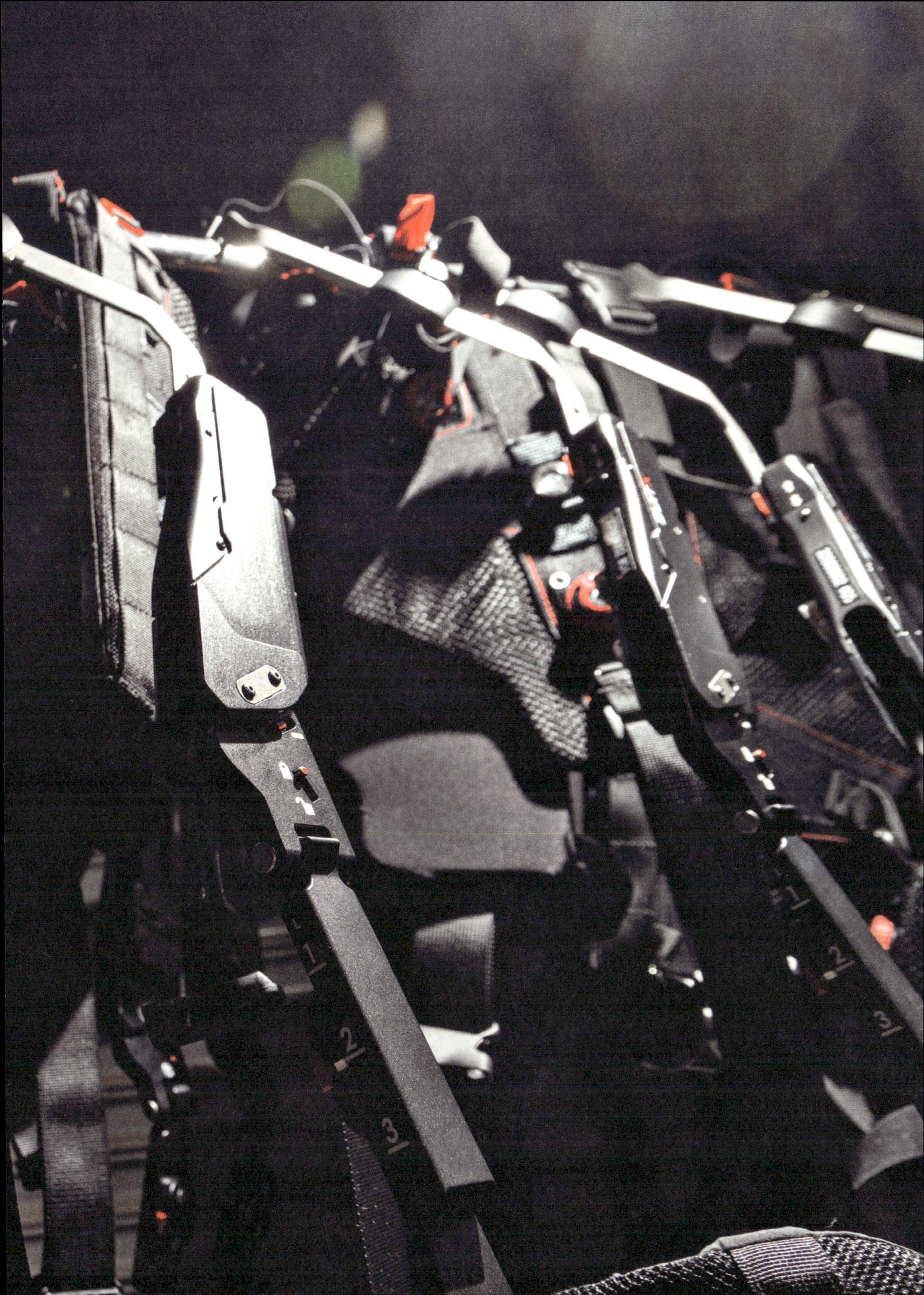

So it's about cultural techniques: gathering information, doing research, and transferring it to industrial products.

Yes. And people. Managing means giving them fish. But I want them to learn how to fish so tomorrow they will continue pushing the technology forward. That's my role. When I carry people with me, I want them to learn that they can do it on their own. They're young, and they need confidence.

They don't have a lot of self-confidence?

They sure do. We all make mistakes; mistakes are the price we pay for future success.

The title of our book, Futuring Human Empowerment, *fits you perfectly. Where do you see the future of human empowerment?*

There are three areas that show our vision for exoskeletons: healthcare, industrial, and recreational. For the next five to ten years, we plan to have a lot more smaller devices, like braces, shoulder braces, or back braces, which we are already selling. As a creator of these devices, we have to consider the price tag, because it is easy to create something expensive. A good friend of mine who works for Honda told me once, "Making a car like a Honda is very difficult. Making a Lamborghini is easy, because you put in the technology that is needed, and customers will pay for it. At Honda, you can't put much in because the price matters." So engineers have to create something out of nothing. We have to make sure our designs are cleverly done so that most people can afford them. I think for the next five or ten years there are going to be small modules that need to cost less and be very clever. After that, the technology would be basically occupied by machine learning. This means that the modules in braces would learn from how users walk in order to assist you even better. So I would say that in five to seven years we will add more intelligence to these devices. The hardware will obviously improve. Cars today are better than they were ten years ago, and in ten years they will be better than now. If we add more intelligence, the machine will adapt and help the person even more. That's the path that exoskeleton technology will go.

You are a very patient person, but you want to go forward.

It makes me happy when somebody teaches me something new. New projects are always exciting since I learn something new. I like to be among people who make and create things out of nothing to change things. It's like giving a blank paper to an artist.

Germans don't like change.

Well, not having changes is good because it means stability.

Don't change a winning team.

Exactly. America is all about changes. There is a big "C" on the top of the hill at the University of California, Berkeley. A good friend of mine at Berkeley—Dr. Richard Lion, Chief Innovation Officer of UC, Berkeley—asked the students at a seminar if they knew what the C stands for, and everybody said, "California." Richard responded "No; C stands for Change." I think he's right. Change is good for all of us. Producing changes through entrepreneurial activities is good for everyone; these changes produce technologies for better quality of life and create jobs.

What I especially like about your products is that they are so stylish, modern, and lifeworld oriented. If you could make it more fashionable, I think young people would love it.

You know what I recommended? Exoskeletons should have vibrant color. I don't think Ottobock would like it. My superiors want Ottobock blue and white colors on all exoskeleton devices. Of course, blue and white are calming colors; however, for industrial settings, brighter colors would be more appropriate. I would like to see Iron Man pink and gold color on our exoskeletons, but I am not sure people here would like that.

Why don't you build a configurator on the internet, like in the automotive industry, so that users can assemble their own colors? It doesn't have to have a million different variants, but it would be great to be able to choose the color. You could cooperate with brands like Nike, and if you have a configurator, users could make an individual product. If young people are your target group, you have to be very individual and fashionable. You have to work with Louis Vuitton!

Yes; we have approached sports equipment companies; I believe the exoskeleton technology is too early for them to adopt it. I was thinking of cooperation with BMW. BMW has created an international image of strength and speed.

Let's go back to the future of human empowerment. What are the next steps in the exoskeleton business?

The first area of improvement is about the increased modularity of the exoskeletons. This modularity and the resulting choice of the modules will lead to a broader use and deployment of the technology. For example, physicians and physical therapists who employ modular exoskeletons will have more choices about what components will be needed for each individual user. For industrial exoskeletons, the modularity allows for suitable choice of exoskeletons for various industrials tasks. The second area of enhancement will include embedded intelligence that requires various sensors and higher computation power. The third area of enhancement is about the communication of the users' performance to other platforms such as AI or machine learning. This will lead to diagnosis and decision-making about the users' total performance.

Connecting the device to the Internet of Things.

For example, imagine that someone is wearing a smart exoskeleton brace and the gait can be recorded in cloud storage. If the user is not walking correctly with the brace, an AI layer will be able to give a summary and perhaps an opinion on the collected data and its relation for further the user's gait improvement. In another example the collected data from a group of industrial athletes in a factory may lead to an understanding of the work setting efficiency and work quality. When I visit my doctor, she makes comments on my health using my vitals that she had taken a few minutes before. Is this enough data for diagnosis? I want my doctor to see a histogram of my vitals. Additionally, a computer model that uses data for a period of time should be able to give my doctor some predictions and suggestions. Further, the model should identify where I am in comparison with a lot of other individuals in a similar group. To me this is a much better way of diagnosis; a diagnosis based on historical data from me and others. The same is true here for an exoskeleton user. The collection of data from many exoskeleton users will be able to give an opinion on the users' performance, both collectively and individually. This is valuable for the users, the caregivers, and the executives. Of course, collection of data alone is not enough to take us to our goal; we still need an intelligent platform to learn the collected data and to suggest improvements.

You could develop a whole digital ecosystem.

Correct. That is the state-of-the-art problem that data scientists are working on. You asked me what I see in the future of the exoskeleton technology? The modularity is about lower costs and focus on the application. Intelligence is about the better performance. And data fusion is about sharing information and performance modification.

In industry, workers who make a wrong movement can be notified to watch the correct movement on a tablet so they can improve it.

Or if there are hundreds of workers in a warehouse moving around, it can establish that the guys are working very hard and tell them that they shouldn't be doing that much. Or, you know, the other option ...

They quantify everything—it's more like an efficiency whip!

Horrible. You're right, some companies go the other way and say, "This guy needs to work harder." But my point is really that it can tell if the workers are not squatting well, bending too much, raising their arms too much, and so on. All this data is useful for diagnostics. So these three elements will come: modularity for lower cost, intelligence for better performance, and data fusion for diagnostic understanding and education. I really believe that recreational devices are coming, and they will be modular. If they have intelligence in there to help me more, that's even better. If they collect data that I can see or direct to my doctor, it will take some time. That's the way bionics and exoskeletons are developing. You see that it's different than with soldiers or in the movies. I haven't really ever thought about making a device for fighting. Science fiction writers write that because they want to sell entertainment. I am inspired by those movies too, but I can't get excited enough to actually make fighting devices. I just don't see it happening—even soldiers don't want it. I talked to soldiers about it, and they told me that when things are critical, they drop everything except their weapon and run. So imagine if they are wearing something that's impeding them. The strength augmentation in movies is inspiring, but it's not real. The real thing is people who need a little bit of an assistance.

I'm curious to know if I can learn to dance with my exoskeleton?

I haven't thought that far. Well, dancing is a complex maneuver.

If you program the exoskeleton to do ten different classical dances, the exoskeleton could show my body how to move.

It's more complex than that. We move sideways, we move backward, and we twist; none of these maneuvers can be fully supported by current exoskeletons due to the lack of actuation and control. We do not have enough space in an exoskeleton to put in that many actuators—at least the actuators that are currently available. We are at a very early stage in making an exoskeleton teach us how to dance.

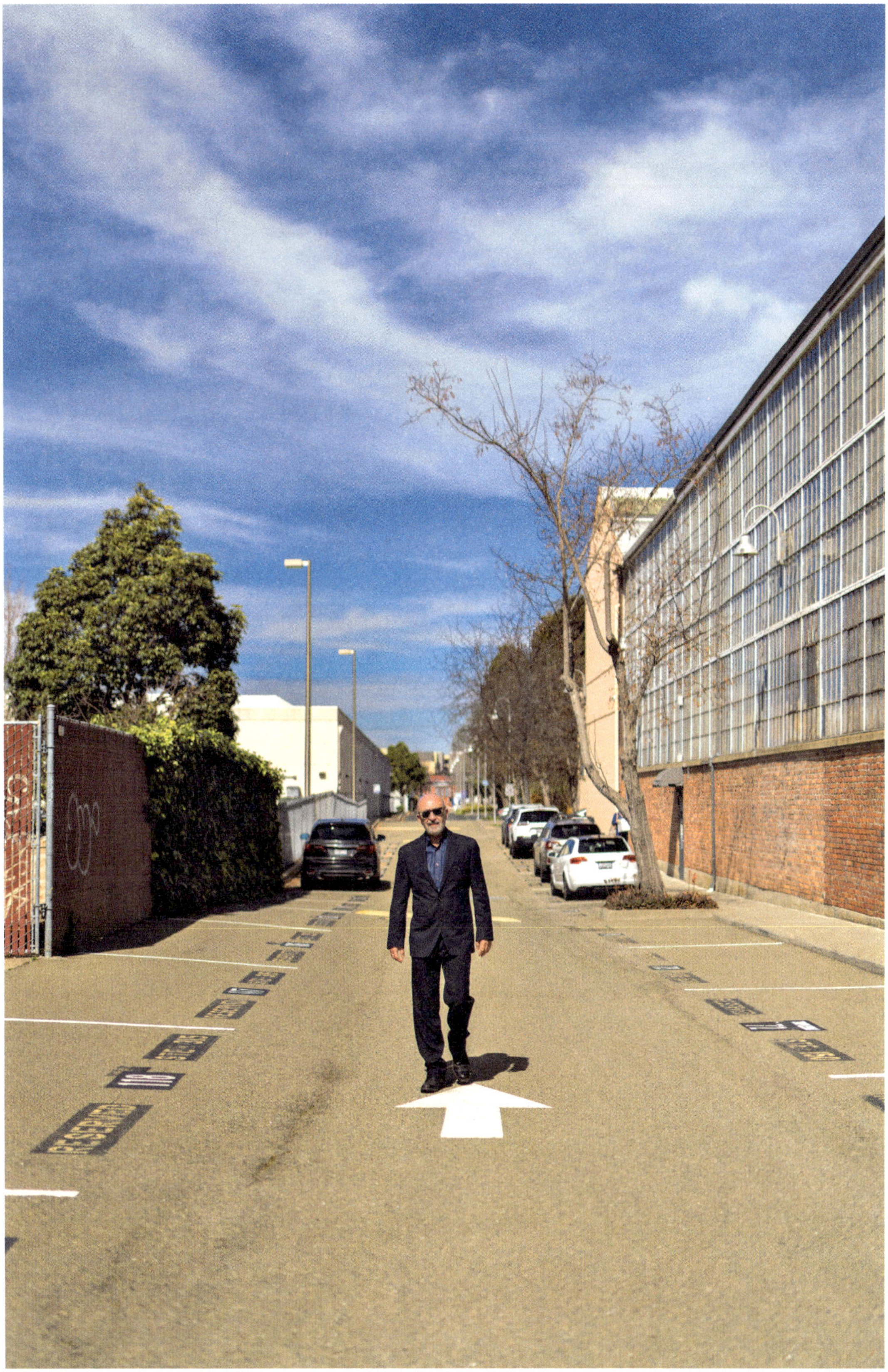

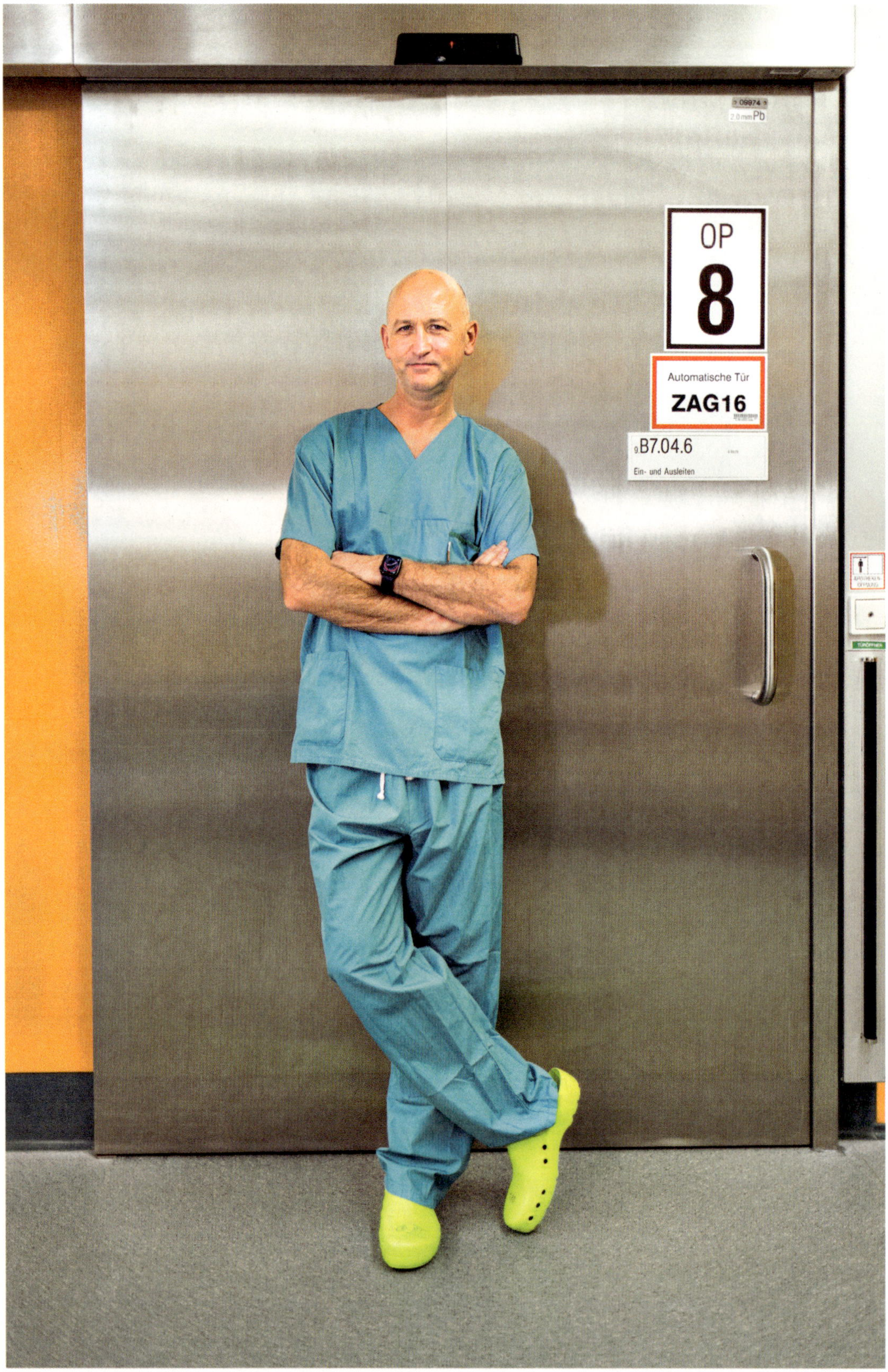
OP
8
Automatische Tür
ZAG16
B7.04.6
Ein- und Ausleiten

OSKAR C. ASZMANN

"You Can Picture the Brain Like a Lasagna"

The successful plastic surgeon works in his "laboratory of corporeality" at the University Hospital in Vienna. He can see the human anatomy with his fingers and works on reconstructing the body at the borderline between man and machine by making use of the brain's perceptual abilities. Sometimes he so skillfully outwits this fascinating neuronal electro-biological machine so that the biceps can suddenly play the piano and a shinbone thinks it is a penis.

Professor Aszmann, during our last discussion we spent a long time talking about Elon Musk. With his company Neuralink, he intends to implant a chip in the human cerebral cortex that will monitor or transmit electrical signals. This will enable him to do things such as bypass nerve damage in combination with a smartphone. Your view, in contrast, is that the quality of information that can be deciphered directly in the brain is far too unspecific. All that one finds there is a nonspecific cloud of data. You propose that the actual quality of information is first generated when the electrical impulse from the brain arrives at the muscle cells to which it is addressed in the extremities, where it is coupled with biological information.

I am one hundred percent convinced that it is impossible to obtain useful information directly at the level of the cerebral cortex. What is frequently not understood: in terms of structure every computer, every iPhone is a processing system of binary information. But our brain doesn't function like this.

So the brain doesn't function with zeros and ones?

No, the brain is a biological system. Memories and information are set down as protein, as a synapse, as a productive process, which is also triggered genetically. The electrical impulse puts a whole biological machine in gear, which then deposits the information. But not as a zero-one system.

What kind of system instead?

If I knew that, I would win a Nobel Prize. Ultimately, it has to do with a neuronal network. For example, when animals don't get certain amino acids, they suffer from memory loss. That shows that memory, both short-term and long-term memory, requires a protein synthesis process.

So the person who found these amino acids could—

This has nothing to do with the individual amino acids. In principle, stored information is the product of an entire neuronal electro-biological process. Protein compounds require the entire biological machinery to build new synapses and thus configure the connectivity of the central nervous system so that a very simple movement, such as changing gears in a car, can be carried out without the brain having to explicitly think about it. You can talk to your fellow passenger without being aware what you are doing with your right hand. That is a memory process.

If you found the key to this mechanism, would it be possible to cure Alzheimer's or Parkinson's?

Unfortunately it is not simply a key that one can place in a lock, but a multilayered phenomenon. Many decades ago the Nobel Prize winner Rita Levi-Montalcini discovered the nerve growth factor, and back then we experimented on this basis, in the hope of finding a molecule, an element, a protein, a pharmacology that one could simply drop onto the nerve so it knew where it should grow. But biology doesn't make things that simple for us.

How should we picture it?

I am not a brain researcher, but you can picture the brain like a lasagna, with six layers of neurons that are distributed across the whole cerebral cortex and networked in a variety of ways. For some of these layers the interconnectivity is simpler, for example, for the cells in the cerebellum. The cerebellum is responsible for the coordination of motor functions and is the executive organ for the more complex motor tasks.

So for the reflexes?

Yes, exactly. In contrast, the cerebral cortex thinks in a cloud. The central nervous system is characterized by its ability to outsource tasks. For example, you can go for a walk in the woods and concentrate on a highly philosophical conversation with a friend, without having to think about your feet. The feet do their work all on their own, because it is an afferent triggered system—

Afferent nerve pathways are those that transmit sensory stimuli to the central nervous system. In contrast, the impulses that are transmitted from the central nervous system to the limbs and organs as motor signals pass through efferent nerve pathways.

Exactly. That is what is so exciting. For example, decerebration experiments have been carried out on cats in which the spinal cord has been severed, cutting off the body from the cerebrum. Despite this, the cats could still take a few steps. Chickens can also keep walking after their head has been cut off, and from historical accounts we know that beheaded people can also walk a few steps.

People can still walk after the guillotine?

Yes, for a few steps. These are afferents. The motor program keeps running, at least for a short while—without motor stimuli from the brain. And so the motor program effectively runs in the background the whole time in living humans. Nobody thinks about their feet when they are walking. It is only when you suddenly trip over a root that you have to take an extra step. You have to interrupt your conversation and think about where and how you will place the next step. The human brain only has to learn everything consciously when first mastering new, unknown movement sequences, for example, driving a car: simultaneously holding the steering wheel, pressing the clutch, and shifting into first gear. This is difficult at first, until it eventually becomes routine. For example, my daughter plays the violin by heart. She doesn't have to remember the notes, instead her fingers know exactly where they have to grip. That is an afferent-driven system. If the fingertips don't know what they are doing, then you can't play piano. Theoretically—once you have learned—you could also play even if you can no longer feel your fingers.

How do you use this knowledge to establish a better connection between prostheses and the human body?

For me the most important finding is: movement is defined through afferents. Movement is a sensory phenomenon. Seven years ago we began to reposition nerves for the control of prostheses. However, in order to do this intelligently we first had to investigate how the

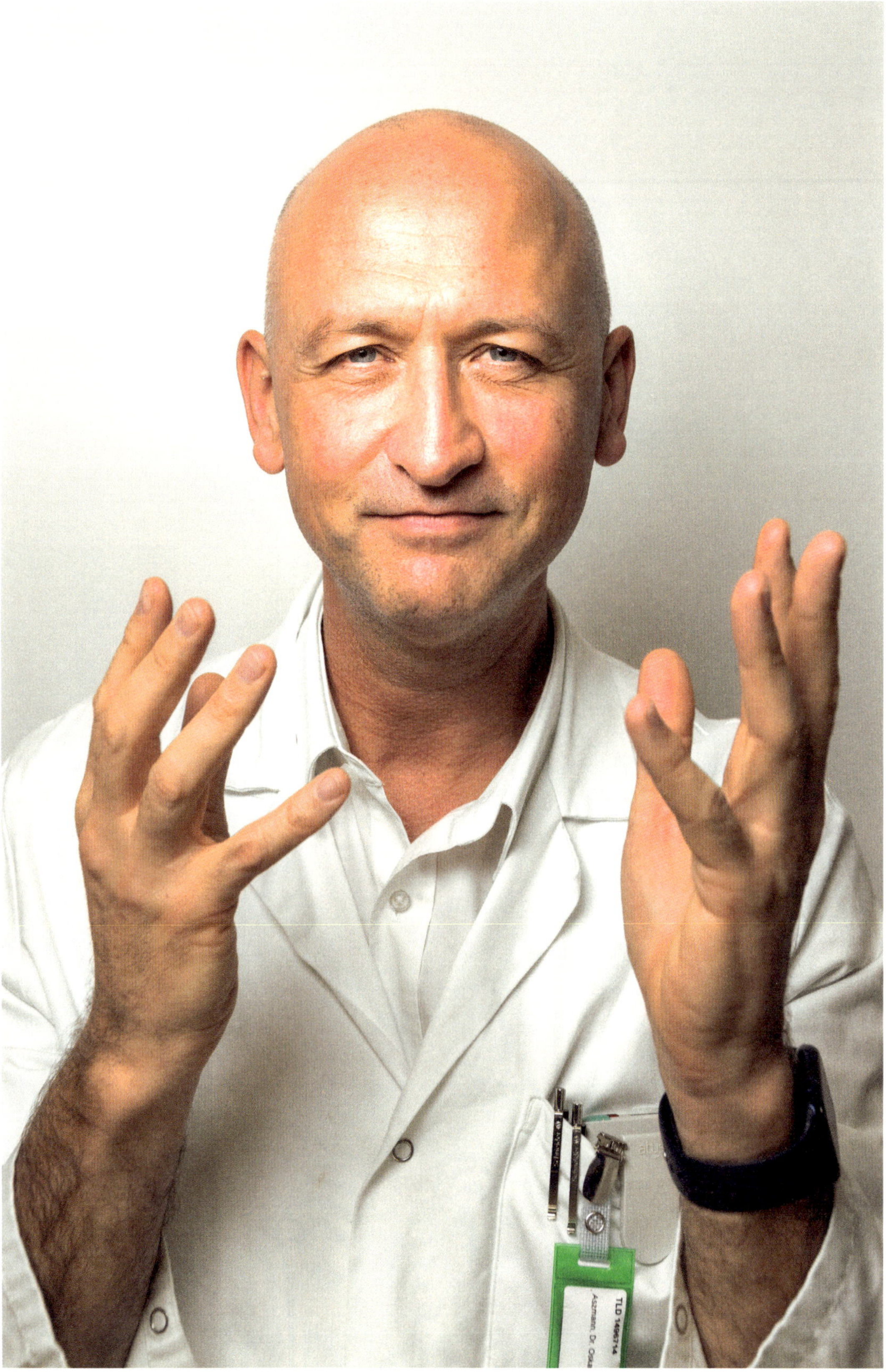

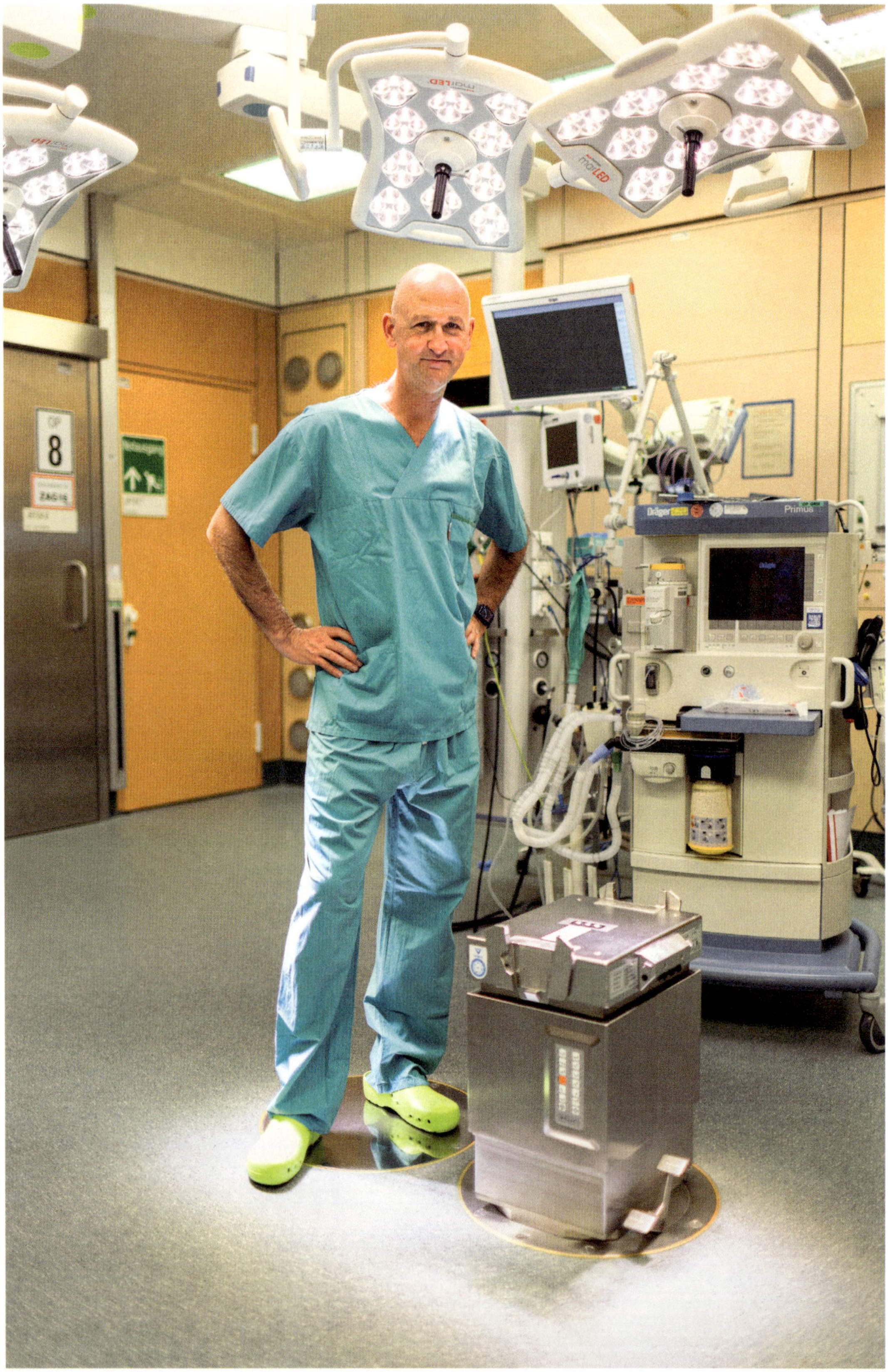
OP
8
Dräger
Primus

nervous system functions in the first place. We have investigated ten nonheartbeating organ donors here in the University Hospital; we exposed the entire nervous system of the arm and looked to see how many afferents and how many efferents there are.

How could you establish this?

We developed dye techniques that dye the human nerve fibers different colors. We were able to establish that ninety percent of the nerve fibers are afferent. That was a minor sensation. That means that ninety percent of all the transported information passes from the periphery to the brain and only ten percent from the brain in the other direction. The vast majority of nerves are used by the brain to listen to the body, and only a very small proportion of the nerves are responsible for the brain's commands. That is completely ergonomic and economical. Our brain activity is best used for meaningful work and not for the routines. And routines include a movement such as walking. When we step onto an escalator, we don't interrupt a conversation. However, a person wearing a prosthesis has to interrupt the conversation and briefly concentrate on his feet because he has to consciously think about what he is going to do.

So that is why the major goal is to perfect things sensorially, so that people actually perceive the prosthesis as part of their body.

The keyword here is *reafferentiation*. We recently equipped our first patient with sensors that transmit sensory feedback to the prosthesis reafferented by this means. When he walked up and down, he said, "That is fantastic!"

What is a reafferent prosthesis, and how does it function exactly?

In this case we mounted four sensors on the residuum of the lower leg. The prosthesis receives a return signal from the surface on which one is walking, indicating when the foot has touched the ground, when you have taken a step, and whether one has lifted the foot again with the heel or the forefoot. The individual only feels this feedback in the form of different vibrations—no more than that. However, with this sensory feedback the brain quickly learns the sequences in the prosthesis and then the brain can outsource it to a more unconscious and automated program.

So in this case it is not necessary to operate on the nerve fibers because the different vibrations are sufficient for the brain?

Yes, that it vibrates here and there on the foot is very quickly grasped.

What does your special area of Targeted Sensory Reinnervation (TSR) involve?

With TSR one severs the original nerve and links it to the remaining nerves of the amputated limb. The resulting electromyographic signals of the target muscle can then be used in the motor system to control a motorized prosthesis. But this also works in the sensory system. I will explain this using an example: a patient had a stroke and as a result of the malfunction of a number of cerebral nerves could no longer properly move or feel his lips. He was always biting his lips until they bled or the soup ran out of his mouth when eating. I relocated a nerve from his shoulder to the nerve that leads to his lips. It took three months before he realized that when he felt a tweak in his shoulder he was actually biting his lip. Six weeks later he knew that when it tweaked there, it was the lip. And now he would say, "That is my lip." The brain simply switches it over.

In another case, a person who had their lower arm amputated, you simply stuck the nerve responsible for the hand movements into the biceps. The biceps is initially irritated by this unknown quality of information. Up until now this muscle had a very simple task: to bend the arm up and down like a bodybuilder. Initially it has no idea what to do with the

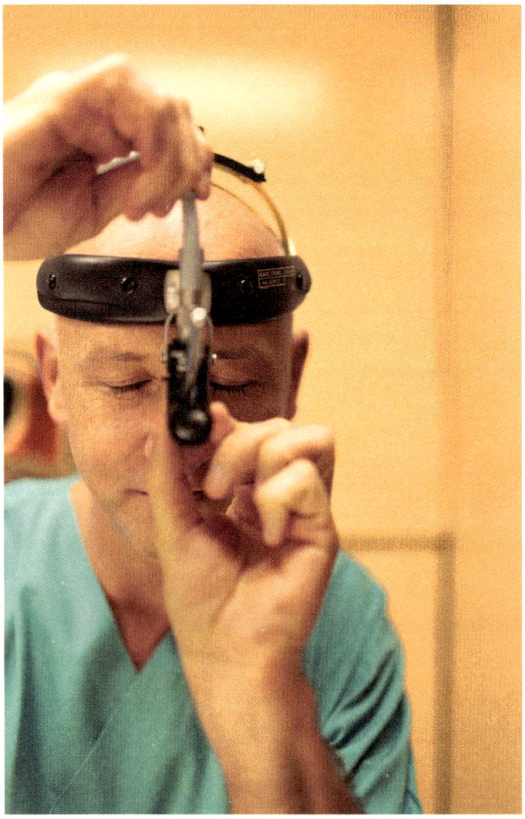

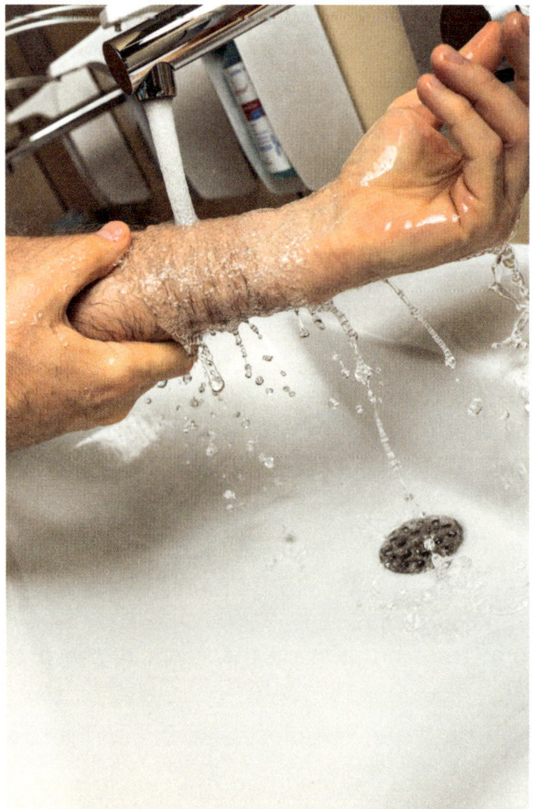

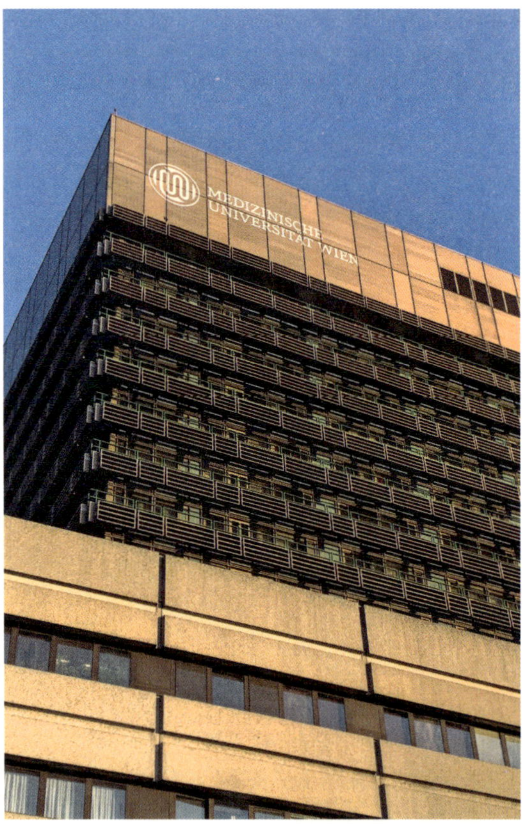
MEDIZINISCHE
UNIVERSITÄT WIEN

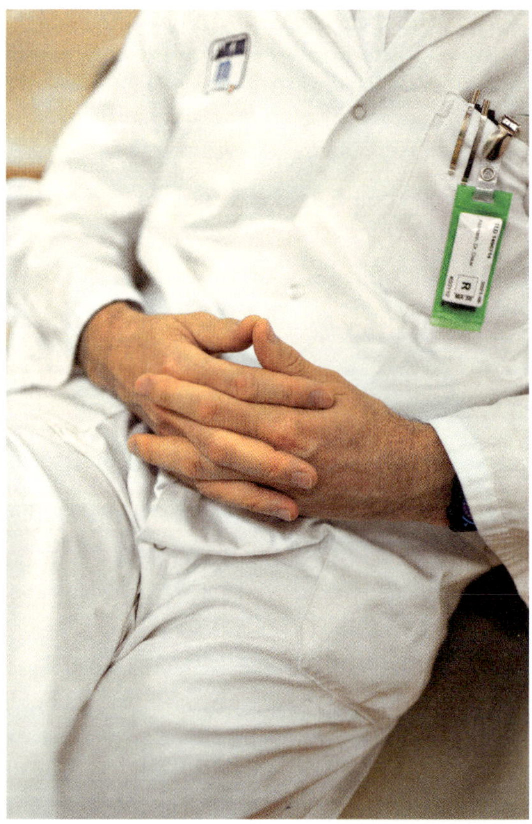

information addressed to the hand. What does this constant message from the brain mean: "Play piano, play piano"?

But then the bicep's muscle cells begin to rearrange themselves and eventually the prosthesis plays piano. In the case of people who have had their lower arm amputated, who no longer have an elbow, the bicep has no function at all. It can't do anything. But when I transfer the *nervus ulnaris*, the nerve that is responsible for the small hand muscles, to the biceps, then at some point the muscle develops the capacity to play the piano. Naturally, we are not this far prosthetically, but theoretically the neuronal information is available. My second example is a patient who suffered an incomplete paraplegia due to a climbing accident. I operated on his legs so that he could walk again. A year later he came back to me and said that everything was great but mentioned to me that since the fall he couldn't feel his penis at all, and asked whether I could also solve this problem. I thought about it somewhat and conducted a thorough examination and discovered that there was one other nerve that he could still feel. The *nervus saphenus* is a long nerve that originates from the thigh and can be felt running over the knee to the ankle because it supplies sensation to the inner thigh and the inside of the lower leg. I then rerouted this nerve into the *nervus dorsalis penis.*

But he still had his legs?

Yes, naturally. The *nervus saphenus* is a pure sensory nerve that is only responsible for the feeling on the calf to the shinbone and ankle. It doesn't do anything else. I met the patient again three months later, and he told me that when he inadvertently pinched his genitals, which he could no longer feel, against the kitchen table, for example, then he could suddenly feel a tweak in his shinbone. Initially he always looked at his leg to check what pinched him until he finally understood that it was the rerouted nerve and that he had actually pinched his penis. Six months later he reported that his shinbone had now become his genitals.

Can he also develop an erection and experience an orgasm?

He can't ejaculate, but he gets a good erection that he is able to penetrate with.

What does he feel when he has sex?

He reports that it is pleasant.

A pleasant tingling of the shinbone?

Yes, so to speak, but no, now it is his genitals, and he is very happy to feel something there again. First of all, tens of thousands of nerve fibers have to grow into the skin, but initially they don't exactly know where they are. That takes its time. Ultimately one has to reconvince the brain, then it reformats itself and can assign the sensory information to the penis. The best rehabilitation program for this patient is to have as much sex as he wants. That is good for the nerve regeneration—not just for the organic regeneration of the nerves themselves, but also for the cognitive-sensory relocation, so he knows that it really is his penis and not his shinbone.

Does he still feel something on the shin?

At first he wouldn't feel anything at the site; it would be numb. But the brain is very ingenious. There are also nerves to the front and back, to the right and left. When we remove nerves, the other nerves begin to sprout into the gap, so that at the end there is hardly any numb area left at all. However, the repositioning of sensory nerves can also be very unpleasant for patients. Many of them don't want this.

If patients are capable of relearning from shinbone to penis, why do we need to conduct a nerve transfer at all?

We recently treated a young man with a congenital dysmelia. He had a shortened leg with a completely twisted lower leg. He had a prosthesis and always suffered from pain in the ankle joint. Naturally we could have carried out a classic lower-leg amputation, however with the disadvantage that he wouldn't be able to end-load it because the bone presses directly onto the soft tissue causing pain. We then filleted the sole from the existing foot together with the nerves, severed the foot guillotine-like, and then placed the neurovascular surface, which in his head is his foot, onto the residuum for load-bearing purposes. Now the effect is such that when he steps on it he can feel his toes and heel. It is like a short foot. He only needs a short shaft and he can walk on it directly. This improves the quality of the lower-leg provision considerably.

That is fantastic. I have never heard of anything like this from an orthopedist.

Yes, there are not many people who have mastered this type of bionic reconstruction. Plastic surgeons tend to think outside the box and have a more creative approach to their discipline.

But what is the holy grail of bionic reconstruction? What would be the most important innovative leap?

The reafferentiation of prostheses.

Didn't we discuss that last year?

Yes, but that is a central theme, especially if movement is actually a sensory phenomenon. With his AMI concept Hugh Herr has established a connection between antagonists and agonists. This is nothing less than an attempt at reafferentiation at a proprioceptive level. *Proprioception* means "self-perception."

In other words, perceiving the movement of one's own hand through space and time with one's eyes closed.

Exactly. In principle that has nothing to do with haptics, that one can feel and perceive things. All it means is that I can say where my extremity is positioned in space. For leg amputees the signal transmission from the sensors and the return signal can function quickly enough, provided one walks slowly. But with the upper extremities, the time until the sensor begins to emit a signal, the signal enters the socket and the actuator—that is, a vibration starts—can be too long. When one requires thirty milliseconds for a reaction, but it actually takes three hundred milliseconds, the egg or the grape has already been squashed and the glass has been shattered. It is not just about the quality of the sensation; the time axis is also relevant. It is very hard to overcome these time spans with external systems. In principle Hugh Herr's AMI concept is very interesting, but it is not relevant for the majority of residua because it is too short. The concept doesn't function there. That is why two years ago I began to investigate whether one could solve the problem without outsourcing it, while retaining its relevance for the man-machine interface. We began analyzing the face's neurobiological system, because the face musculature is the most complex neuromuscular system that we know. A wink can start a relationship, raising a corner of the mouth can start an argument. The smallest movements of the facial expression are meaningful. And so you have to know what is happening in the face. But the astonishing thing is, the facial musculature doesn't have any proprioception.

So no spatial feeling?

Every muscle normally has a neuromuscular spindle with gamma afferents, which can tell me at lightning speed what state of contraction my muscle is currently in. But the face doesn't have that. Over recent months we have conducted a large number of animal experiments in order to get to the bottom of this. The most exciting thing is that proprioception in the face only seems to operate via the skin. When you receive an injection at the dentist's in order to anaesthetize the *nervus alveolaris inferior*, the nerve for the lower jaw, you can't feel your lip anymore and want to keep looking in the mirror because you are worried you look like Quasimodo with your face hanging down. But the face is completely normal, and you can hardly believe it. So the facial musculature receives its feedback via the feeling in the face. We now know that the finest movements of the fingers, for example, when drawing, proceed via the sensitivity of the skin over the finger joints, not via the extrinsic musculature. So when the proprioception in the most sensitive system, namely the face, is outsourced to the sensibility, then we will attempt to do this with prosthetics too. Our test animals, rats, can finely control their vibrissae. When the animal is exploring its environment it raises its hairs to the front so it can sense what is taking place there. In animal experiments we severed the nerve, after which the hairs lie completely flat. After approximately six weeks the nerves have grown back, the hairs stand erect again, and one can see the start of reinnervation.

How could you prove that in the case of the rats this also occurred via the sensitivity of the skin?

You can see the connectivity of these nerves via retrograde dyeing of the brain. The idea is that in the case of a patient with amputated arms, for example, we carry out two classic nerve transfers in the biceps for the efference—in other words for the control—but also transplant a piece of skin, for example, from the palm of the hand, onto the muscle. The transplanted skin stimulates the nerves, and the brain perceives it as the hand. At least we had this effect in the animal experiments.

But it is only the skin without any nerves—

Yes, the nerve grows into the skin.

—and then it recognizes the molecular structure of the muscle cells and knows that it is the hand?

It is a little more complicated than that. For example, if you compress or stretch the skin over your PIP joint, then a movement is simulated that is perceived by the brain as movement, although no movement has taken place.

Because I can simulate it via the skin?

That's the way it is. One can perceive and simulate movements just by means of the skin. If someone has lost an arm, I can trick the brain into thinking a movement has occurred through the stimulation of an inserted skin transplant, even though the hand is no longer there. This also proceeds a lot quicker because it occurs within one's own biology. Nevertheless, one can also influence this system from the outside with small implanted magnets, using them as interfaces for prostheses.

So you no longer need to fiddle about with the nerves at all?

I don't need to operate this axis down to the fingers because this is incredibly complicated, and ultimately it is not even necessary because it can be achieved a lot easier with the piece of skin.

Then you have made great progress within the space of a year.

Yes, after all we are a laboratory!

On top of this, you are also a gifted plastic surgeon. You see patients that no one else is able to help, and you always find a solution that makes their lives easier.

I have learned a great deal in my life and compiled a substantial portfolio of adventurous surgeries. I have almost exclusively special patients. Every patient has a different needs profile, for every patient you have to find a solution, and this obviously is a creative process. Drawing a line between that which you can solve biologically and what not, is also exciting. For example, I had a thirteen-year-old patient with a malignant tumor in the arm. The oncologists asked me how one should amputate the arm for the best possible prosthesis. But his hand was still healthy. Amputating a thirteen-year-old boy above the elbow is a crime, if not absolutely necessary. So we cut above and below the entire diseased segment and sewed the hand back on. Now he has a shorter arm, but he still has his own hand with which he can do everything.

So is it possible that bionic reconstruction could eventually make prosthetics superfluous?

No. I have a great deal of faith in bionic reconstruction, but sometimes one reaches the limits of biology. As long as I have the possibility of reconstructing someone biologically, then naturally I make use of this. However, there are patients whom one can only help prosthetically. Fifteen years ago I saw a patient who had lost both arms. At the time I didn't know Ottobock, but I did some research on the internet to help the patient with Targeted Muscle Reinnervation and came across Hans Dietl. Another patient was bitten by a dog and contracted a sepsis with a Waterhouse-Friderichsen syndrome and fell into an acute coma. Both legs had to be amputated at the lower leg, he lost all his fingers and his nose. I reconstructed his nose with skin and bone from his right hand, and with his hands we thought about how we could remedy this with prosthetics. In the end we carried out a metacarpal bone transfer, so that he had a thumb again, which he is very happy with as he can hold a bottle with it, brush his teeth, or even write. He is still sitting in a wheelchair, but he will soon be fitted with prostheses and will be able to walk about on two legs. I work at precisely the interface between bionics and prosthetics and always try to find the best solution for everyone. It is about the reconstruction of the body and the border between man and machine.

How is it that you know the body so well, for finding such unconventional solutions?

We have the great advantage that we have a center for anatomy here in Vienna. We have always had a good supply of body donations here in Austria. During my medical studies I had the privilege of dissecting hundreds of corpses, and as a result have gained a thorough understanding of the human matrix. That is my piano. I know every detail by name and know where it is situated. But not because I have read it in the textbook, but because I have explored it on dead bodies with my own hands. You don't just look with the eyes, but always with the hands too. The surgical finger is actually an endoscope. You also say, "Let me take a quick look," when reaching into the pocket. The fingers are capable of stereognosis, that is, the ability to sense the surface of a part and recognize it blindfolded. You look with the fingers. A lot of the things you do as a surgeon is haptic. And after having dissected so many corpses, my fingers know a great deal about the anatomy of the human body.

So you are a pianist who has all his skill in the tips of his fingers?

Yes, that is true. The surgeon can't work without sensitivity. The human body is a mystery. And the confrontation with the worst injuries teaches you a certain humility. In my job you develop a high regard for your own corporeality.

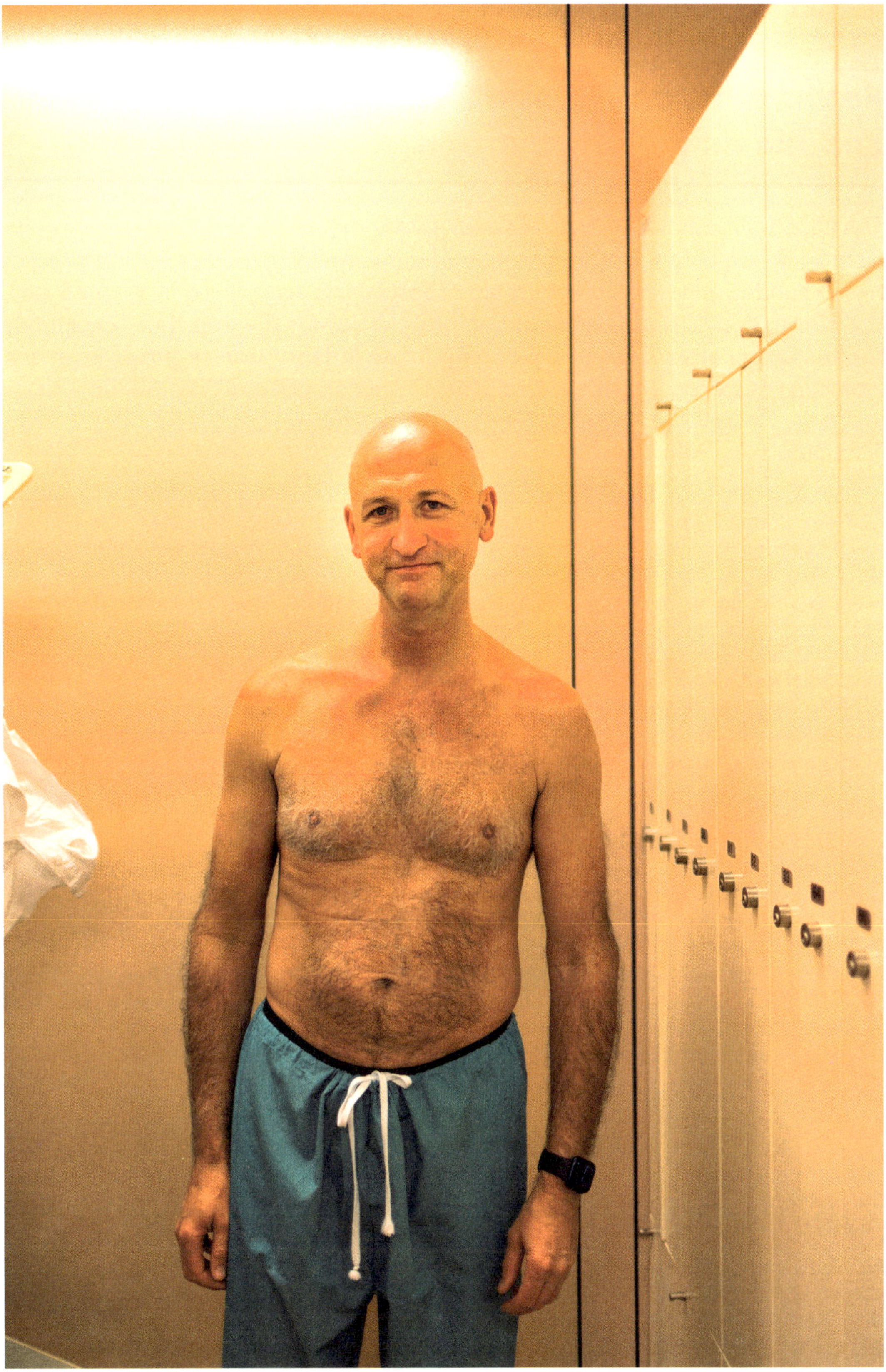

MARTIN BÖHM, HANS DIETL, ANDREAS GOPPELT,
BERNHARD GRAIMANN, AND GÜNGÖR KARA

Reflections on Wearable Robotics and Bionic Reconstruction

The Ottobock experts discuss the growth opportunities that open up through Ottobock's strategic focus on human empowerment. The acquisition of suitX has expanded the range of wearable robotics and exoskeletons. Beyond this, the company plans to develop active prosthetics for the first time. Enormous technological advances are also being made in laboratories throughout the world in the core competence of bionic reconstruction.

Andreas Goppelt: The chapter on wearable robotics and bionic reconstruction encompasses Ottobock's core competences. On the one hand, with the acquisition of suitX in November 2021, our company has become one of the world's leading suppliers of exoskeletons for industrial workplaces. The close cooperation with the founder Homayoon Kazerooni will push forward the development of wearable robots and, most of all, provide access to younger target groups. "Kaz" visited us just a few days ago in Europe, and we have identified a multitude of common projects. This example shows how Ottobock's orientation on human empowerment is opening up new growth markets. In our more classical core business of prosthetics there are also encouraging developments worldwide with bionic reconstruction. For example, Hugh Herr, thanks to the cooperation with MIT and the Brigham and Women's Hospital, has already provided over twenty patients with AMI, or agonist-antagonist myoneural interface. The clinical tests have proven that through the reestablishment of the natural proprioception in the neuromuscular system of the amputation residuum it is possible to create a high-performance myoelectrical interface. This can be used to very effectively control prostheses, and patients report that it feels as if they have got their leg back. That is a decisive step in the direction of embodiment, in other words, the acceptance of the prosthesis as part of oneself. In Vienna Oskar Aszmann has made great progress with Targeted Sensory Reinnervation (TSR), which is extremely valuable for Ottobock's technological development. With TSR the original nerve is severed and linked to the remaining nerves of the amputated limb. The resulting electromyographic signals of the target muscle can then be used in the motor system to control a motorized prosthesis.

BG: In Utah, Tommaso Lenzi is building active prostheses—and the human-machine interface is also quite important for him. The intention behind his technology also begins with the user. In fact, Lenzi is not that far away from our prostheses for the lower extremities. The difference is that he builds active systems with motors which, for example, support the user when standing up. In contrast we have a passive system. With us, the force must be exerted by the user of the prosthesis themselves.

AG: Provocatively, Lenzi stated that the algorithms of his wearable robots control things autonomously, and the user is only given the illusion of control. But Lenzi knows how patients tick: they want to maintain control at all costs. The intention and the control impulse must still come from the patient. Everything else can be a downstream intelligent automatism.

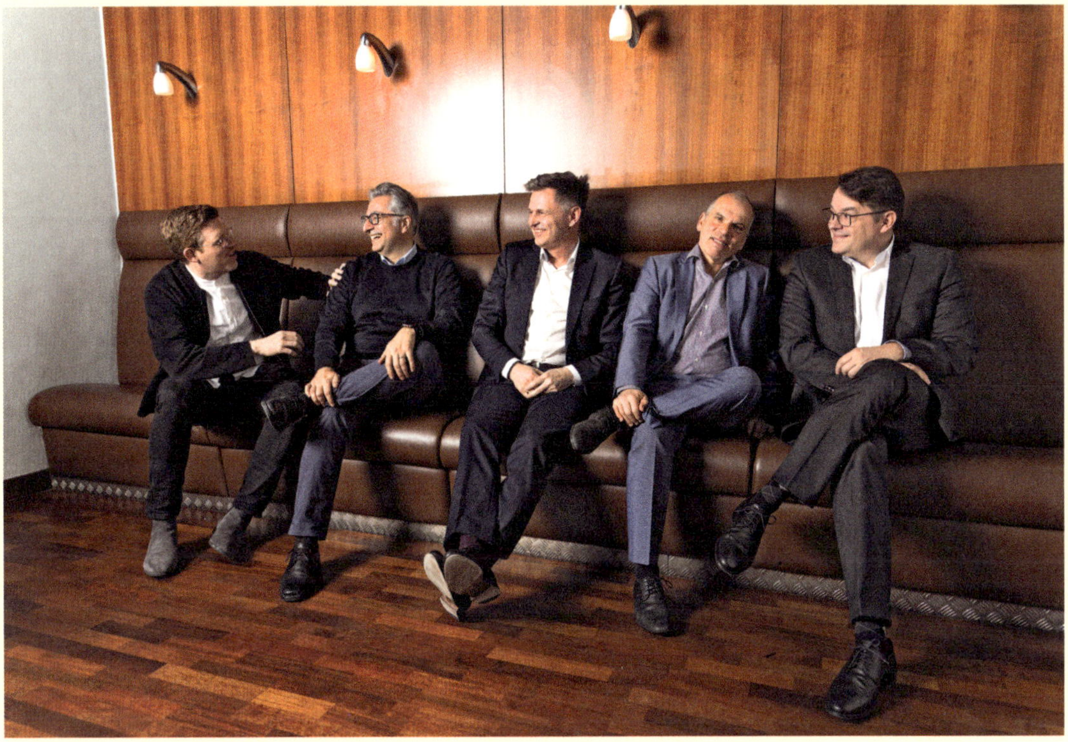

Martin Böhm, Güngör Kara, Andreas Goppelt, Hans Dietl, and Bernhard Graimann (from left to right)

BG: Lenzi's system simultaneously enables intuitive control and precise placement, as the motors actively support the control. That means that the user can place the leg more intuitively, because a control algorithm assesses the situation and carries out the movement more precisely than a passive system. The motor-driven assistance when climbing steps is a very popular feature, as it relieves the patient of the physical exertion.

AG: Naturally, this raises the question why Ottobock hasn't offered any active prostheses to date. Ottobock is actually not the first on the market with such systems. There are already active prostheses from other providers, although they are sometimes very heavy and loud. The existing advantages such as assistance when standing up, when climbing steps, and during long periods of walking on the level are outweighed by the disadvantages.

Hans Dietl: That was a longstanding problem with the earlier versions of the power foot developed by Hugh Herr: you could hear it long before you saw it. We know from clinical user studies that the noise was one of the main reasons why partners and families didn't accept the active prosthesis. These prostheses function wonderfully, but they are simply too loud.

AG: Ottobock will not be putting any active prostheses on the market until we have reduced these disadvantages to a minimum. Our aim is to keep weight and noise level to a minimum. Very early on Lenzi had the goal of designing an active knee of the size and, most importantly, low weight of our C-Leg. While he has succeeded, the motor noises are still a problem. We have been working on active knees for years, and there are numerous research projects in which we are following a number of approaches, including our collaboration with Lenzi. These cooperations are very important for us.

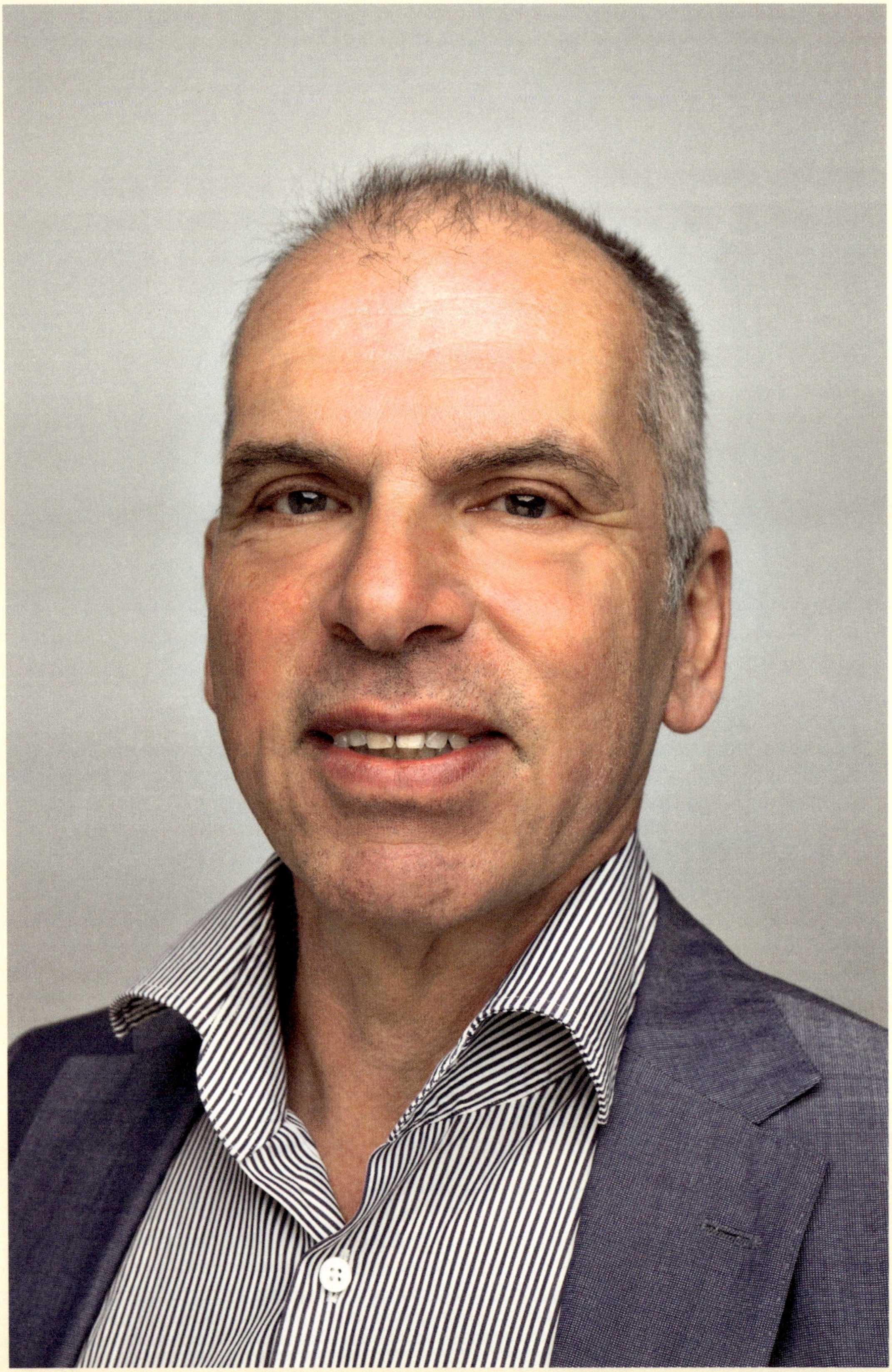

Dr. Hans Dietl was Chief Technology Officer at Ottobock for over thirty years.

BG: While passive technology has a number of limitations, it can satisfy the patients' most important needs to a large extent. Nevertheless, the metabolic costs are high, for example, when standing up or walking. There are studies—including those by Tommaso Lenzi—that show that the exertion and the metabolic parameters can be many times higher than with a healthy leg. In contrast, the three value propositions of the active system are: firstly, lower metabolic costs, secondly, better positioning during the movement sequence, and thirdly, a better and more intuitive control of the prostheses.

AG: But just because the technology is active, it doesn't necessarily result in better control.

BG: Correct. The prosthesis itself must provide the basis so it can be intuitively controlled.

AG: You can also formulate it differently: the active system allows more degrees of freedom, and because it offers these degrees of freedom, you can ultimately produce a better result—but you also have to know how. A major competitive advantage of Ottobock is that we are very good at translating biomechanical knowledge into intelligent controllers. For example, the algorithms are coded into the knee's fit component and coupled to the corresponding sensor system and can anticipate the patient's next steps. Our passive systems are also adjusted via the dampening so they are extremely comfortable to wear. If in the future we extend our system with the "active" degree of freedom and still maintain our existing advantages, we will be almost unbeatable on the market.

BG: You have to see things in perspective: bionic reconstruction is one possibility for improving the human-machine interface and providing better control. In contrast to a passive system an active system can be controlled in a variety of ways. This combination of a powerful human-machine interface and an active system would be especially effective.

AG: That is why Ottobock will continue to invest in active prostheses. The future is active!

The Human
Empowerment
Company

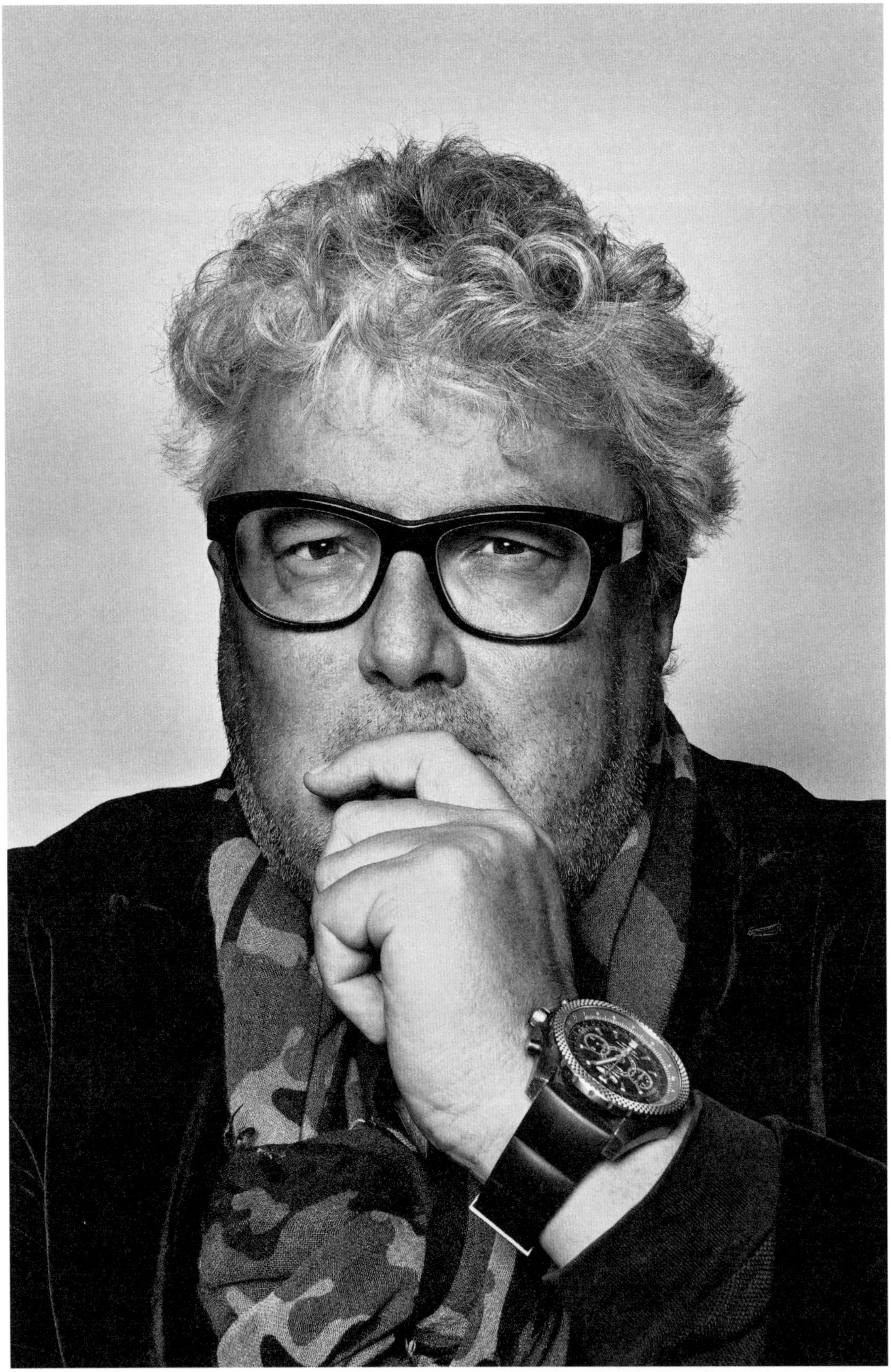

HANS GEORG NÄDER

The Fourth Revolution

The founding of Ottobock over one hundred years ago was a small revolution in itself. My grandfather, Otto Bock, had the trailblazing idea of how to better meet the demand for prostheses needed by the countless people disabled in World War I. By breaking down the production process, previously organized along artisanal lines, and dividing the prostheses into components that they could be manufactured in industrial serial production, he laid the foundation for the orthopedic industry. My father, Max Näder, was also driven by curiosity and always searching for new innovations. He revolutionized orthopedic technology with the Ottobock modular system, which continues to set the global standard. Furthermore, he also invented myoelectrical prosthetics, developed new materials, and introduced plastics into orthopedics. Under my leadership Ottobock began producing the world's first microprocessor-controlled knee joint, the C-Leg, twenty-five years ago. It revolutionized orthopedic technology for the third time, considerably expanding value creation for the whole industry. To this was recently added the C-Brace orthosis system, which enables patients with paralysis to be treated using a sensor technology that electronically regulates the leg so that a wide range of patients can get out of their wheelchairs and walk again.

The history of the company is thus a sequence of three revolutions in orthopedic technology, which has allowed Ottobock to rise to the position of unchallenged innovation leader. What is at stake today is nothing less than the fourth revolution—the complete digitization of orthopedic technology. While the transformation of a traditional craft in the digital world is everyone's goal—from opticians to hearing-aid acousticians—we are taking the lead. Digitization is decisive for the future of every company that claims leadership in technology and innovation. At Ottobock the digitization of the entire value creation chain, from customer relations to production, is the order of the day. We are moving away from plaster and resin toward a fully digitized supply chain with scanners, 3D printing, software-based development and planning processes, and artificial intelligence. This digitization enables us to find solutions that are much better, far more individualized, and can also be produced more quickly, more effectively, and more sustainably.

The digitization of Ottobock has two directions: the digitization of customer relations as well as product development and manufacturing. The digitization of customer relations is the fundamental precondition for improving the user experience. The central credo of user experience is placing a greater focus on the user as the heart of all corporate processes. For users to have all-around affirmative experiences with Ottobock's products and services, the company first needs to gain a better understanding of what users really need. Our goal is to optimize the patient journey, that is, a patient's path from the initial contact with Ottobock and their decision to use a product to aftercare.

The digitization of product development will not function without the digitization of customer relations. It involves scanning and digital modeling. The new Digital-to-Walk platform plays a central role in this. In medical supply stores of the future, the user's body will be measured with a new, handheld scanner the size of an electric razor that Ottobock will make available to the medical supply stores. The orthopedic technicians can make a 3D scan of body parts without the need for additional technical equipment. This generates a highly precise data model of the body, a digital twin. In the next step, the appropriate 3D models of the appropriate counterparts are downloaded from the cloud using algorithms. Prior to the start of the actual manufacturing, these digital prototype models from the digital cloud library are virtually tested in different variants on the future prothesis. The protheses can then be manufactured in a digitally networked 3D printing center that is located nearby. Ottobock's digital platform makes it possible to standardize the quality specifications in the virtual medical supply store of the future. The orthopedic technicians, equipped with virtual tools such as augmented reality glasses, can be guided by the expert database in the cloud during all steps of the procedure for the assembly and adjustment of the device.

The digital platform will enable Ottobock to overcome the existing hurdles in supplying quickly growing, emerging economies. It will finally be possible to offer substantial numbers of high-quality orthopedic engineering solutions in Africa, South America, and Asia without having to establish complex workshops with highly qualified employees over many years. The more expert knowledge is collected in the digital database for worldwide access via augmented reality and other technologies, the less specialist orthopedic knowledge is required on the ground. China, for example, still does not provide treatment that is comparable to that of the German health system, and to date all attempts to train orthopedic technicians there have failed. Despite this, the Chinese middle class now has enormous purchasing power. With the digitization of customer relations and production we can leverage this growth potential.

Our digitization of the manufacturing process is changing our craft so fundamentally that it is opening up completely new growth opportunities worldwide. With our wealth of ideas and our work, we are radically changing the life of users. That was and remains the highly emotional and highly empathetic mission of our company to this day. Ottobock is human and technological—and in the future it will also be digital.

Two Paths of Growth for Ottobock

This book provides a comprehensive overview of the main scientific disciplines that are working on the optimization of the human body. It also describes a variety of the growth paths that the company is pursuing with great attention and varying degrees of intensity. In the coming years, Ottobock will continue to invest in diverse innovations, thus consolidating its position as the world's leader in technology. We will focus on growth fields that contribute to our strategic corporate goal as the human empowerment company. As a pioneer we will restore the powers and abilities of the human body and will continually enhance them in the future by translating intriguing technologies into solutions that empower people to transcend themselves.

Ottobock will increase its corporate value with further innovation leaps, which, as in all high-tech companies, are generated at the interface between artificial intelligence and new technologies. Ottobock's growth strategy focuses on two main areas: the further development of innovations in the traditional business sector of restoring human mobility and the creation of new growth by strategically reorienting our company on human empowerment.

Human Mobility

In restoring human mobility, Ottobock generates not only sales and profit, but also makes a significant contribution to society. We help our users and patients improve their mobility and return to everyday life. While individuals benefit greatly from this, it also serves our social system and society at large when patients can be gradually reintegrated into the working world and earn their own money. In our quest to restore human mobility we are now going one step further to increasingly develop products that help people whose mobility is significantly limited although they have not suffered an amputation. To this end we will use our technological know-how and innovation leadership in the field of prosthetics and in segments of orthotics to develop products for people suffering from paralysis that results from specific conditions such as diabetes, multiple sclerosis, paraplegia, and strokes. This increases the number of possible applications of our technology, and we can significantly improve the treatment of many people with such conditions. This is why Ottobock has created a new division called NeuroMobility, which provides solutions for neuro-orthotics as well as the rehabilitation and wheelchair sector. By clustering our competencies and services, we offer overall solutions for patients with neurological disorders. Our complete takeover of the Swedish start-up Exoneural Network in 2021 also points in this direction. The company developed the innovative Exopulse Mollii Suit, which is equipped with fifty-eight electrodes and relaxes muscle cramps using electrical impulses. This will enable people with multiple sclerosis and cerebral palsy to lead considerably more independent lives.

Human Empowerment

In 2021 Ottobock expanded its strategic corporate objective to include human empowerment. This means that in the future, in addition to helping people with limited mobility, we will also aid users who wish to increase and strengthen their physical powers in the workplace or in their leisure time with wearable robotics. This has the potential to become an enormous mass market that is just at the beginning of its development. The demand for wearable robotics that can ease the physical activities of workers is currently on the rise in industry. This complies with the values of Ottobock: if we can go beyond just helping patients and shape work society on a more sustained basis, the company will develop beyond medical technology and establish itself in the field of prevention. With our technologies—especially bionic exoskeletons—we can significantly reduce the stress of hard physical labor. That is why in late 2021 we expanded our exoskeleton portfolio through our acquisition of the American company suitX, which specializes in research and development of exoskeletons in occupational and medical environments (see the interview with the founder of suitX, Homayoon Kazerooni, on page 291).

Technological Progress in Seven Disciplines

While Ottobock's growth strategy in the two fields of human mobility and human empowerment focuses on pinpointing new areas of application, new target groups, and new markets, innovative technologies take the center stage as a drivers of growth. With varying intensity, our main concentration lies on the seven technological and research areas that are presented in this book: Digitization, Quantum Technologies, Advanced Materials / 3D Printing, Bionic Regeneration, Artificial Intelligence, Human-Machine Interface, and Wearable Robotics / Bionic Reconstruction. With its comprehensive, in-depth interviews, this book has opened the door to the laboratories of the world and simultaneously enabled us to enormously expand our horizon. We are precisely monitoring the progress in all the relevant research areas described in these pages to quickly translate the advances into our core competence of human mobility.

This strategy is also pursued in this book by the five Ottobock experts, who currently accompany Ottobock on this journey or have done so in the past and are representative of the numerous employees enriching the innovative power of Ottobock worldwide with their cultural diversity and brave ideas. In each chapter of this book the five experts establish how relevant the advances in the respective disciplines are for Ottobock's overall growth strategy. Throughout we have always had a realistic timeline in mind, along which the technological advances can be translated into short-term, medium-term, and long-term product innovations.

The focus of this book is the overarching vision of the complete *Digitization* of our century-old craft. We will make fundamental advances in this area in the near future. In contrast, in the field of *Quantum Technologies* the tangible results tend to be in the distant future. While the maturity of the respective technology that we invest in is normally more advanced, we also have the task of better understanding emerging technologies. We can best do this as participating observers in concrete research projects such as the project "Quantum-Human-Machine-Interface Based on Quantum Sensors" in the Cluster4Future QSens, in which we are participating as a research partner. We are interested in fundamentally improving the intelligent control of prostheses and orthoses using quantum sensors and soon anticipate the first applications on and in the human body.

Ottobock has considerable experience in the field of *Advanced Materials* and *3D Printing*. Innovative materials have always played a decisive role in our pioneering history. We are currently experimenting with the use of soft polymers for rapid liquid printing. These are soft polymer structures that are fabricated in a gel. In this revolutionary technology for all applications, the soft polymers are anatomically formed or individually tailored. This represents a major breakthrough for Ottobock. Our technology, which we have already patented worldwide, enables a diversity of applications extending well beyond our industry.

The future-oriented field of *Bionic Regeneration* is intensively searching for a formula for a longer life. The central focus is the issue of how the self-healing powers that are already present in humans can be amplified and how biomedical techniques can help in the fight against cancer and other incurable diseases. It is already possible to artificially grow stem cells and muscle cells on an industrial scale; the first artificial organs are already being printed on 3D bioprinters. In the field of bionic regeneration we can benefit from the growing convergence between the engineering and life sciences. We have many points of contact with biotechnologies in the company. Artificial skin and tissue are already being printed in a number of research laboratories using 3D printers today, and it will not be long before there are numerous applications for tissue engineering, including on an industrial scale. In principle, bionic regeneration means perfect healing of wounds through the mobilization of the human body's hidden self-healing powers. This could soon become reality for certain reconstructive tasks with a small cell volume, providing interesting application possibilities for Ottobock, such as considerably improving the integration of prostheses in the body and the interface between residuum and socket. This represents a major step forward in terms of the quality of life of our users.

As in all high-tech companies, *Artificial Intelligence* already plays a key role at Ottobock. Examples are the intuitive mind control of prostheses, baby helmets that help correct deformed head shapes, and the evaluation of the data of millions of patients in compliance with data protection laws. The absolute strength of artificial intelligence is in machine learning and pattern recognition, which we employ in numerous technical solutions. AI will play an increasingly strong role in Ottobock's smart, intuitive supply solutions and intelligent processes.

The Human-Machine Interface is at the heart of Ottobock's research. The optimization of the connection between man and machine is not only decisive for technological progress, but particularly for user acceptance of wearable robotics and prostheses. Our overarching goal is to increasingly fuse technical aids with the human body so that they are perceived as part of the user's own body. In the long term Ottobock will continue to extend technological solutions in its core area of orthotics and prosthetics in the direction of the consumer and make greater use of all its experience to develop applications that extend far beyond the repair of the human body and are interesting for far larger target groups. This will open up unimagined market opportunities.

The chapter *Wearable Robotics / Bionic Reconstruction* encompasses Ottobock's core competences. In our more tradition core business of bionic reconstruction there are encouraging developments worldwide in the laboratories of Hugh Herr, Tommaso Lenzi, and Homayoon Kazerooni as well as in Oskar Aszmann's operating theater that are quite valuable for the technological development of Ottobock. Our further advance in robotics shows how Ottobock's new strategic orientation on human empowerment is opening up new growth markets and strengthening our competence for the development of active prostheses. Ottobock will not be putting any active prostheses on the market until we have reduced the disadvantages of this technology to a minimum. We have been working on active knees for the last years and are following the numerous approaches of various research projects. Ottobock's major competitive advantage is that we are very good at translating biomechanical knowledge into intelligent controllers. As soon as we succeed in extending our system with the "active" degree of freedom, we will enter a new dimension. A combination of our current powerful human-machine interface and an actively controlled system would be especially effective. That is why Ottobock will continue to invest in active prostheses in the future and thus generate new growth.

This vision—starting from my family's values—is in the company's genes. Ottobock will offer people technological solutions for all the challenges of everyday life, so that they can do what they want to do for longer periods—all as intuitively as possible and at an affordable price. We are now in the process of increasing the body's own strengths with the aid of innovative technology. Our goal is to empower people. We are the human empowerment company.

Biographies

Prof. Dr. Jens Anders is the director of the Institute of Smart Sensors and Theoretical Electrical Engineering at the University of Stuttgart and spokesperson for Cluster4future QSens. One focus of his research is quantum technology, particularly how to make quantum sensors useful in everyday life. For example, he has developed the first wearable electron spin resonance spectrometer that will fundamentally change the detection of free radicals, since measurements can currently only be made using surrogate parameters. The exchange of free radicals and antitoxins has a direct bearing on things such as a person's medical condition and the quality of new energy materials that could be applied in modern solar cells.

Matt Angle, PhD, is a scientist-entrepreneur and the founder and CEO of Paradromics, a venture-backed company that is building brain-computer interfaces (BCI) to address unmet medical needs. Together with a team of unparalleled domain experts and advisors in the BCI field, he aims to develop a new generation of neural interfaces that will be used in clinical applications. Chips will act as modems for the brain, allowing direct data exchange with a computer, changing the way we treat and analyze diseases. Paradromics is creating the first high-datarate interface between brains and computers, essentially transforming neural data into medicine.

Prof. Dr. Oskar C. Aszmann is a specialist in plastic and reconstructive surgery. He is the deputy director of the University Clinic for Plastic, Reconstructive, and Aesthetic Surgery at the University Hospital Vienna, where he is the director of the outpatient clinic for reconstruction of peripheral nerves and the special outpatient clinic for bionic reconstruction of extremities. Aszmann's research focuses on all aspects of reconstructive surgery, from both a clinical perspective and with respect to basic research. Patients with complex losses in the function of their extremities are at the core of his work involving a wide array of biological and reconstructive measures as well as modern prosthetic systems.

Anthony Atala, MD, is the G. Link Professor and Director of the Wake Forest Institute for Regenerative Medicine and the W. Boyce Professor and Chair of the Department of Urology at Wake Forest School of Medicine. His work focuses on growing human cells, tissues, and organs. Fifteen applications of technologies developed in Dr. Atala's laboratory have been used clinically in human patients. Atala and his team of 450 researchers developed the first lab-grown organ (a bladder) to be implanted into a human. He is also developing experimental technology that can 3D print human tissue on demand.

Martin Böhm is the first Chief Experience Officer at Ottobock, a position that he has held since June 2021. He was previously Chief Digital Officer at Beiersdorf, and he plans to combine the marketing and IT departments at Ottobock. As director of this field, he will focus on developing the digital user experience, particularly in the areas of social media, e-commerce, and customer relations management (CRM). With his teams, he is creating a data-driven platform for users and customers that offer personalized solutions for everyday life.

Dr. Hans Dietl was Chief Technology Officer at Ottobock for over thirty years, advancing research and development with his expertise of prostheses, ortheses, and neuroprostheses. He is still involved in the company as a consultant. Among his many innovations, he designed a new generation of microprocessor-driven prosthetic knees (C-Leg) and innovative myoelectical systems for the rehabilitation of limited upper extremities. He studied engineering at the Technical University in Vienna and completed his dissertation in the Department of Bio Fluid Mechanics and Magnetic Hydrodyamics at the Institute of Fluid Mechanics and Heat Transfer.

Prof. Dr. Claus Emmelmann has been a professor at the Technical University of Hamburg since 2001. He is the director of the Institute of Lasers and System Technologies. In his research and collaboration with the industry, his goal is to industrialize digital and additive research. From 1991 to 2001, before accepting the offer to work at the University of Hamburg, he worked for ROFIN-SINAR, one of the world's largest laser producers in the international application market of lasers, where he was responsible for the business sector of the solid-state laser, which was new at the time. He is the director of Laser Zentrum Nord in Hamburg, and until 2020 he was the director of the Fraunhofer Institution for Additive Production Technologies IAPT.

Prof. Dr. Dario Farina was a full professor at Aalborg University in Denmark and the University Medical Center Göttingen, where he founded and directed the Institute of Neurorehabilitation Systems, until he moved to Imperial College London as Professor and Chair in Neurorehabilitation Engineering. His research focuses on biomedical signal processing, neurorehabilitation technology, and neural control of movement. Among other achievements, he pioneered noninvasive devices for neural interfacing, wearable neural interfaces that allow a direct noninvasive connection with the human nervous system and are intended to be used in a variety of human-machine interfacing application scenarios. The developed noninvasive device for neural interfacing breaks the traditional information-invasiveness bottleneck observed in earlier neural interfacing approaches.

Martin Ford is a futurist, Silicon Valley entrepreneur, and author focusing on artificial intelligence and robotics, and the impact of accelerating technology on workplaces, the economy, and society. In 2009 Ford published his first book *The Lights in the Tunnel: Automation, Accelerating Technology, and the Economy of the Future*, making a strong argument that advances in robotics and artificial intelligence would make a large fraction of human workforce obsolete. He also predicted that artificial intelligence will be the next Killer App. He is most known for his book *Rise of the Robots: Technology and the Threat of a Jobless Future*, which won the 2015 FT & McKinsey Business Book of the Year Award and was a New York Times Bestseller. In Ford's most recent book, *Rule of the Robots: How Artificial Intelligence Will Transform Everything* (2021), he argues that AI is a systemic, general-purpose technology that will ultimately compare to electricity in terms of its impact on the economy and society.

Prof. Dr. Benjamin M. Friedrich is Heisenberg Professor for Biological Algorithms at the cluster of excellence Physics of Life at the Technical University of Dresden. As an applied mathematician and theoretical physicist, he collaborates closely with experimental biologists. He is affiliated with the Center for Molecular and Cellular Bioengineering (CMCB), the Center for Advancing Electronics Dresden (cfaed), and the Institute of Theoretical Physics (ITP). In his cross-disciplinary research, he aims to understand nonlinear control loops in biological systems and their robustness in the presence of strong fluctuations, using mathematical modeling, image and data analysis, and numerical simulations. The mission of Friedrich's Biological Algorithms Group is to identify simple paradigms of robust mobility control and pattern formation in complex biological systems.

Dr. Andreas Goppelt was Chief Technology Officer (CTO) in the area of research and development at Ottobock from 2017 until 2022. Prior to that, he was CEO of Otto Bock Mobility Solutions GmbH, the wheelchair business. He came to the Ottobock consortium in 2015 as CEO of nstim NeuroBionics GmbH. Before that, he was director of the area R&D for the business segment BioSurgery and Regenerative Medicine at Baxter. He began his business career as the founder and Chief Scientific Officer of Switch Biotech AG, which is specialized in disorders in wound healing and skin disease. Originally from Erlangen, Germany, he studied chemistry at Ludwig Maximilian University (LMU) in Munich and has a doctorate in biochemistry. He did postdoc work at the Gene Center of the LMU and at the Max Planck Institute for Biochemistry in Martinsried.

Dr. Bernhard Graimann completed his doctorate in biomedical engineering in 2002 and received his postdoctoral qualification in applied computer science in 2006, both at the Technical University of Graz, Austria. He then moved to the Institute for Automatization Technology at the University of Bremen, Germany, where he worked on controlling robots with brain-computer interfaces. He has worked for Ottobock since 2008, first in Strategic Technology Management, then as Global Expert in Neurotechnology, and since 2012 he has been the director of Transatlantic Research and Knowledge Management. He is Vice President Global Research & Development.

Hugh Herr, PhD, is a previously elite rock climber, biophysicist, and professor of Media Arts and Sciences at the MIT Media Lab, where he directs the Biomechatronic Research Group and codirects the K. Lisa Yang Center for Bionics at MIT. *Time* magazine coined Hugh Herr the "Leader of the Bionic Age" in 2011, because of his revolutionary work in the emerging field of Biomechatronic technology that marries human physiology with electromechanics. A double amputee himself, he is responsible for breakthrough advances in bionic limbs that provide greater mobility and new hope to those with physical disabilities. As founder of BionX Medical Technologies he produced the powered-ankle-foot prothesis called emPOWER, which emulates the action of a biological leg and, for the first time, provides amputees with a natural gait. Ottobock acquired the company in 2017.

Thomas Huber founded his agency semanticom GmbH in Berlin in 2006. Since 2019 he has also been Associate Partner in the strategy consultancy Prenew. The graduate of the German School of Journalism in Munich has been responsible for the conception and editorial direction of a number of books in close cooperation with Steidl, including *August Bebel* (2013, ed. Manfred Bissinger), *ZukunftsWerte* (2018, Otto Group), *Digital Shift* (2019, Deutsche Telekom), *Futuring Human Mobility*, (2019, Ottobock), and *Begegnungen* (2021, Hans Georg Näder). Before starting his agency, he was Head of Public Relations at Gruner+Jahr, director of communication at BBDO Germany, and editor of corporate issues at Daimler-Benz, where he wrote speeches for CEO Jürgen E. Schrempp.

Güngör Kara, a digital expert, has been Chief Digital Officer at Ottobock since 2019. Previous to that he held a similar position at EOS Electro Optical Systems, which is specialized in industrial 3D production. Kara has developed and implemented innovation centers, Industry 4.0 solutions, and industrial 3D printing. At Ottobock he develops ambitious digital strategies, establishes business models, directs the internal start-up teams, and refines collaboration with external start-ups to promote innovations that contribute to the digital transformation of the company.

Prof. Dr. Homayoon Kazerooni is a roboticist and professor of mechanical engineering at the University of California, Berkeley, where he also serves as the director of the Berkeley Robotics and Human Engineering Laboratory—known as KAZ LAB. Kazerooni holds a doctorate degree from MIT, and he is a leading expert in robotics, control sciences, exoskeletons, human augmentation, mechatronics, and intelligent assist devices. In addition to his teaching and research experience in academia, Kazerooni is also an entrepreneur. In 2005 he founded Ekso Bionics, which went on to become a publicly owned company in 2014 to market medical exoskeletons. He is also the founder of suitX, which designs and manufactures industrial, medical, and recreational exoskeletons. Ottobock acquired suitX in November 2021, and Kaz remains with the company as the company's chief scientist.

Taavi Kotka is an entrepreneur and IT visionary who became the first CIO for the government of Estonia in 2013 to oversee the country's development as an advanced digital nation. He is the pioneer of the leading e-residency program e-Estonia, which has placed Estonia at the epicenter of the global tech community, and innovations such as data embassies, country-as-a-service (CAAS), the no-legacy policy, and VAT fraud detection. As a result of his work in the public sector, Kotka was named European CIO of the Year 2014 and received Estonia's prestigious White Star III Class Order in 2016. He is now back in the private sector, helping start-ups develop and consulting large enterprises and governments on digital transformation. He was also a special advisor to European Commission Vice President Andrus Ansip on European Digital Single Market.

Mark R. Kotter, MD PhD, is a doctor, scientist, and serial entrepreneur. As a neurosurgeon, he treats patients with spinal cord injury. Kotter is best known for discovering the importance of macrophages for brain regeneration, which led to the first regenerative medicine trial for degenerative cervical myelopathy, and for developing opti-ox, a gene targeting approach that enables faithful execution of genetic information in pluripotent stem cells and demonstrated that this is necessary and sufficient to deterministically induce a new cellular identity. These findings challenge the theory that cell reprogramming depends on stochastically determined permissive states and enable the production of any human cell within days at purities approaching one hundred percent. He is the founder of bit.bio, cofounder of cultured meat start-up Meatable and of rejuvenation company clock.bio, and co-founder and trustee of Myelopathy.org, the first charity dedicated to a common yet often overseen condition causing a "slow motion spinal cord injury."

Prof. Robert Langer, ScD, is an American chemical engineer, scientist, entrepreneur, inventor, and one of the twelve Institute Professors at the Massachusetts Institute of Technology (MIT). Langer's research laboratory at MIT is the largest academic biomedical engineering lab in the world. Langer is a widely recognized and cited researcher in biotechnology, especially in the fields of drug delivery systems and tissue engineering. He is also a prolific entrepreneur, having participated in the founding of over forty biotechnology companies including Moderna. The most-cited engineer in human history, he is considered a pioneer of many new technologies, including controlled release systems and transdermal delivery systems, which allow the administration of drugs or extraction of analytes from the body through the skin without needles or other invasive methods. Langer is regarded as the founder of tissue engineering in regenerative medicine. He and his team have made advances in tissue engineering, such as the creation of engineered blood vessels and vascularized engineered muscle tissue.

Dr. Tommaso Lenzi is Assistant Professor in the Department of Mechanical Engineering at University of Utah, and a Core Faculty in the Utah Robotics Center, where he directs the newly founded Bionic Engineering Lab, sponsored by the Department of Defense, the National Science Foundation, and the National Institute of Health. His main research interests include robotics, mechatronics, and rehabilitation medicine with a major emphasis on the design and control of wearable robots for human assistance and rehabilitation. As a scientist, engineer, and inventor he is interested in bridging the gap between robotics, controls, design, prosthetics, biomechanics, and rehabilitation and creates bionic wearable technologies that help people walk and move better.

Prof. Christian Lüscher is a neurobiologist and professor at the Department of Basic Neurosciences of the University of Geneva who also maintains clinical activity at the clinic of Neurology of the Geneva University Hospital. His work aims to understand the neural mechanisms underlying addiction and other behavioral diseases. Lüscher is known for his contributions in the field of addiction, particularly for establishing links of causality between the drug-evoked synaptic plasticity and adaptive behavior in mice. Lüscher studies how addictive drugs alter synaptic transmission in the reward system of the mouse brain. In his most recent work, he explores the molecular basis of individual vulnerability to addiction and launched several translational projects. He has initiated the OptoDBS conference series, seeking to emulate optogenetic innervations with deep brain stimulation.

Dr. Francesca Moresco is research group leader of the Single Molecule Machines Group at the Center for Advancing Electronics Dresden (cfaed). She joined the Technical University of Dresden at the end of 2009, after an experience in the semiconductor industry. She headed the Molecular Manipulation and Nanomachines Group at the Institute of Material Science until 2017, studying electronic and mechanical properties of single molecules at surfaces by scanning tunneling microscopy. Since October 2017 Moresco coordinates the FET Open European project "Mechanics with Molecules" (MEMO) with the objective of understanding the transmission of motion from molecule to molecule toward the construction of atomic scale mechanical calculators.

Prof. Hans Georg Näder was born in 1961 in Duderstadt to Maria Näder, née Bock, and Max Näder. He studied business administration and took over Ottobock with approximately one thousand employees at the age of twenty-nine in 1990. Under his leadership, the company has become the world market leader in prosthetics and major milestones such as the first microprocessor-controlled leg prosthesis C-Leg have been launched. Näder is a passionate entrepreneur, futurist, art collector, sailor, and philanthropist.

Christoph Neumann, born in 1982 in Lüdenscheid, Germany, lives and works in Berlin. He works for major German publishers and advertising agencies, and has been the recipient of many prestigious prizes such as the C/O Berlin Close-Up Award (gold), the Design Prize of the Federal Republic of Germany (silver), and the Art Directors Club Germany Prize (bronze). He became known through his photo column in the German magazine *ART*. Neumann was assisted by **Sascha Boldt** in his work on the photographs for this book. Boldt was born in 1972 in Bremen, Germany, and lives and works in Berlin. His focus is on digital collages, painting, and photography. He has won various art prizes, including a grant to study abroad from the City of Frankfurt am Main. His artwork has been included in many exhibitions around the world.

Prof. Dr. Christian Ospelkaus is member of the executive board of the Institute of Quantum Optics at Leibniz University Hannover and leads a joint research group of Leibniz University Hannover and the Physikalisch-Technische Bundesanstalt (PTB) Braunschweig, focusing on trapped ions for quantum information processing and precision measurements. He is coordinator of BMBF ATIQ Quantum Computer Demonstration Setups and brings together major ion-trap QC groups in Germany with QC users and technology developers to push the limit of ion-trap QC.

Dr. Viola Priesemann is a physicist who is the director of the research group Theory of Neuron Systems at the Max Planck Institute in Göttingen. The focus of Priesemann's research is the modeling of complex systems. She is especially interested in the self-organization and emergence of information distribution in living and artificial neuronal networks. Priesemann would like to improve the autonomous learning and efficiency of neural artificial neuronal networks as well as investigate how information distribution of a complex environment occurs in the brain. In addition, she is exploring how activity spreads in neuronal networks and how this can be quantified, even if only a tiny part of all neurons can be measured. During the COVID-19 pandemic she has also focused on modeling the spread of infection in order to study which strategies of pandemic containment are most effective.

Dr. Tatiana Sandoval-Guzmán is the leader of the research group Vertebrate Tissue Repair and Regeneration at the Medical Faculty of the Technical University of Dresden. The vertebrate limb contains various tissue types including bone, skin, nerves, muscle, and blood vessels. After an injury these tissues respond to a plethora of signals driving repair and, in few animal species, regeneration. The main goal is to understand how individual tissues respond to an injury to further understand their interaction during axolotl limb regeneration. The fine-tuned coordination of both, the individual tissue regeneration and interaction with other tissue types could be the key for successful appendage regeneration.

Prof. Dr. Piet Oliver Schmidt is a physicist and professor at the Institute of Quantum Optics at Leibniz University Hannover (LUH) who investigates applications of quantum physics in metrology. He is the director of the Institute of Experimental Quantum Metrology of the research group Quantum Logic Spectroscopy. The institute was founded in 2009 in a collaboration between the Physikalische-Technische Bundesanstalt (PTB) in Braunschweig, Germany's national metrology institute, and the LUH, with the goal of promoting basic research in the area of time and frequency metrology and developing new kinds of quantum technology with saved ions. He is cospokesperson of the cluster of excellence QuantumFrontiers—Light and Material at the Quantum Frontier and spokesperson of the special research area of Designed Quantum States for Matter—Production, Manipulation, and Detection of Metrological Applications and Tests of Fundamental Physics.

First edition published in 2022

Concept and editorial staff: semanticom, Berlin
Thomas Huber, Angela Christ, Johann von Schubert,
Dr. Tas Skorupa (New York / Berlin)

Translations: Colin Shepherd, Berlin

Photographs: Christoph Neumann, Berlin
in collaboration with Sascha Boldt, Berlin

Ottobock team:
Merle Florstedt, Marc C. Schneider, Sebastian Zavelberg

Book design: Gerhard Steidl, Holger Feroudj, Matthias Langner
Separations by Steidl image department
Production and printing: Steidl, Göttingen

Steidl
Düstere Str. 4 / 37073 Göttingen, Germany
Phone +49 551 49 60 60
mail@steidl.de
steidl.de

ISBN 978-3-96999-113-8
Printed in Germany by Steidl